Lecture Notes in Computer Science 11810

More information about this series at http://www.springer.com/series/7409

Markus Krötzsch · Daria Stepanova (Eds.)

Reasoning Web

Explainable Artificial Intelligence

15th International Summer School 2019
Bolzano, Italy, September 20–24, 2019
Tutorial Lectures

Springer

Editors
Markus Krötzsch ⓘ
Technische Universität Dresden
Dresden, Sachsen, Germany

Daria Stepanova ⓘ
Bosch Center for Artificial Intelligence
Renningen, Germany

ISSN 0302-9743   ISSN 1611-3349 (electronic)
Lecture Notes in Computer Science
ISBN 978-3-030-31422-4   ISBN 978-3-030-31423-1 (eBook)
https://doi.org/10.1007/978-3-030-31423-1

LNCS Sublibrary: SL3 – Information Systems and Applications, incl. Internet/Web, and HCI

This Springer imprint is published by the registered company Springer Nature Switzerland AG
The registered company address is: Gewerbestrasse 11, 6330 Cham, Switzerland

Preface

Artificial Intelligence has become the defining technology of our time. Impressive and effective though it may be, it often defies our understanding, and hence our control. Nevertheless, today's main concern is not that AI might become too smart - it is far from it - but that it often surprises us by being dangerously dumb, prejudiced, credulous, or unreliable. Fortunately, AI is an extremely rich research area, which keeps developing in many different directions that outline ways of constructing intelligent systems that can be understood, validated, and controlled. This is the promise of ongoing works towards Explainable AI.

With Explainable AI as its focus topic, the 15th Reasoning Web Summer School explored a variety of ideas being developed in this research. The lectures, given in Bolzano, Italy, during September 20–24, 2019, brought together leading researchers in AI and related areas. Topics ranged from explainable forms of machine learning, over data mining, knowledge representation, and query answering, to the explanation of complex software systems using formal methods.

These topics are not only highly relevant and timely, but also a logical next step in Reasoning Web's continuing evolution. Since its inception as a summer school on reasoning in (Semantic) Web applications in 2005, Reasoning Web has developed into a prime educational event covering many aspects of intelligent systems and attracting young and established researchers. The 2019 edition, hosted by the Free University of Bozen-Bolzano, marked the school's 15th anniversary. A highlight of Reasoning Web 2019 was the integration into the Bolzano Rules and Artificial Intelligence Summit (BRAIN), including the International Joint Conference on Rules and Reasoning (RuleML+RR 2019), the Global Conference on Artificial Intelligence (GCAI 2019), and DecisionCAMP 2019.

These proceedings compile tutorial papers that complement the lectures given in Bolzano. The tutorials include in-depth surveys as well as shorter extended abstracts that point to existing works. All papers have been written as accompanying material for the students of the summer school, in order to deepen their understanding and to serve as a reference for further detailed study. This volume contains the following tutorial papers, each accompanying a lecture:

- "Classical Algorithms for Reasoning and Explanation in Description Logics," in which Yevgeny Kazakov (presenter) and Birte Glimm give a detailed introduction to explainable reasoning in description logics, with the focus on tableau-based inference algorithms and the computation of justifications.
- "Explanation-friendly Query Answering Under Uncertainty," in which Maria Vanina Martinez (presenter) and Gerardo I. Simari discuss their work on knowledge representation with probabilistic and inconsistency-tolerant rule languages, and show ways of explaining reasoning in each case.
- "Provenance in Databases: Principles and Applications," in which Pierre Senellart provides a short overview of provenance as a key concept for explaining answers to

database queries, as well as references to works that explore this concept for various data models and query languages.

- "Knowledge Representation and Rule Mining in Entity-Centric Knowledge Bases," in which Fabian M. Suchanek (presenter), Jonathan Lajus, Armand Boschin, and Gerhard Weikum introduce graph-like knowledge bases, overview rule mining, and neural knowledge-base completion methods.
- "Explaining Data with Formal Concept Analysis," in which Bernhard Ganter, Sebastian Rudolph (presenter), and Gerd Stumme introduce the mathematical foundations and algorithmic basics of Formal Concept Analysis, which can in particular be applied in data mining.
- "Logic-Based Learning of Answer Set Programs," in which Mark Law (presenter), Alessandra Russo (presenter), and Krysia Broda discuss inductive logic programming approaches that apply to the popular ASP formalism.
- "Constraint Learning: An Appetizer," in which Stefano Teso gives a brief introduction to the field of constraint learning, some of its core methods, and its relationship with other areas of machine learning.
- "A Modest Markov Automata Tutorial," in which Arnd Hartmanns and Holger Hermanns (presenter) give a first introduction into modeling of (systems of) software systems, including several forms of non-deterministic behavior, and show how to analyze such models using the MODEST toolset.
- "Explainable AI Planning (XAIP): Overview and the Case of Contrastive Explanation," in which Jörg Hoffmann (presenter) and Daniele Magazzeni briefly discuss various existing types of explanations within AI planning.

Many people have contributed to the realization of this event. First and foremost, we would like to thank the lecturers and their co-authors for their contributions, which are the very essence of this school. Furthermore, we would like to thank the general chair of BRAIN 2019, Diego Calvanese, and his team: Evellin Cardoso, Ana Ozaki, and Nicolas Troquard provided support for speakers and accommodation; Julien Corman and Andrey Rivkin coordinated the social program; Rafael Peñaloza acted as a sponsorship chair; and Paolo Felli managed the web pages. Special thanks are also due to Kati Domann, who organized speaker refunds, and to Hannes Hell, who coordinated student scholarships.

We also gratefully acknowledge the support of our sponsors. In particular, we thank the Emerald Sponsors Free University of Bozen-Bolzano and Center for Perspicuous Computing (CPEC), and the Platinum Sponsor Bosch. Their generous contributions allowed us to financially support a record number of participants and to cover the cost of participation for most lecturers. We also thank the Gold Sponsor *Artificial Intelligence Journal*, Silver Sponsor Hotel Greif, and Bronze Sponsor Ontopic for their valuable contributions. Moreover, we appreciated the assistance and professional service provided by the Springer LNCS publishing team. Finally, thanks are due to all participants of Reasoning Web 2019; we hope that their stay in Bolzano was most profitable and enjoyable.

September 2019   Markus Krötzsch
                                                                 Daria Stepanova

Organization

General Chair

Diego Calvanese — University of Bozen-Bolzano, Italy

Reasoning Web Chairs

Markus Krötzsch — TU Dresden, Germany
Daria Stepanova — Bosch Center for Artificial Intelligence, Germany

Publicity Chairs

Livia Predoiu — Free University of Bozen-Bolzano, Italy
Guohui Xiao — Free University of Bozen-Bolzano, Italy

Speaker Support and Accommodation

Evellin Cardoso — Free University of Bozen-Bolzano, Italy
Ana Ozaki — Free University of Bozen-Bolzano, Italy
Nicolas Troquard — Free University of Bozen-Bolzano, Italy

Sponsorship Chair

Rafael Peñaloza — Università degli Studi di Milano Bicocca, Italy

Social Program

Julien Corman — Free University of Bozen-Bolzano, Italy
Andrey Rivkin — Free University of Bozen-Bolzano, Italy

Webmaster and Design

Paolo Felli — Free University of Bozen-Bolzano, Italy

Program Committee

Stefan Borgwardt — TU Dresden, Germany
David Carral — TU Dresden, Germany
Sergey Chubanov — Bosch Center for Artificial Intelligence, Germany
Tom Hanika — University of Kassel, Germany
Evgeny Kharlamov — University of Oslo, Sweden
Sascha Klüppelholz — TU Dresden, Germany

Maximilian Marx	TU Dresden, Germany
Pauli Miettinen	University of Eastern Finland, Finland
Sophie Tourret	Max Planck Institute for Informatics, Germany

Reasoning Web 2019 Sponsors

Emerald Sponsors

Free University of Bozen-Bolzano
https://www.unibz.it/

DFG CRC/Transregio 248 "CPEC"
https://www.perspicuous-computing.
science/

Platinum Sponsor

Bosch
https://www.bosch.com/

Gold Sponsor

Artificial Intelligence Journal (AIJ)
https://www.journals.elsevier.com/
artificial-intelligence

Silver Sponsor

Hotel Greif
https://www.greif.it/en/hotel-bolzano/1-0.
html

Bronze Sponsor

Ontopic
http://ontopic.biz/

Contents

Classical Algorithms for Reasoning and Explanation in Description Logics

Birte Glimm and Yevgeny Kazakov[✉]

The University of Ulm, Ulm, Germany
{birte.glimm,yevgeny.kazakov}@uni-ulm.de

Abstract. Description Logics (DLs) are a family of languages designed to represent conceptual knowledge in a formal way as a set of ontological axioms. DLs provide a formal foundation of the ontology language OWL, which is a W3C standardized language to represent information in Web applications. The main computational problem in DLs is finding relevant consequences of the information stored in ontologies, e.g., to answer user queries. Unlike related techniques based on keyword search or machine learning, the notion of a consequence is well-defined using a formal logic-based semantics. This course provides an in-depth description and analysis of the main reasoning and explanation methods for ontologies: tableau procedures and axiom pinpointing algorithms.

1 Introduction

It often happens that one needs to find some specific information on the Web. Information can be of different kinds: a local weather report, shop opening hours, a cooking recipe, or some encyclopedic information like the birth place of Albert Einstein. To search for this information, one usually enters some keywords into Web search engines and inspects Web pages that contain such keywords in the hope that the relevant information can be found there. The modern search engines are a little more advanced: they can also search for Web pages that use synonyms or related keywords, they use vocabularies of terms to structure information on the Web (e.g. schema.org[1]) or to disambiguate search results, they use some machine learning techniques to rank Web pages according to their likeliness of containing the relevant information, they use facts or knowledge (e.g., extracted from Wikipedia or from internal sources such as Google's Knowledge Graph[2]) to answer some queries, and they give direct answers for certain queries, e.g., Google.com directly answers the question "Who was US president in 2015?" with a knowledge panel for Barack Obama. In general, however, there is *no guarantee* that the searched piece of information can be found, even if a corresponding Web page exists.

Although an average Web user can live with such limitations, there are some critical applications in which incorrect or missed results cannot be tolerated. For

---

[1] https://schema.org/.
[2] https://developers.google.com/knowledge-graph/.

© Springer Nature Switzerland AG 2019
M. Krötzsch and D. Stepanova (Eds.): Reasoning Web 2019, LNCS 11810, pp. 1–64, 2019.
https://doi.org/10.1007/978-3-030-31423-1_1

example, if a medical patient is misdiagnosed, i.e., a correct diagnosis based on the description of the patient's symptoms is not found, the consequences can be severe. Several similar examples can be given for other domains ranging from banking to autonomous driving. Supporting such applications requires representing the knowledge in a *precise* and *unambiguous* way so that it can be correctly processed by *automated tools.*

Ontology languages, such as OWL [44], offer a solution to this problem by defining the formal syntax and semantics to describe and interpret expert knowledge. The basic principle of ontologies is similar to Wikipedia: instead of extracting the knowledge from general sources, such as Web pages, the knowledge is described and curated in one place by *domain experts.* The main difference between Wikipedia and ontologies is in the way how the knowledge is described. Wikipedia pages provide mostly textual (natural language) descriptions, which are easy to understand by humans, but difficult to process by computers. Today Wikipedia is also partly fed from Wikidata,[3] which provides a knowledge base of facts. The Wikidata project was founded in 2012 and contains, at the time of writing, facts about roughly 60 million entities. Ontologies go beyond a knowledge base of facts and provide often complex descriptions by means of *formulas;* each formula can use a limited set of constructors with well-defined meaning.

The main benefit of *formal* ontology languages such as OWL is that it is possible to answer questions by combining several sources of information. For example, if an ontology knows that 'Albert Einstein was a physicist who was born in Ulm,' and that 'Ulm is a city in Germany,' the ontology can answer questions like 'Which physicists were born in Germany?' by returning 'Albert Einstein' as one of the answers. That is, an answer to a question may be obtained not only from the *explicitly stated* information, but also from the (*implicit*) consequences.

This course is concerned with logic-based languages called *Description Logics* (short: DLs) that provide the formal foundation of the ontology language OWL. DLs are not just one language but a whole *family* of languages, designed to offer a great variety of choices for knowledge-based applications. Each language defines its own set of *constructors* that can be used to build the ontological formulas, called *axioms.* Each axiom describes a particular aspect of the real world; for example an axiom saying that 'Ulm is a city in Germany' describes a relation between the entities 'Ulm' and 'Germany'. The goal of the ontology is to provide axioms that describe (a part of) the real world as accurately as needed for an application. Since the real world is extremely complex, each of these descriptions are necessarily *incomplete*, which means that they can be also *satisfied* in situations that are different from the real world. The semantics of DL defines an abstract notion of a *model* to represent such situations. If an axiom holds in all such models, it is said to be a *logical consequence* of the ontology.

Of course, ontologies cannot answer questions on their own; they require special programs that can analyze and combine axioms to obtain the answers, called *ontology reasoners.* Ontology reasoners are usually able to solve several types of *reasoning problems*, such as checking if there are logical contradictions

---

[3] https://www.wikidata.org.

in the ontology or finding logical consequences of a certain form. To be practically useful, the *reasoning algorithms* implemented in ontology reasoners need to possess certain formal properties. An algorithm is *sound* if any answer that it returns is correct. If the algorithm returns every correct answer, it is *complete*. An algorithm might not return any answer if it does not *terminate*. If an algorithm always terminates, it is also useful to know how long one needs to wait for the answer. This can be measured using a notion of *algorithmic complexity*. Generally, algorithms with lower complexity should be preferred.

Modern ontology reasoners use very sophisticated and highly optimized algorithms for obtaining reasoning results, and often such results are counterintuitive or hard to understand for humans. Reasoning results may also be incorrect, which indicates that some axioms in the ontology must have errors. In order to improve the user experience when working with ontologies, and, in particular, facilitate ontology *debugging*, ontology development tools include capabilities for *explaining* reasoning results.

Almost every year since its inception, the Reasoning Web Summer School offered lectures focused on different topics of reasoning in DLs ranging from overview courses [2, 49, 50, 65] to more specialized topics such as lightweight DLs [64], query answering [10, 36, 43], and non-standard reasoning problems [9, 12, 59]. The purpose of this course is to provide a deeper understanding of the key reasoning and explanation algorithms used in DLs. We provide a detailed account on *tableau procedures*, which are the most popular reasoning procedures for DLs, including questions such as soundness, completeness, termination, and complexity. For explaining the reasoning results we look into general-purpose *axiom pinpointing procedures* that can efficiently identify some or all subsets of axioms responsible for a reasoning result. Some of these procedures can be also used to *repair* unintended entailments by identifying possible subsets of axioms whose removal breaks the entailment.

This paper is (partly) based on the course "Algorithms for Knowledge Representation" given at the University of Ulm, and includes full proofs, detailed examples, and simple exercises. The material should be accessible to students of the general university (bachelor) level in technical subjects such as, but not limited to, computer science. All relevant background, such as the basics of the computational complexity theory is introduced as needed. The course should be of particular interest to those who are interested in developing and implementing (DL) reasoning procedures. Since the duration of this course is limited to two lectures, we mainly focus on the *basic description logic* $\mathcal{ALC}$, to nevertheless provide a detailed account on the topics of DL reasoning and explanation. For this reason, this course does not provide a comprehensive literature survey. For the latter, we refer the reader to previous overview courses [2, 49, 50, 65], DL textbooks [4, 6], PhD theses [26, 60] and some recent overviews [45].

2   Description Logics

Description Logics (DLs) are specialized logic-based languages designed to represent conceptual knowledge in a machine-readable form so that this information can be processed by automated tools. Most DLs correspond to decidable fragments of first-order logic, which is a very expressive general-purpose language, however, with undecidable standard reasoning problems. Decidability has been one of the key requirements for the development of DL languages; to achieve decidability, the languages are often restricted to contain the features most essential for knowledge representation. For example, in natural language, one rarely speaks about more than two objects at a time. For this reason, DLs usually restrict the syntax to only unary and binary relations and to constants. Unary relations usually specify types (or classes) of objects and are called *concepts* in DLs (and *classes* in OWL). Binary relations specify how objects are related to each other, and are called *roles* in DLs (and *properties* in OWL). Constants refer to particular objects by their names. In DLs (and OWL) they are called *individuals*. In this paper, we mainly focus on the *basic description logic $\mathcal{ALC}$* [51], which is regarded by some as the smallest sensible language for knowledge representation. This language traditionally serves not only as the basis of more expressive languages, but also as a relatively simple example on which the main ideas about reasoning in DLs can be explained.

2.1 Syntax

The *vocabulary* of the DL $\mathcal{ALC}$ consists of countably-infinite sets N_C of *concept names* (or *atomic concepts*), N_R of *role names* (or *atomic roles*), N_I of *individual names* (or *individuals*), *logical symbols*: $\top$, $\bot$, $\neg$, $\sqcap$, $\sqcup$, $\forall$, $\exists$, $\sqsubseteq$, $\equiv$, and *structural symbols*: '(', ')', '.'. These symbols are used to construct formulas that are called DL *axioms*.

Intuitively, (atomic) concepts are used to describe sets of objects. For example, we may introduce the following atomic concepts representing the respective sets of objects:

Human — the set of all human beings,
Male — the set of all male (not necessarily human) beings,
Country — the set of all countries.

Likewise, (atomic) roles represent binary relations between objects:

hasChild — the parent-child relation between objects,
hasLocation — a relation between objects and their (physical) locations.

Individuals represent some concrete object, for example:

germany — the country of Germany,
john — the person John.

In our examples, we usually use a *convention* for writing (atomic) concepts, roles, and individuals so that one can easily tell them apart: concepts are written starting with capital letters, while role and individual names start with a lower case letter. In addition, we reserve single letters (possibly with decorations) A, B for atomic concepts, r and s for (atomic) roles, a, b, c for individuals, and C, D, E for complex concepts, which are introduced next.

The logical symbols $\top, \bot, \neg, \sqcap, \sqcup, \forall$, and $\exists$ are used for constructing *complex concepts* (or just *concepts*). Just like for atomic concepts, complex concepts represent sets of objects, but these sets are uniquely determined by the *subconcepts* from which they are constructed. The set of $\mathcal{ALC}$ *concepts* can be defined using the grammar definition:

$$C, D ::= A \mid \top \mid \bot \mid C \sqcap D \mid C \sqcup D \mid \neg C \mid \exists r.C \mid \forall r.C. \tag{1}$$

This definition means that the set of concepts (which are named by C and D) is recursively constructed starting from atomic concepts A, *top concept* $\top$, *bottom concept* $\bot$, by applying *conjunction* $C \sqcap D$, *disjunction* $C \sqcup D$, *negation* $\neg C$, *existential restriction* $\exists r.C$, and *universal restriction* $\forall r.C$. Intuitively, $\top$ represents the set of all objects of the modeled domain, $\bot$ the empty set of objects, $C \sqcap D$ the set of common objects of C and D, $C \sqcup D$ the union of objects in C and D, $\neg C$ all objects that are not in C, $\exists r.C$ all object that are related by r to *some* object in C, $\forall r.C$ all objects that are related by r to *only* objects in C. For example, one can construct the following $\mathcal{ALC}$-concepts:

Male $\sqcap$ Human	– the set of all male humans,
Dead $\sqcup$ Alive	– the union of all dead and all alive things,
$\neg$Male	– the set of all non-male things,
($\neg$Male) $\sqcap$ Human	– the set of all non-male humans,
$\exists$hasChild.Male	– all things that have a male child,
$\forall$hasChild.Female	– all things that have only female children,
Male $\sqcap$ ($\forall$hasChild.$\neg$Male)	– all male things all of whose children are not male.

Once complex concepts are constructed, they can be used to describe various properties by writing *axioms*. In the DL $\mathcal{ALC}$ we consider four possible types of axioms: a *concept inclusion* $C \sqsubseteq D$ states that every object of the concept C must be an object of the concept D, a *concept equivalence* $C \equiv D$ states that the concepts C and D must contain exactly the same objects, a *concept assertion* $C(a)$ states that the object represented by the individual a is an object of the concept C, and a *role assertion* $r(a, b)$ states that the objects represented by the individuals a and b are connected by the relation represented by the role r. Here are some examples of these axioms:

Human $\sqsubseteq$ Dead $\sqcup$ Alive	– every human is either dead or alive,
Parent $\equiv$ $\exists$hasChild.$\top$	– parents are exactly those that have some child,

Male(john)   – John is a male,
bornIn(einstein, ulm)   – Albert Einstein was born in Ulm.

Axioms are usually grouped together to form *knowledge bases* (or *ontologies*). An $\mathcal{ALC}$ ontology $\mathcal{O}$ is simply a (possibly empty) set of $\mathcal{ALC}$ axioms. The axioms of an ontology are usually split into two parts: the *terminological part* (short: *TBox*) contains only concept inclusion and concept equivalence axioms, the *assertional part* (short: *ABox*) contains only concept and role assertion axioms. This distinction is often used to simplify the analysis of algorithms. For example, to answer questions about concepts, in many cases it is not necessary to consider the ABox, which is usually the larger part of an ontology.

Example 1. Consider the ontology $\mathcal{O}$ consisting of the following axioms:

1.   Parent $\equiv \exists$hasChild.$\top$,
2.   GrandParent $\equiv \exists$hasChild.Parent,
3.   hasChild(john, mary).

Then the TBox of $\mathcal{O}$ consists of the first two axioms, and the ABox of $\mathcal{O}$ consists of the last axiom.

The main application of ontologies is to extract new information from the information explicitly stated in the ontologies. For example, from the first two axioms of the ontology $\mathcal{O}$ from Example 1 it follows that each grandparent must be a parent because each grandparent has a child (who happens to be a parent). This new information can be formalized using a concept inclusion axiom GrandParent $\sqsubseteq$ Parent. Likewise, from the first and the last axiom of $\mathcal{O}$ one can conclude that the object represented by the individual john must be a parent because he has a child (mary). This piece of information can be formalized using a concept assertion axiom Parent(john). The two new axioms are said to be *logical consequences* of the ontology $\mathcal{O}$.

2.2   Semantics

To be able to calculate (preferably automatically) which axioms are logical consequences of ontologies and which are not, we need to define the *semantics* of ontologies. So far we have defined the *syntax* of ontologies, which describes how axioms in the ontologies can be constructed from various symbols. This information is not enough to understand the *meaning* of concepts and axioms. In fact, in $\mathcal{ALC}$ the same information can be described in many different ways. For example, the concept Male $\sqcap$ Human describes exactly the same set of objects as the concept Human $\sqcap$ Male. The axiom Human $\sqsubseteq$ Dead $\sqcup$ Alive describes exactly the same situation as the axiom Human $\sqcap (\neg$Dead$) \sqsubseteq$ Alive. The formal semantics describes how to determine the meaning of concepts and axioms, while abstracting from the particular syntactic ways in which they are written down.

Like in many other logic-based formalisms (including propositional and first-order logic), the semantics of description logics is defined using (Tarski-style set-theoretic) interpretations. Intuitively, an interpretation describes a possible state of the world modeled by the ontology. Formally, an *interpretation* is a pair $\mathcal{I} = (\Delta^{\mathcal{I}}, \cdot^{\mathcal{I}})$ where $\Delta^{\mathcal{I}}$ is a non-empty set called the *domain* of $\mathcal{I}$ and $\cdot^{\mathcal{I}}$ is an *interpretation function* that assigns to each atomic concept $A \in N_C$ a set $A^{\mathcal{I}} \subseteq \Delta^{\mathcal{I}}$, to each atomic role $r \in N_R$ a binary relation $r^{\mathcal{I}} \subseteq \Delta^{\mathcal{I}} \times \Delta^{\mathcal{I}}$, and to each individual $a \in N_I$ an element $a^{\mathcal{I}} \in \Delta^{\mathcal{I}}$. Intuitively, the domain $\Delta^{\mathcal{I}}$ represents the objects that can be part of the modeled world; this can be an infinite (and even an uncountable) set, but it must contain at least one element because otherwise it is not possible to assign $a^{\mathcal{I}} \in \Delta^{\mathcal{I}}$ for $a \in N_I$. Although the interpretation function requires an assignment for every symbol of the vocabulary (and there are infinitely many available symbols in N_C, N_R, and N_I), when defining interpretations for ontologies, we usually provide the values only for the symbols present in the ontology, assuming that all other symbols are interpreted in an arbitrary way.

Example 2. We can define an interpretation $\mathcal{I} = (\Delta^{\mathcal{I}}, \cdot^{\mathcal{I}})$ of the symbols appearing in the ontology $\mathcal{O}$ from Example 1, for example, as follows:

- $\Delta^{\mathcal{I}} = \{a, b, c\}$,
- $\mathsf{Parent}^{\mathcal{I}} = \{a, b\}$, $\mathsf{GrandParent}^{\mathcal{I}} = \{a\}$,
- $\mathsf{hasChild}^{\mathcal{I}} = \{\langle a, b \rangle, \langle b, c \rangle\}$,
- $\mathsf{john}^{\mathcal{I}} = a$, $\mathsf{mary}^{\mathcal{I}} = b$.

Once the interpretation is fixed, it can be *recursively extended* to complex $\mathcal{ALC}$ concepts according to the following rules that match the respective cases of the grammar definition (1). Assuming that the values of $C^{\mathcal{I}} \subseteq \Delta^{\mathcal{I}}$ and $D^{\mathcal{I}} \subseteq \Delta^{\mathcal{I}}$ for concepts C and D have already been determined, the interpretations of concepts build from C and D can be computed as follows:

- $\top^{\mathcal{I}} = \Delta^{\mathcal{I}}$,
- $\bot^{\mathcal{I}} = \emptyset$,
- $(C \sqcap D)^{\mathcal{I}} = C^{\mathcal{I}} \cap D^{\mathcal{I}}$,
- $(C \sqcup D)^{\mathcal{I}} = C^{\mathcal{I}} \cup D^{\mathcal{I}}$,
- $(\neg C)^{\mathcal{I}} = \Delta^{\mathcal{I}} \setminus C^{\mathcal{I}}$,
- $(\exists r.C)^{\mathcal{I}} = \{x \in \Delta^{\mathcal{I}} \mid \exists y : \langle x, y \rangle \in r^{\mathcal{I}} \ \& \ y \in C^{\mathcal{I}}\}$,
- $(\forall r.C)^{\mathcal{I}} = \{x \in \Delta^{\mathcal{I}} \mid \forall y : \langle x, y \rangle \in r^{\mathcal{I}} \Rightarrow y \in C^{\mathcal{I}}\}$.

The last two cases of this definition probably require some further clarifications. The interpretation of $\exists r.C$ contains exactly those elements $x \in \Delta^{\mathcal{I}}$ that are connected to *some* element y by a binary relation $r^{\mathcal{I}}$ (i.e., $\langle x, y \rangle \in r^{\mathcal{I}}$) such that $y \in C^{\mathcal{I}}$. The interpretation of $\forall r.C$ contains exactly those elements $x \in \Delta^{\mathcal{I}}$ such that *every* element y connected from x by $r^{\mathcal{I}}$ (i.e., for which $\langle x, y \rangle \in r^{\mathcal{I}}$ holds), is a member of $C^{\mathcal{I}}$ (i.e., $y \in C^{\mathcal{I}}$). In other words, x is $r^{\mathcal{I}}$-connected to *only* elements of $C^{\mathcal{I}}$. Importantly, if an element x does not have any $r^{\mathcal{I}}$-successor, i.e., $\langle x, y \rangle \in r^{\mathcal{I}}$ holds for no $y \in \Delta^{\mathcal{I}}$, then $x \notin (\exists r.C)^{\mathcal{I}}$ but $x \in (\forall r.C)^{\mathcal{I}}$ for every concept C.

Example 3. Consider the interpretation $\mathcal{I} = (\Delta^{\mathcal{I}}, \cdot^{\mathcal{I}})$ from Example 2. Then:

- $\top^{\mathcal{I}} = \{a, b, c\}$,
- $(\mathsf{Parent} \sqcap \mathsf{GrandParent})^{\mathcal{I}} = \{a\}$,
- $(\mathsf{Parent} \sqcup \mathsf{GrandParent})^{\mathcal{I}} = \{a, b\}$,
- $(\neg\mathsf{GrandParent})^{\mathcal{I}} = \{b, c\}$,
- $(\mathsf{Parent} \sqcap \neg\mathsf{GrandParent})^{\mathcal{I}} = \{b\}$,

- $(\exists\mathsf{hasChild}.\top)^{\mathcal{I}} = \{a, b\}$,
- $(\exists\mathsf{hasChild}.\mathsf{Parent})^{\mathcal{I}} = \{a\}$,
- $(\forall\mathsf{hasChild}.\mathsf{Parent})^{\mathcal{I}} = \{a, c\}$,   (!!!)
- $(\forall\mathsf{hasChild}.\mathsf{GrandParent})^{\mathcal{I}} = \{c\}$, (!!!)
- $(\forall\mathsf{hasChild}.\forall\mathsf{hasChild}.\bot)^{\mathcal{I}} = \{b, c\}$.

We next define how to interpret $\mathcal{ALC}$ axioms. The purpose of axioms in an ontology is to describe the characteristics of concepts, roles, and individuals involved in these axioms. These properties hold in some interpretations and are violated in other interpretations. For an interpretation $\mathcal{I}$ and an axiom α we write $\mathcal{I} \models \alpha$ if α *holds* (or *is satisfied*) in $\mathcal{I}$, defined as follows:

- $\mathcal{I} \models C \sqsubseteq D$ if and only if $C^{\mathcal{I}} \subseteq D^{\mathcal{I}}$,
- $\mathcal{I} \models C \equiv D$ if and only if $C^{\mathcal{I}} = D^{\mathcal{I}}$,
- $\mathcal{I} \models C(a)$ if and only if $a^{\mathcal{I}} \in C^{\mathcal{I}}$,
- $\mathcal{I} \models r(a, b)$ if and only if $\langle a^{\mathcal{I}}, b^{\mathcal{I}}\rangle \in r^{\mathcal{I}}$.

If it is not the case that $\mathcal{I} \models \alpha$, we write $\mathcal{I} \not\models \alpha$ and say that α is *violated* (or *not satisfied*) in $\mathcal{I}$. Table 1 summarizes the syntax and semantics of $\mathcal{ALC}$.

Table 1. The summary of syntax and semantics of the DL $\mathcal{ALC}$

	Syntax	Semantics	
Roles:			
atomic role	r	$r^{\mathcal{I}} \subseteq \Delta^{\mathcal{I}} \times \Delta^{\mathcal{I}}$	(given)
Concepts:			
atomic concept	A	$A^{\mathcal{I}} \subseteq \Delta^{\mathcal{I}}$	(given)
top	$\top$	$\Delta^{\mathcal{I}}$	
bottom	$\bot$	$\emptyset$	
conjunction	$C \sqcap D$	$C^{\mathcal{I}} \cap D^{\mathcal{I}}$	
disjunction	$C \sqcup D$	$C^{\mathcal{I}} \cup D^{\mathcal{I}}$	
negation	$\neg C$	$\Delta^{\mathcal{I}} \setminus C^{\mathcal{I}}$	
existential restriction	$\exists r.C$	$\{x \mid \exists y : \langle x, y\rangle \in r^{\mathcal{I}} \ \& \ y \in C^{\mathcal{I}}\}$	
universal restriction	$\forall r.C$	$\{x \mid \forall y : \langle x, y\rangle \in r^{\mathcal{I}} \Rightarrow y \in C^{\mathcal{I}}\}$	
Individuals:			
individual	a	$a^{\mathcal{I}} \in \Delta^{\mathcal{I}}$	(given)
Axioms:			
concept inclusion	$C \sqsubseteq D$	$C^{\mathcal{I}} \subseteq D^{\mathcal{I}}$	
concept equivalence	$C \equiv D$	$C^{\mathcal{I}} = D^{\mathcal{I}}$	
concept assertion	$C(a)$	$a^{\mathcal{I}} \in C^{\mathcal{I}}$	
role assertion	$r(a, b)$	$\langle a^{\mathcal{I}}, b^{\mathcal{I}}\rangle \in r^{\mathcal{I}}$	

Example 4. Continuing Example 3, we can determine the interpretation of the following axioms in the defined $\mathcal{I}$:

- $\mathcal{I} \models \mathsf{GrandParent} \sqsubseteq \mathsf{Parent}$: $\mathsf{GrandParent}^{\mathcal{I}} = \{a\} \subseteq \mathsf{Parent}^{\mathcal{I}} = \{a, b\}$,
- $\mathcal{I} \not\models \mathsf{Parent} \sqsubseteq \mathsf{GrandParent}$: $\mathsf{Parent}^{\mathcal{I}} = \{a, b\} \not\subseteq \mathsf{GrandParent}^{\mathcal{I}} = \{a\}$,
- $\mathcal{I} \models \exists\mathsf{hasChild}.\mathsf{GrandParent} \equiv \bot : (\exists\mathsf{hasChild}.\mathsf{GrandParent})^{\mathcal{I}} = \emptyset = \bot^{\mathcal{I}}$
- $\mathcal{I} \models (\exists\mathsf{hasChild}.\mathsf{Parent})(\mathsf{john})$: $\mathsf{john}^{\mathcal{I}} = a \in (\exists\mathsf{hasChild}.\mathsf{Parent})^{\mathcal{I}} = \{a\}$,
- $\mathcal{I} \not\models \mathsf{hasChild}(\mathsf{mary}, \mathsf{john})$: $\langle \mathsf{mary}^{\mathcal{I}}, \mathsf{john}^{\mathcal{I}} \rangle = \langle b, a \rangle \notin \mathsf{hasChild}^{\mathcal{I}} = \{\langle a, b \rangle\}$,
- $\mathcal{I} \models (\forall\mathsf{hasChild}.\neg\mathsf{Parent})(\mathsf{mary})$: $\mathsf{mary}^{\mathcal{I}} = b \in (\forall\mathsf{hasChild}.\neg\mathsf{Parent})^{\mathcal{I}} = \{b, c\}$.

As mentioned earlier, an axiom may hold in one interpretation, but may be violated in another interpretation. For example, the axiom $A \sqsubseteq B$ holds in $\mathcal{I} = (\Delta^{\mathcal{I}}, \cdot^{\mathcal{I}})$ with $\Delta^{\mathcal{I}} = \{a\}$, $A^{\mathcal{I}} = B^{\mathcal{I}} = \emptyset$, but is violated in $\mathcal{J} = (\Delta^{\mathcal{J}}, \cdot^{\mathcal{J}})$ with $\Delta^{\mathcal{J}} = \{a\}$, $A^{\mathcal{J}} = \{a\}$, and $B^{\mathcal{J}} = \emptyset$. There are, however, axioms that hold in *every* interpretation. We call such axioms *tautologies*.

Example 5. The following $\mathcal{ALC}$ axioms are tautologies because they hold in every interpretation $\mathcal{I} = (\Delta^{\mathcal{I}}, \cdot^{\mathcal{I}})$:

- $C \sqsubseteq C$:   because $C^{\mathcal{I}} \subseteq C^{\mathcal{I}}$,
- $C \sqsubseteq \top$:   because $C^{\mathcal{I}} \subseteq \Delta^{\mathcal{I}} = \top^{\mathcal{I}}$,
- $C \sqcap D \sqsubseteq C$:   because $(C \sqcap D)^{\mathcal{I}} = C^{\mathcal{I}} \cap D^{\mathcal{I}} \subseteq C^{\mathcal{I}}$,
- $\forall r.\top \equiv \top$:   because

$$(\forall r.\top)^{\mathcal{I}} = \{x \in \Delta^{\mathcal{I}} \mid \forall y : \langle x, y \rangle \in r^{\mathcal{I}} \Rightarrow y \in \top^{\mathcal{I}} = \Delta^{\mathcal{I}}\} = \Delta^{\mathcal{I}} = \top^{\mathcal{I}}.$$

- $\exists r.C \sqcap \forall r.D \sqsubseteq \exists r.(C \sqcap D)$:   because

$$
\begin{aligned}
(\exists r.C \sqcap \forall r.D)^{\mathcal{I}} &= (\exists r.C)^{\mathcal{I}} \cap (\forall r.D)^{\mathcal{I}} \\
&= \{x \in \Delta^{\mathcal{I}} \mid \exists y : \langle x, y \rangle \in r^{\mathcal{I}} \ \& \ y \in C^{\mathcal{I}}\} \cap \\
&\quad \{x \in \Delta^{\mathcal{I}} \mid \forall y : \langle x, y \rangle \in r^{\mathcal{I}} \Rightarrow y \in D^{\mathcal{I}}\} \\
&\subseteq \{x \in \Delta^{\mathcal{I}} \mid \exists y : \langle x, y \rangle \in r^{\mathcal{I}} \ \& \ y \in C^{\mathcal{I}} \ \& \ y \in D^{\mathcal{I}}\} \\
&= \{x \in \Delta^{\mathcal{I}} \mid \exists y : \langle x, y \rangle \in r^{\mathcal{I}} \ \& \ y \in C^{\mathcal{I}} \cap D^{\mathcal{I}} = (C \sqcap D)^{\mathcal{I}}\} \\
&= (\exists r.(C \sqcap D))^{\mathcal{I}}.
\end{aligned}
$$

The following $\mathcal{ALC}$ axioms are not tautologies as they do not hold in at least one interpretation $\mathcal{I} = (\Delta^{\mathcal{I}}, \cdot^{\mathcal{I}})$:

- $C \sqsubseteq C \sqcap D$:   Take $\Delta^{\mathcal{I}} = \{a\}$, $C^{\mathcal{I}} = \{a\}$, and $D^{\mathcal{I}} = \emptyset$. Then $C^{\mathcal{I}} = \{a\} \not\subseteq \emptyset = \{a\} \cap \emptyset = C^{\mathcal{I}} \cap D^{\mathcal{I}} = (C \sqcap D)^{\mathcal{I}}$.
- $\forall r.C \sqsubseteq \exists r.C$:   Take $\Delta^{\mathcal{I}} = \{a\}$ and $C^{\mathcal{I}} = r^{\mathcal{I}} = \emptyset$. Then $(\forall r.C)^{\mathcal{I}} = \{a\} \not\subseteq \emptyset = (\exists r.C)^{\mathcal{I}}$.

– $\exists r.C \sqcap \exists r.D \sqsubseteq \exists r.(C \sqcap D)$: Take $\Delta^{\mathcal{I}} = \{a, c, d\}$, $C^{\mathcal{I}} = \{c\}$, $D^{\mathcal{I}} = \{d\}$, and $r^{\mathcal{I}} = \{\langle a, c\rangle, \langle a, d\rangle\}$. Then $a \in (\exists r.C)^{\mathcal{I}}$ because $\langle a, c\rangle \in r^{\mathcal{I}}$ and $c \in C^{\mathcal{I}}$. Similarly, $a \in (\exists r.D)^{\mathcal{I}}$ since $\langle a, d\rangle \in r^{\mathcal{I}}$ and $d \in D^{\mathcal{I}}$. But $(C \sqcap D)^{\mathcal{I}} = C^{\mathcal{I}} \cap D^{\mathcal{I}} = \{c\} \cap \{d\} = \emptyset$. Thus, $(\exists r.(C \sqcap D))^{\mathcal{I}} = \emptyset$. Hence, $(\exists r.C \sqcap \exists r.D)^{\mathcal{I}} = (\exists r.C)^{\mathcal{I}} \cap (\exists r.D)^{\mathcal{I}} = \{a\} \cap \{a\} = \{a\} \not\sqsubseteq \emptyset = (\exists r.(C \sqcap D))^{\mathcal{I}}$.

Interpretations that satisfy the axioms in an ontology will be of special interest to us, because these interpretations agree with the requirements imposed by the axioms. These interpretations are called models. Formally, an interpretation $\mathcal{I}$ is a *model* of an ontology $\mathcal{O}$ (in symbols: $\mathcal{I} \models \mathcal{O}$) if $\mathcal{I} \models \alpha$ for every $\alpha \in \mathcal{O}$. We say that $\mathcal{O}$ is *satisfiable* if $\mathcal{O}$ has at least one model, i.e., if $\mathcal{I} \models \mathcal{O}$ holds for at least one interpretation $\mathcal{I}$. Otherwise, we say that $\mathcal{O}$ is *unsatisfiable*.

Example 6. Consider the ontology $\mathcal{O}$ containing the first two axioms from Example 1:

1. $\qquad\qquad\qquad$ Parent $\equiv \exists$hasChild.$\top$,
2. $\qquad\qquad\qquad$ GrandParent $\equiv \exists$hasChild.Parent.

We can prove that $\mathcal{O}$ is satisfiable by presenting a simple model $\mathcal{I} = (\Delta^{\mathcal{I}}, \cdot^{\mathcal{I}})$ of $\mathcal{O}$:

– $\Delta^{\mathcal{I}} = \{a\}$,
– Parent$^{\mathcal{I}}$ = GrandParent$^{\mathcal{I}}$ = hasChild$^{\mathcal{I}}$ = $\emptyset$,
– john$^{\mathcal{I}}$ = mary$^{\mathcal{I}}$ = a.

Note that $(\exists$hasChild.$\top)^{\mathcal{I}} = \emptyset$ and $(\exists$hasChild.Parent$)^{\mathcal{I}} = \emptyset$ since hasChild$^{\mathcal{I}} = \emptyset$. Thus, Parent$^{\mathcal{I}} = (\exists$hasChild.$\top)^{\mathcal{I}}$ and GrandParent$^{\mathcal{I}} = (\exists$hasChild.Parent$)^{\mathcal{I}}$, which implies that $\mathcal{I}$ satisfies both axioms in $\mathcal{O}$.

Let us now extend $\mathcal{O}$ with the third axiom from Example 1:

3. $\qquad\qquad\qquad$ hasChild(john, mary).

The previous interpretation $\mathcal{I}$ is no longer a model of $\mathcal{O}$ since $\langle$john$^{\mathcal{I}}$, mary$^{\mathcal{I}}\rangle = \langle a, a\rangle \notin \emptyset = $ hasChild$^{\mathcal{I}}$. This does not, however, mean that the ontology $\mathcal{O}$ is unsatisfiable since we can find another interpretation $\mathcal{J} = (\Delta^{\mathcal{J}}, \cdot^{\mathcal{J}})$ that satisfies all three axioms:

– $\Delta^{\mathcal{J}} = \{a\}$,
– Parent$^{\mathcal{J}}$ = GrandParent$^{\mathcal{J}}$ = $\{a\}$, hasChild$^{\mathcal{J}}$ = $\{\langle a, a\rangle\}$,
– john$^{\mathcal{J}}$ = mary$^{\mathcal{J}}$ = a.

It is easy to verify that $(\exists$hasChild.$\top)^{\mathcal{J}} = (\exists$hasChild.Parent$)^{\mathcal{J}} = \{a\}$, which proves that $\mathcal{J}$ still satisfies the first two axioms. Since $\langle$john$^{\mathcal{J}}$, mary$^{\mathcal{J}}\rangle = \langle a, a\rangle \in \{\langle a, a\rangle\} = $ hasChild$^{\mathcal{J}}$, $\mathcal{J}$ now also satisfies the third axiom.

Besides the notion of satisfiability of ontologies, in description logics one also considers a notion of satisfiability of concepts. We say that a *concept C is satisfiable* if there exists an interpretation $\mathcal{I}$ such that $C^{\mathcal{I}} \neq \emptyset$. For example, the concept $\forall r.\bot$ is satisfiable because $(\forall r.\bot)^{\mathcal{I}} = \Delta^{\mathcal{I}} \neq \emptyset$ for every interpretation $\mathcal{I}$ such that $r^{\mathcal{I}} = \emptyset$ (and there is certainly at least one such interpretation). On the other hand, the concept $\exists r.\bot$ is not satisfiable since $\bot^{\mathcal{I}} = \emptyset$ and, consequently, $(\exists r.\bot)^{\mathcal{I}} = \emptyset$ for every $\mathcal{I}$.

Sometimes the interpretation that should satisfy the concept is constrained to be a model of a given ontology. We say that a *concept C is satisfiable with respect to an ontology* $\mathcal{O}$ if $C^{\mathcal{I}} \neq \emptyset$ for some $\mathcal{I}$ such that $\mathcal{I} \models \mathcal{O}$. Note that if the ontology $\mathcal{O}$ is not satisfiable, then no concept C is satisfiable with respect to this ontology.

Example 7. Let $\mathcal{O} = \{A \sqsubseteq \neg A\}$. Note that $\mathcal{O}$ is satisfiable in any interpretation $\mathcal{I}$ such that $A^{\mathcal{I}} = \emptyset$ since $A^{\mathcal{I}} = \emptyset \subseteq (\neg A)^{\mathcal{I}} = \Delta^{\mathcal{I}} \setminus \emptyset = \Delta^{\mathcal{I}}$. However, the concept A is not satisfiable w.r.t. $\mathcal{O}$. Indeed, assume that $A^{\mathcal{I}} \neq \emptyset$ for some $\mathcal{I}$. Then $a \in A^{\mathcal{I}}$ for some domain element $a \in \Delta^{\mathcal{I}}$. But then $a \notin \Delta^{\mathcal{I}} \setminus A^{\mathcal{I}} = (\neg A)^{\mathcal{I}}$. Hence, $A^{\mathcal{I}} \not\subseteq (\neg A)^{\mathcal{I}}$. Thus, $\mathcal{I} \not\models A \sqsubseteq \neg A$. Hence, for each $\mathcal{I}$ such that $A^{\mathcal{I}} \neq \emptyset$, we have $\mathcal{I} \not\models \mathcal{O}$.

We are finally ready to formally define the notion of logical entailment. We say that $\mathcal{O}$ *entails* an axiom α (written $\mathcal{O} \models \alpha$), if every model of $\mathcal{O}$ satisfies α. Intuitively this means that the axiom α should hold in every situation that agrees with the restrictions imposed by $\mathcal{O}$. Note that according to this definition, if $\mathcal{O}$ is unsatisfiable then the entailment $\mathcal{O} \models \alpha$ holds for every axiom α.

Example 8. Consider $\mathcal{O} = \{C \sqsubseteq \exists r.D, D \sqsubseteq E\}$. We prove that $\mathcal{O} \models C \sqsubseteq \exists r.E$. Take any $\mathcal{I}$ such that $\mathcal{I} \models \mathcal{O}$. We show that $\mathcal{I} \models C \sqsubseteq \exists r.E$ or, equivalently, $C^{\mathcal{I}} \subseteq (\exists r.E)^{\mathcal{I}}$. Suppose that $x \in C^{\mathcal{I}}$. Since $\mathcal{I} \models C \sqsubseteq \exists r.D$, we have $C^{\mathcal{I}} \subseteq (\exists r.D)^{\mathcal{I}}$, so $x \in (\exists r.D)^{\mathcal{I}}$. Then there exists some $y \in \Delta^{\mathcal{I}}$ such that $\langle x, y \rangle \in r^{\mathcal{I}}$ and $y \in D^{\mathcal{I}}$. Since $\mathcal{I} \models D \sqsubseteq E$, we have $y \in D^{\mathcal{I}} \subseteq E^{\mathcal{I}}$. Therefore, since $\langle x, y \rangle \in r^{\mathcal{I}}$ and $y \in E^{\mathcal{I}}$ we obtain $x \in (\exists r.E)^{\mathcal{I}}$. Thus, for each $x \in C^{\mathcal{I}}$, we have $x \in (\exists r.E)^{\mathcal{I}}$, which means that $C^{\mathcal{I}} \subseteq (\exists r.E)^{\mathcal{I}}$ or, equivalently, $\mathcal{I} \models C \sqsubseteq \exists r.E$.

2.3 Reasoning Problems

There are many situations in which one is interested in performing logical operations with ontologies, such as checking consistency or verifying entailments. Just like computer software, ontologies are usually developed manually by humans, and humans tend to make mistakes. For programming languages, dedicated software tools, such as syntax checkers, compilers, debuggers, testing frameworks, and static analysis tools help preventing and finding errors. Similar tools also exist for ontology development.

Just like for programming languages, one usually distinguishes several types of errors in ontologies. *Syntax errors* usually happen when the syntax rules for constructing concepts and axioms described in Sect. 2.1 are not used correctly. For example $C\neg D$ is not a correct concept according to grammar (1) since the negation operation is unary. Syntax errors also include situations where an atomic role is used in the position of an atomic concept or when the parentheses are not balanced. Syntax errors are relatively easy to find using *parsers*, which can verify that the ontology is well-formed.

It can be that the ontology is syntactically well-formed, but some of its axioms do not make sense. For example, an ontology may contain the axiom Father $\equiv$ Male $\sqcup$ Parent, in which, clearly a disjunction was accidentally used instead of a conjunction. Although for a human the problem seems obvious, it is hard to detect such an error using automated tools since computers do not know the meaning of the words involved. From the computer point of view, this axiom looks like $C \equiv D \sqcup E$, which is a legitimate axiom. These kinds of errors are usually called *semantic* or *modeling errors*.

Although it is not possible to automatically detect modeling errors in general, there are some common *symptoms* for such errors. For example, an incorrectly formulated axiom may cause a logical contradiction with other axioms in the ontology, which makes the whole ontology unsatisfiable. Another common symptom is unsatisfiability of *atomic* concepts with respect to an ontology. Each atomic concept is usually introduced to capture a certain non-empty subset of objects in the modeled domain. For example, the concept Parent was introduced to capture the individuals who are parents in the real world. If an atomic concept is unsatisfiable, this indicates that the modeled domain cannot correspond to any model of the ontology. Note that, as shown in Example 7, an atomic concept can be unsatisfiable even with respect to a satisfiable ontology.

A modeling error may also result in incorrect entailments of the ontology, which are sometimes easier to detect than the error itself. For example, the erroneous axiom Father $\equiv$ Male $\sqcup$ Parent entails the simpler concept inclusions Male $\sqsubseteq$ Father and Parent $\sqsubseteq$ Father, which are also incorrect from the modeling point of view. By observing the entailed concept inclusions $A \sqsubseteq B$ between atomic concepts appearing in the ontology, an ontology developer can usually quickly identify those incorrect entailments. When the entailment $\mathcal{O} \models C \sqsubseteq D$ holds, it is often said that the concept C is *subsumed by* the concept D (or the concept D *subsumes* the concept C) w.r.t. $\mathcal{O}$. For detecting problems involving individuals, one can similarly inspect the entailed concept assertions $A(a)$ between atomic concepts A and individuals a appearing in the ontology. When $\mathcal{O} \models C(a)$, it is often said that a is an *instance* of C (or C is a *type* of a) w.r.t. $\mathcal{O}$. Checking subsumptions and instances is not only useful for finding modeling errors, but also for answering *queries*, which is usually the main purpose of ontologies in applications. For example, given a (complex) concept C, it is possible to query for all atomic concepts A for which the subsumption $\mathcal{O} \models C \sqsubseteq A$ holds or to query for all individuals a which are instances of C.

Thus, one can distinguish several *standard reasoning problems* that are of interest in ontology-based applications:

1. *Ontology satisfiability checking*:
   - Given: an ontology $\mathcal{O}$,
   - Return: *yes* if $\mathcal{O}$ is satisfiable and *no* otherwise.
2. *Concept satisfiability checking*:

- Given: an ontology $\mathcal{O}$ and a concept C,
- Return: *yes* if C is satisfiable w.r.t. $\mathcal{O}$ and *no* otherwise.
3. *Concept subsumption checking*:
    - Given: an ontology $\mathcal{O}$ and a concept inclusion $C \sqsubseteq D$,
    - Return: *yes* if $\mathcal{O} \models C \sqsubseteq D$ and *no* otherwise.
4. *Instance checking*:
    - Given: an ontology $\mathcal{O}$ and a concept assertion $C(a)$,
    - Return: *yes* if $\mathcal{O} \models C(a)$ and *no* otherwise.

Example 9. Consider the ontology $\mathcal{O}$ from Example 1:

1. $$\mathsf{Parent} \equiv \exists \mathsf{hasChild}.\top,$$
2. $$\mathsf{GrandParent} \equiv \exists \mathsf{hasChild}.\mathsf{Parent},$$
3. $$\mathsf{hasChild}(\mathsf{john}, \mathsf{mary}).$$

As was shown in Example 6, this ontology has a model $\mathcal{J} = (\Delta^{\mathcal{J}}, \cdot^{\mathcal{J}})$ with

- $\Delta^{\mathcal{J}} = \{a\}$,
- $\mathsf{Parent}^{\mathcal{J}} = \mathsf{GrandParent}^{\mathcal{J}} = \{a\}$, $\mathsf{hasChild}^{\mathcal{I}} = \{\langle a, a \rangle\}$,
- $\mathsf{john}^{\mathcal{J}} = \mathsf{mary}^{\mathcal{J}} = a$.

Therefore, the answer to the ontology satisfiability checking problem for $\mathcal{O}$ is *yes*.

The answer to the concept satisfiability checking problem for $\mathcal{O}$ and concept Parent is also *yes* because $\mathcal{J} \models \mathcal{O}$ and $\mathsf{Parent}^{\mathcal{J}} = \{a\} \neq \emptyset$. The same answer is also obtained for the inputs $\mathcal{O}$ and GrandParent.

We next check which subsumptions hold between these concepts. The subsumption $\mathsf{Parent} \sqsubseteq \mathsf{GrandParent}$ holds in $\mathcal{J}$ since $\mathsf{Parent}^{\mathcal{J}} = \{a\} \subseteq \{a\} = \mathsf{GrandParent}^{\mathcal{J}}$, but there is another model $\mathcal{I} = (\Delta^{\mathcal{I}}, \cdot^{\mathcal{I}})$ of $\mathcal{O}$ in which this subsumption does not hold:

- $\Delta^{\mathcal{I}} = \{a, b\}$,
- $\mathsf{Parent}^{\mathcal{I}} = \{a\}$, $\mathsf{GrandParent}^{\mathcal{I}} = \emptyset$, $\mathsf{hasChild}^{\mathcal{I}} = \{\langle a, b \rangle\}$,
- $\mathsf{john}^{\mathcal{I}} = a$, $\mathsf{mary}^{\mathcal{I}} = b$.

Indeed, $\mathsf{Parent}^{\mathcal{I}} = \{a\} = (\exists \mathsf{hasChild}.\top)^{\mathcal{I}}$. Therefore, $\mathcal{I} \models \mathsf{Parent} \equiv \exists \mathsf{hasChild}.\top$. We further have $\mathcal{I} \models \mathsf{GrandParent} \equiv \exists \mathsf{hasChild}.\mathsf{Parent}$ since $\mathsf{GrandParent}^{\mathcal{I}} = \emptyset = (\exists \mathsf{hasChild}.\mathsf{Parent})^{\mathcal{I}}$. Since $\langle \mathsf{john}^{\mathcal{I}}, \mathsf{mary}^{\mathcal{I}} \rangle = \langle a, b \rangle \subseteq \{\langle a, b \rangle\} = \mathsf{hasChild}^{\mathcal{I}}$, we have $\mathcal{I} \models \mathsf{hasChild}(\mathsf{john}, \mathsf{mary})$. Therefore, $\mathcal{I} \models \mathcal{O}$. However, $\mathsf{Parent}^{\mathcal{I}} = \{a\} \not\subseteq \emptyset = \mathsf{GrandParent}^{\mathcal{I}}$. Therefore, $\mathcal{I} \not\models \mathsf{Parent} \sqsubseteq \mathsf{GrandParent}$. Since we have found a model $\mathcal{I}$ of $\mathcal{O}$ for which the subsumption $\mathsf{Parent} \sqsubseteq \mathsf{GrandParent}$ does not hold, we have proved that $\mathcal{O} \not\models \mathsf{Parent} \sqsubseteq \mathsf{GrandParent}$. Therefore, the answer to the concept subsumption checking problem for $\mathcal{O}$ and $\mathsf{Parent} \sqsubseteq \mathsf{GrandParent}$ is *no*.

The subsumption $\mathsf{GrandParent} \sqsubseteq \mathsf{Parent}$ holds in both $\mathcal{J}$ and $\mathcal{I}$, and in fact, in all models of $\mathcal{O}$. Indeed, assume that $\mathcal{I} \models \mathcal{O}$. We will show that $\mathsf{GrandParent}^{\mathcal{I}} \subseteq \mathsf{Parent}^{\mathcal{I}}$. To do this, take any $x \in \mathsf{GrandParent}^{\mathcal{I}}$. If there is no such x then, trivially, $\mathsf{GrandParent}^{\mathcal{I}} = \emptyset \subseteq \mathsf{Parent}^{\mathcal{I}}$. Since $\mathcal{I} \models \mathsf{GrandParent} \equiv$

$\exists$hasChild.Parent, we have $x \in$ GrandParent$^{\mathcal{I}} = (\exists$hasChild.Parent$)^{\mathcal{I}}$. Hence, there exists some y such that $\langle x, y \rangle \in$ hasChild$^{\mathcal{I}}$ and $y \in$ Parent$^{\mathcal{I}} \subseteq \Delta^{\mathcal{I}} = \top^{\mathcal{I}}$. Hence $x \in (\exists$hasChild.$\top)^{\mathcal{I}}$. Since $\mathcal{I} \models$ Parent $\equiv \exists$hasChild.$\top$, we have Parent$^{\mathcal{I}} = (\exists$hasChild.$\top)^{\mathcal{I}}$. Hence, $x \in$ Parent$^{\mathcal{I}}$. Since $x \in$ GrandParent$^{\mathcal{I}}$ was arbitrary, we proved that GrandParent$^{\mathcal{I}} \subseteq$ Parent$^{\mathcal{I}}$, that is, $\mathcal{I} \models$ GrandParent $\sqsubseteq$ Parent and so, $\mathcal{O} \models$ GrandParent $\sqsubseteq$ Parent.

Finally, we check which of the individuals john and mary appearing in $\mathcal{O}$ are instances of the atomic concepts Parent and GrandParent. For the model $\mathcal{I}$ defined above, GrandParent$^{\mathcal{I}} = \emptyset$, hence, $\mathcal{I} \not\models$ GandParent(john) and $\mathcal{I} \not\models$ GandParent(mary) since john$^{\mathcal{I}} = a \notin \emptyset =$ GandParent$^{\mathcal{I}}$ and mary$^{\mathcal{I}} = b \notin \emptyset =$ GandParent$^{\mathcal{I}}$. Also $\mathcal{I} \not\models$ Parent(mary) since mary $= b \notin \{a\} =$ Parent$^{\mathcal{I}}$. Hence, the answer to the instance checking problem for $\mathcal{O}$ and each concept assertion GandParent(john), GandParent(mary), and Parent(mary) is *no*.

The answer to the instance checking problem for $\mathcal{O}$ and the fourth concept assertion Parent(john) is *yes* since $\mathcal{I} \models$ Parent(john) for each $\mathcal{I} \models \mathcal{O}$. Indeed, let $x =$ john$^{\mathcal{I}}$ and $y =$ mary$^{\mathcal{I}}$. Since $\mathcal{I} \models$ hasChild(john, mary), we have $\langle x, y \rangle \in$ hasChild$^{\mathcal{I}}$. Trivially, $y \in \Delta^{\mathcal{I}} = \top^{\mathcal{I}}$. Hence $x \in (\exists$hasChild.$\top)^{\mathcal{I}}$. Since $\mathcal{I} \models$ Parent $\equiv \exists$hasChild.$\top$, we have Parent$^{\mathcal{I}} = (\exists$hasChild.$\top)^{\mathcal{I}}$. Therefore, $x \in$ Parent$^{\mathcal{I}}$. Since $x =$ john$^{\mathcal{I}}$, we proved that $\mathcal{I} \models$ Parent(john) and, since $\mathcal{I}$ was an arbitrary model of $\mathcal{O}$, we proved that $\mathcal{O} \models$ Parent(john).

Exercise 1. Determine which of the individuals john and mary are instances of the *negated* concepts $\neg$Parent and $\neg$GrandParent for the ontology $\mathcal{O}$ in Example 9. Are there any surprises? Can you explain the unexpected answers you obtained?

In Example 9 we have solved all reasoning problems "by hand" by either providing counter-models for entailments or proving that entailments hold for all models. Of course, in practice, it is not expected that the ontology developers or anybody else is going to solve these tasks manually. It is expected that these tasks are solved by computers *automatically* and, preferably, *quickly*. The main focus of the research in DLs, therefore, was development and analysis of algorithms for solving reasoning problems.

We have listed four standard reasoning tasks for ontologies: (1) ontology satisfiability checking, (2) concept satisfiability checking, (3) concept subsumption checking, and (4) instance checking. Developing *separate* algorithms for solving each of these problems would be too time consuming. Fortunately, it turns out, as soon as we find an algorithm for solving one of these problems, we can solve the remaining three too using simple modifications of this algorithm.

2.4   Reductions Between Reasoning Problems

In the remainder of this course, we are interested in measuring the *computational complexity* of problems and algorithms. We also develop *polynomial reductions* between the four standard reasoning problems mentioned above. Appendix A.1 provides additional material for readers who first want to refresh their knowledge about these notions.

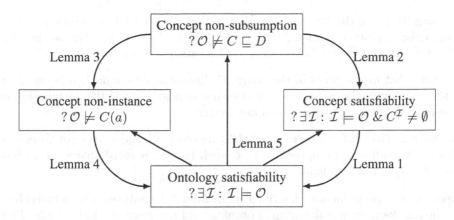

Fig. 1. An overview of reductions between the standard DL reasoning problems

Notice the similarities in the formulations of the reasoning problems 1–4 given earlier. All problems are decision problems, i.e., they get some objects as inputs and are expected to produce either *yes* or *no* as the output. We next apply the common approach of formulating polynomial reductions between these reasoning problems: We first prove that all problems can be reduced to the ontology satisfiability problem and then show how to reduce the ontology satisfiability problem to all other problems. The overview of the reductions is shown in Fig. 1. Note that for the concept subsumption and instance checking problems, we provide reductions for their *complementary* problems, i.e., where the answers *yes* and *no* are swapped.

The following lemma shows how to reduce the concept satisfiability problem to the ontology satisfiability problem. Intuitively, to check if a concept C is satisfiable w.r.t. $\mathcal{O}$, it is sufficient to extend $\mathcal{O}$ with a new concept assertion $C(a)$ and check satisfiability of the resulting ontology. Clearly, the reduction $R(\langle \mathcal{O}, C \rangle) = \{\mathcal{O} \cup C(a)\}$ can be computed in polynomial time in the size of $\mathcal{O}$ plus C.

Lemma 1. *Let $\mathcal{O}$ be an ontology, C a concept, and a an individual not appearing in $\mathcal{O}$. Then C is satisfiable w.r.t. $\mathcal{O}$ if and only if $\mathcal{O} \cup \{C(a)\}$ is satisfiable.*

Proof. ($\Rightarrow$): To prove the "only if" direction, assume that C is satisfiable w.r.t. $\mathcal{O}$. Then there exists a model $\mathcal{I} \models \mathcal{O}$ such that $C^{\mathcal{I}} \neq \emptyset$. That is, there exists some $x \in C^{\mathcal{I}} \subseteq \Delta^{\mathcal{I}}$. Let $\mathcal{J} = (\Delta^{\mathcal{J}}, \cdot^{\mathcal{J}})$ be a new interpretation defined as follows:

- $\Delta^{\mathcal{J}} = \Delta^{\mathcal{I}}$,
- $A^{\mathcal{J}} = A^{\mathcal{I}}$ and $r^{\mathcal{J}} = r^{\mathcal{I}}$ for each atomic concept A and atomic role r,
- $b^{\mathcal{J}} = b^{\mathcal{I}}$ for every individual $b \neq a$,
- $a^{\mathcal{J}} = x \in C^{\mathcal{I}} \subseteq \Delta^{\mathcal{I}} = \Delta^{\mathcal{J}}$.

Clearly, $\mathcal{J} \models \mathcal{O}$ since the interpretation of every symbol in $\mathcal{O}$ remained unchanged, and $\mathcal{J} \models C(a)$ since $a^{\mathcal{J}} = x \in C^{\mathcal{I}} = C^{\mathcal{J}}$. Hence, $\mathcal{O} \cup \{C(a)\}$ is satisfiable.

($\Leftarrow$): To prove the "if" directly, assume that $\mathcal{O} \cup \{C(a)\}$ is satisfiable. Then there exists an interpretation $\mathcal{I}$ such that $\mathcal{I} \models \mathcal{O}$ and $\mathcal{I} \models C(a)$. The last implies $a^{\mathcal{I}} \in C^{\mathcal{I}}$. Hence $C^{\mathcal{I}} \neq \emptyset$. Consequently, C is satisfiable w.r.t. $\mathcal{O}$. $\square$

Note that in the proof of the "only if" direction of Lemma 1, it is essential that the individual a is *fresh*, i.e., it does not appear in $\mathcal{O}$. If this assumption is dropped, the lemma does not hold any longer.

Exercise 2. Give an example of an $\mathcal{ALC}$ ontology $\mathcal{O}$ and a concept assertion $C(a)$, with individual a appearing in $\mathcal{O}$ such that C is satisfiable w.r.t. $\mathcal{O}$ but $\mathcal{O} \cup \{C(a)\}$ is not satisfiable.

Exercise 3. The reduction described in Lemma 1 introduces a new individual to the ontology, even if the original ontology did not have any individuals. This may be undesirable if the algorithm for checking ontology satisfiability can work only with TBoxes. Formulate a different reduction from concept satisfiability to ontology satisfiability that does not introduce any individuals. Prove that this reduction is correct like in Lemma 1.

Hint 1: Using a concept assertion $C(a)$ one forces the interpretation of C to be nonempty. Using which other axioms one can force non-emptiness of concepts? Which concepts are always interpreted by nonempty sets? Hint 2: Similar to fresh individuals, the reduction can use fresh atomic concepts and roles.

We next show how to reduce the problem of checking concept non-subsumption to the problem of concept satisfiability, which, in turn, as shown by Lemma 1, can be reduced to checking ontology satisfiability.

Lemma 2. *Let $\mathcal{O}$ be an ontology and C, D concepts. Then $\mathcal{O} \not\models C \sqsubseteq D$ if and only if $C \sqcap \neg D$ is satisfiable w.r.t. $\mathcal{O}$.*

Proof. It is easy to see that the following statements are equivalent:

1. $\mathcal{O} \not\models C \sqsubseteq D$,
2. There exists a model $\mathcal{I} \models \mathcal{O}$ such that $C^{\mathcal{I}} \not\subseteq D^{\mathcal{I}}$,
3. There exists a model $\mathcal{I} \models \mathcal{O}$ such $C^{\mathcal{I}} \setminus D^{\mathcal{I}} \neq \emptyset$,
4. There exists a model $\mathcal{I} \models \mathcal{O}$ such $(C \sqcap \neg D)^{\mathcal{I}} \neq \emptyset$,
5. $C \sqcap \neg D$ is satisfiable w.r.t. $\mathcal{O}$.

In particular, 3 and 4 are equivalent because:
$(C \sqcap \neg D)^{\mathcal{I}} = C^{\mathcal{I}} \cap (\neg D)^{\mathcal{I}} = C^{\mathcal{I}} \cap (\Delta^{\mathcal{I}} \setminus D^{\mathcal{I}}) = C^{\mathcal{I}} \setminus D^{\mathcal{I}}$. $\square$

The problem of checking concept (non-)subsumption can alternatively be reduced to checking (non-)entailment of concept instances. The following lemma proves that to check $\mathcal{O} \models C \sqsubseteq D$, one can extend $\mathcal{O}$ with a concept assertion $C(a)$ for a fresh individual a, and check if the resulting ontology entails the concept instance $D(a)$.

Lemma 3. *Let $\mathcal{O}$ be an ontology, C, D concepts, and a an individual not appearing in $\mathcal{O}$. Then $\mathcal{O} \models C \sqsubseteq D$ if and only if $\mathcal{O} \cup \{C(a)\} \models D(a)$.*

Proof. ($\Rightarrow$): To prove the "only if" direction, assume that $\mathcal{O} \models C \sqsubseteq D$. To show that $\mathcal{O} \cup \{C(a)\} \models D(a)$, take any model $\mathcal{I}$ of $\mathcal{O} \cup \{C(a)\}$. Since $\mathcal{I} \models \mathcal{O}$ and $\mathcal{O} \models C \sqsubseteq D$, we have $C^{\mathcal{I}} \subseteq D^{\mathcal{I}}$. Then, since $\mathcal{I} \models C(a)$, we have $a^{\mathcal{I}} \in C^{\mathcal{I}} \subseteq D^{\mathcal{I}}$. Hence, $\mathcal{I} \models D(a)$. Since $\mathcal{I}$ was an arbitrary model such that $\mathcal{I} \models \mathcal{O} \cup \{C(a)\}$, we have shown that $\mathcal{O} \cup \{C(a)\} \models D(a)$.

($\Leftarrow$): We prove the "if" direction by showing the contrapositive: if $\mathcal{O} \not\models C \sqsubseteq D$ then $\mathcal{O} \cup \{C(a)\} \not\models D(a)$. Assume that $\mathcal{O} \not\models C \sqsubseteq D$. Then there exists a model $\mathcal{I}$ of $\mathcal{O}$ such that $\mathcal{I} \not\models C \sqsubseteq D$, or, equivalently, $C^{\mathcal{I}} \not\subseteq D^{\mathcal{I}}$. This means that there exists $x \in C^{\mathcal{I}}$ such that $x \notin D^{\mathcal{I}}$. Define another interpretation $\mathcal{J} = (\Delta^{\mathcal{J}}, \cdot^{\mathcal{J}})$, which is identical to $\mathcal{I}$ on all symbols, except for the interpretation of the individual a:

- $\Delta^{\mathcal{J}} = \Delta^{\mathcal{I}}$,
- $A^{\mathcal{J}} = A^{\mathcal{I}}$ and $r^{\mathcal{J}} = r^{\mathcal{I}}$ for each atomic concept A and atomic role r,
- $b^{\mathcal{J}} = b^{\mathcal{I}}$ for every individual $b \neq a$,
- $a^{\mathcal{J}} = x \in C^{\mathcal{I}} \setminus D^{\mathcal{I}} \subseteq \Delta^{\mathcal{I}} = \Delta^{\mathcal{J}}$.

Clearly, $\mathcal{J} \models \mathcal{O}$ since the interpretation of every symbol in $\mathcal{O}$ remained unchanged. Since $a^{\mathcal{J}} \in C^{\mathcal{I}} \setminus D^{\mathcal{I}} = C^{\mathcal{J}} \setminus D^{\mathcal{J}}$, we have $\mathcal{J} \models C(a)$ and $\mathcal{J} \not\models D(a)$. Thus, we found $\mathcal{J} \models \mathcal{O} \cup \{C(a)\}$ such that $\mathcal{J} \not\models D(a)$, which proves that $\mathcal{O} \cup \{C(a)\} \not\models D(a)$, as required. □

Exercise 4. Similarly to Exercise 2, show that the condition that the individual a used in Lemma 3 is fresh cannot be dropped. Give an example where the statement of the lemma is not true without this condition.

As the next lemma shows, the concept instance checking problem can easily be reduced to checking ontology satisfiability.

Lemma 4. *Let $\mathcal{O}$ be an ontology, C a concept, and a an individual. Then $\mathcal{O} \not\models C(a)$ if and only if $\mathcal{O} \cup \{(\neg C)(a)\}$ is satisfiable.*

Proof. It is easy to see that the following statements are equivalent:

1. $\mathcal{O} \not\models C(a)$,
2. There exists a model $\mathcal{I} \models \mathcal{O}$ such that $a^{\mathcal{I}} \notin C^{\mathcal{I}}$,
3. There exists a model $\mathcal{I} \models \mathcal{O}$ such that $a^{\mathcal{I}} \in (\neg C)^{\mathcal{I}} = \Delta^{\mathcal{I}} \setminus C^{\mathcal{I}}$,
4. $\mathcal{O} \cup \{(\neg C)(a)\}$ is satisfiable. □

Finally, we show that the ontology satisfiability problem can easily be reduced to the other three problems. Specifically, to check if an ontology $\mathcal{O}$ is satisfiable, one can check if the concept $\top$ is satisfiable with respect to $\mathcal{O}$ (thus, reducing to the concept satisfiability problem), or check if $\mathcal{O}$ does not entail the subsumption $\top \sqsubseteq \bot$ (thus reducing to the concept non-subsumption problem), or check if $\mathcal{O}$ does not entail an instance $\top(a)$ for some individual a (thus, reducing to the concept non-instance problem).

Lemma 5. *Let $\mathcal{O}$ be an ontology. Then the following conditions are equivalent:*

1. $\mathcal{O}$ *is satisfiable,*
2. $\top$ *is satisfiable with respect to* $\mathcal{O}$,
3. $\mathcal{O} \not\models \top \sqsubseteq \bot$,
4. $\mathcal{O} \not\models \bot(a)$ *for every individual* a,
5. $\mathcal{O} \not\models \bot(a)$ *for some individual* a.

Proof. Case $1 \Rightarrow 2$: If $\mathcal{O}$ is satisfiable then $\mathcal{I} \models \mathcal{O}$ for some $\mathcal{I} = (\Delta^{\mathcal{I}}, \cdot^{\mathcal{I}})$, then $\top^{\mathcal{I}} = \Delta^{\mathcal{I}} \neq \emptyset$ for some $\mathcal{I} \models \mathcal{O}$, then $\top$ is satisfiable with respect to $\mathcal{O}$.

Case $2 \Rightarrow 3$: If $\top$ is satisfiable with respect to $\mathcal{O}$ then there exists $\mathcal{I} \models \mathcal{O}$ such that $\top^{\mathcal{I}} \neq \emptyset$, then $\top^{\mathcal{I}} \not\sqsubseteq \emptyset = \bot^{\mathcal{I}}$, then $\mathcal{I} \not\models \top \sqsubseteq \bot$, then $\mathcal{O} \not\models \top \sqsubseteq \bot$ since $\mathcal{I} \models \mathcal{O}$.

Case $3 \Rightarrow 4$: If $\mathcal{O} \not\models \top \sqsubseteq \bot$, then there exists $\mathcal{I} \models \mathcal{O}$ (such that $\top^{\mathcal{I}} \not\sqsubseteq \bot^{\mathcal{I}}$) then, for every individual a: $a^{\mathcal{I}} \notin \emptyset = \bot^{\mathcal{I}}$, hence $\mathcal{I} \not\models \bot(a)$, hence $\mathcal{O} \not\models \bot(a)$ since $\mathcal{I} \models \mathcal{O}$.

Case $4 \Rightarrow 5$: If $\mathcal{O} \not\models \bot(a)$ for every individual a then, trivially, $\mathcal{O} \not\models \bot(a)$ for some individual a since the set of individuals N_I is nonempty.

Case $5 \Rightarrow 1$: If $\mathcal{O} \not\models \bot(a)$ for some individual a, then there exists $\mathcal{I} \models \mathcal{O}$ (such that $a^{\mathcal{I}} \notin \bot^{\mathcal{I}}$), then $\mathcal{O}$ is satisfiable.   □

3   Tableau Procedures

In this section, we introduce the so-called *tableau procedures*, which are the most popular procedures for reasoning in DLs, particularly, for very expressive languages. Tableau procedures or variants thereof have been implemented in many *ontology reasoners*, such as HermiT [42], FacT++ [62], Konclude [58], and Pellet [56]. Intuitively, tableau procedures work by trying to construct ontology models of a particular shape, called the *tree* models. To simplify our exposition, in this section we mainly focus on tableau procedures for TBox reasoning, i.e., we assume that our ontologies do not contain concept and role assertions.

The construction of a model is governed by a number of rules that incrementally expand the model by adding new domain elements and requirements that they need to satisfy (e.g., be instances of particular concepts). To describe this process in a convenient way, in tableau procedures one works with a different representation of interpretations, which is called a tableau.

Definition 1. *A* tableau *is a tuple* $T = (V, L)$, *where*

- V *is a nonempty set of* tableau nodes *of* T,
- L *is a* labeling function *that assigns:*
  - *to every node* $v \in V$ *a subset* $L(v)$ *of concepts,*
  - *to every pair of nodes* $\langle v, w \rangle \in V \times V$ *a subset* $L(v, w)$ *of roles.*

A tableau T is usually drawn as a *labeled graph* with the set of vertices V and the set of (directed) edges $E = \{\langle v, w \rangle \in V \times V \mid L(v, w) \neq \emptyset\}$, in which every node $v \in V$ is labeled with $L(v)$ and every edge $\langle v, w \rangle \in E$ is labeled with $L(v, w)$. In what follows we assume that if $L(v)$ or $L(v, w)$ were not explicitly assigned for some nodes $\{v, w\} \subseteq V$, then $L(v) = L(v, w) = \emptyset$.

Example 10. Consider the interpretation $\mathcal{I} = (\Delta^{\mathcal{I}}, \cdot^{\mathcal{I}})$ from Example 2 extended with $\text{Human}^{\mathcal{I}} = \{a, b\}$:

- $\Delta^{\mathcal{I}} = \{a, b\}$,
- $\text{Human}^{\mathcal{I}} = \{a, b\}$, $\text{Parent}^{\mathcal{I}} = \{a\}$, $\text{GrandParent}^{\mathcal{I}} = \emptyset$,
- $\text{hasChild}^{\mathcal{I}} = \{\langle a, b \rangle\}$,
- $\text{john}^{\mathcal{I}} = a$, $\text{mary}^{\mathcal{I}} = b$.

This interpretation can be equivalently represented by a tableau $T = (V, L)$ with:

- $V = \{a, b\}$,
- $L(a) = \{\text{Human}, \text{Parent}\}$,
- $L(b) = \{\text{Human}\}$,
- $L(a, b) = \{\text{hasChild}\}$.

This tableau is graphically illustrated on the right.

Note that according to Definition 1, tableau nodes can be labeled with arbitrary concepts, not necessarily with atomic ones like in Example 10. This is in contrast to interpretations, which define only the values for atomic concepts and roles. For interpretations, it is not necessary to define how complex concepts are interpreted, since these values can always be *calculated* according to the rules given in Sect. 2.2. For the tableau procedures, the information $C \in L(v)$ represents a *requirement* that $v \in C^{\mathcal{I}}$ should hold for the constructed model $\mathcal{I}$. The tableau should subsequently be expanded to satisfy all such requirements. For example, if $C \sqcap D \in L(v)$ then $L(v)$ should be expanded by adding the concepts C and D, since $v \in (C \sqcap D)^{\mathcal{I}}$ implies $v \in C^{\mathcal{I}}$ and $v \in D^{\mathcal{I}}$. This expansion process is governed using dedicated *tableau expansion rules*.

As we have seen in Sect. 2.3, all standard reasoning problems can be reduced to each other in polynomial time. Therefore, an algorithm for solving any of these problems can easily be modified to solve the other problems. In the next sections, therefore, we use tableau procedures for solving one of these problems: concept satisfiability.

3.1 Deciding Concept Satisfiability

In this section, we formulate a simplified version of the procedure for checking concept satisfiability that works without considering the axioms in the ontology. That is, given an $\mathcal{ALC}$ concept C we need to check if there exists an interpretation $\mathcal{I}$ such that $C^{\mathcal{I}} \neq \emptyset$. Although this problem is not of much use in ontology-based applications, it allows us to illustrate the main principles of tableau procedures.

To check satisfiability of a given concept C, we first transform it into a special *normal form*, which is easier to work with.

Definition 2. *An $\mathcal{ALC}$ concept C is in* negation normal form *(short: NNF) if negation can appear in C only in the form of $\neg A$, where A is an atomic concept.*

Table 2. Tableau expansion rules for checking satisfiability of $\mathcal{ALC}$ concepts

Rule	Conditions	Expansions
⊓-Rule	$D \sqcap E \in L(x)$, $\{D, E\} \not\subseteq L(x)$	Set $L(x) := L(x) \cup \{D, E\}$
⊔-Rule	$D \sqcup E \in L(x)$, $\{D, E\} \cap L(x) = \emptyset$	Set $L(x) := L(x) \cup \{D\}$ or $L(x) := L(x) \cup \{E\}$
∃-Rule	$\exists r.D \in L(x)$ and there is no $y \in V$ such that $r \in L(x, y)$ and $D \in L(y)$	Extend $V := V \cup \{y\}$ for a new y, set $L(x, y) := \{r\}$ and $L(y) := \{D\}$
∀-Rule	$\forall r.D \in L(x)$, $r \in L(x, y)$, $D \notin L(y)$	Set $L(y) := L(y) \cup \{D\}$
⊥-Rule	$\{A, \neg A\} \subseteq L(x)$, $\bot \notin L(x)$	Set $L(x) := L(x) \cup \{\bot\}$

In other words, to construct a concept in NNF, it is permitted to apply negation only to atomic concepts. Thus, $\mathcal{ALC}$ concepts in NNF can be defined by the grammar:

$$C, D ::= A \mid \top \mid \bot \mid C \sqcap D \mid C \sqcup D \mid \neg A \mid \exists r.C \mid \forall r.C. \tag{6}$$

Example 11. The concept $\forall r.(\neg A \sqcup \exists S.\neg B)$ is in NNF; the concepts $\neg\exists r.A$, $\forall r.\neg(A \sqcap B)$, and $A \sqcap \exists r.\neg\top$ on the other hand are *not* in NNF.

Each $\mathcal{ALC}$ concept C can be converted to an equivalent concept in NNF by applying simple rules to "push negation inwards" that are reminiscent of De Morgan's Laws:

$$\neg(C \sqcap D) \quad \Rightarrow \quad (\neg C) \sqcup (\neg D), \qquad\qquad \neg\neg C \quad \Rightarrow \quad C,$$
$$\neg(C \sqcup D) \quad \Rightarrow \quad (\neg C) \sqcap (\neg D), \qquad\qquad \neg\top \quad \Rightarrow \quad \bot,$$
$$\neg(\exists r.C) \quad \Rightarrow \quad \forall r.(\neg C), \qquad\qquad\qquad \neg\bot \quad \Rightarrow \quad \top.$$
$$\neg(\forall r.C) \quad \Rightarrow \quad \exists r.(\neg C),$$

Example 12. Consider the $\mathcal{ALC}$ concept $(\exists r.A) \sqcap \neg((\exists r.A) \sqcap \neg B)$. This concept can be converted to NNF as follows:

$$(\exists r.A) \sqcap \neg((\exists r.A) \sqcap \neg B) \quad \Rightarrow \quad (\exists r.A) \sqcap (\neg(\exists r.A) \sqcup \neg\neg B)$$
$$\Rightarrow \quad (\exists r.A) \sqcap (\forall r.(\neg A) \sqcup B).$$

Exercise 5. Show that the transformation of concepts to NNF described above preserves satisfiability of concepts. That is, the input concept is satisfiable if and only if its NNF is satisfiable. Hint: show for each transformation step $C \Rightarrow D$ that $C^{\mathcal{I}} = D^{\mathcal{I}}$ holds for every interpretation $\mathcal{I}$.

To check satisfiability of an $\mathcal{ALC}$ concept C in NNF, we create a new Tableau $T = (V, L)$ with $V = \{v_0\}$ and $L(v_0) = \{C\}$, and apply the *tableau expansion* rules from Table 2. A rule is *applicable* if all *conditions* of the rule are satisfied in the current tableau T for certain choices of *rule parameters*, such as the values of x, y or the matching concepts and roles in the labels. For example, the ⊓-Rule

$$(\exists r.A) \sqcap (\forall r.(\neg A) \sqcup B),$$

$v_0 \bullet \quad \exists r.A, \quad \forall r.(\neg A) \sqcup B,$

$r \quad\Big\downarrow \quad \forall r.(\neg A)$

$v_1 \bullet \quad A, \neg A, \bot$

$$(\exists r.A) \sqcap (\forall r.(\neg A) \sqcup B),$$

$v_0 \bullet \quad \exists r.A, \quad \forall r.(\neg A) \sqcup B,$

$r \quad\Big\downarrow \quad B$

$v_1 \bullet \quad A$

Fig. 2. Two possible tableau expansions for the concept $C = (\exists r.A) \sqcap (\forall r.(\neg A) \sqcup B)$ due to the non-deterministic $\sqcup$-Rule

is applicable to a node $x \in V$ if some conjunction $D \sqcap E$ belongs to the label $L(x)$ of this node, but at least one of the conjuncts D or E does not belong to $L(x)$. In this case, T is *expanded* by adding new nodes or labels as specified in the *expansions* part of the rules. In this case we say that the rule is *applied* (for the specific choice of the rule parameters). For example, the $\sqcap$-Rule is applied by adding the conjuncts D and E to the label $L(x)$ of the node x. Note that after applying each rule in Table 2, the rule is no longer applicable for the same choices of rule parameters. The tableau expansion rules are applied until no rule is applicable any longer. In this case we say that the T is *fully expanded*.

Example 13. Consider the concept $C = (\exists r.A) \sqcap (\forall r.(\neg A) \sqcup B)$ obtained by the conversion to NNF in Example 12. We check satisfiability of C by applying the tableau expansion rules from Table 2. Consider first the left-hand side of Fig. 2. We initialize $T = (V, L)$ by setting $V = \{v_0\}$ and $L(v_0) = \{C\}$. Since C is a conjunction, the conditions of the $\sqcap$-Rule are satisfied for $x = v_0$, $D = \exists r.A$, and $E = \forall r.(\neg A) \sqcup B$. Applying this rule adds the conjuncts $\exists r.A$ and $\forall r.(\neg A) \sqcup B$ to $L(v_0)$. Now the conditions of the $\exists$-Rule are satisfied for $x = v_0$ and $\exists r.A \in L(v_0)$: note that there is no $v \in V$ such that $r \in L(v_0, v)$. Applying this rule creates a new node v_1, and sets $L(v_0, v_1) = \{r\}$ and $L(v_1) = \{A\}$. Similarly, since $\forall r.(\neg A) \sqcup B \in L(v_0)$, but neither $\forall r.(\neg A) \in L(v_0)$ nor $B \in L(v_0)$, the $\sqcup$-Rule is applicable. There are two ways this rule can be applied to T: either we add the first disjunct $\forall r.(\neg A)$ to $L(v_0)$, or we add the second disjunct B to $L(v_0)$. It is not necessary to add both of them. Let us chose the first disjunct and see what happens. After we apply the $\sqcup$-Rule in this way, we obtain $\forall r.(\neg A) \in L(v_0)$. Since $r \in L(v_0, v_1)$ and $\neg A \notin L(v_1)$, the $\forall$-Rule is now applicable for $x = v_0$, $y = v_1$, and $\forall r.(\neg A) \in L(v_0)$. The application of this rule adds $\neg A$ to $L(v_1)$. Now we have both A and $\neg A$ in $L(v_1)$, which satisfies the conditions of the $\bot$-Rule since $\bot \notin L(v_1)$. The application of the $\bot$-Rule, therefore, adds $\bot$ to $L(v_1)$. After applying this rule, no further rule is applicable, so the tableau is fully expanded.

If during the application of the $\sqcup$-Rule to $\forall r.(\neg A) \sqcup B \in L(v_0)$ we, alternatively, choose to add the second disjunct B to $L(v_0)$, we obtain another fully expanded tableau without $\bot \in L(v_1)$ shown in the right-hand side of Fig. 2.

Remark 1. Note that the tableau edges in Example 13 were labeled by just a single role r. Although Definition 1 allows for arbitrary *sets* of roles in edge labels, the tableau rules for $\mathcal{ALC}$, can only create *singleton* sets of roles. Indeed, it is easy to see from Table 2, that the $\exists$-Rule is the only rule that can modify edge labels, and can only set them to singleton role sets $\{r\}$. More expressive languages, such as the DL $\mathcal{ALCH}$ to be considered in Exercise 10, can have additional rules that can *extend* edge labels similarly to node labels, thus resulting in edge labels that contain multiple roles.

As seen from Example 13, the result of applying the tableau expansion rules is not uniquely determined. If we choose to apply the $\sqcup$-Rule by adding the first disjunct to the label of v_0, we eventually obtain a *clash* $\bot \in L(v_1)$. We say that a tableau $T = (V, L)$ *contains a clash* if $\bot \in L(v)$ for some $v \in V$. Otherwise, we say that T is *clash-free*. A clash means that the tableau cannot correspond to an interpretation since $\bot \in L(v)$ corresponds to the requirement $v \in \bot^{\mathcal{I}} = \emptyset$, which cannot be fulfilled. In our example, the clash was obtained as a result of the "wrong choice" in the application of the $\sqcup$-Rule. When, instead, we choose the second disjunct, a clash-free tableau can be produced. We show next how to construct an interpretation from such a tableau.

Remark 2. Note that a clash $\bot \in L(v)$ may be produced by other rules than the $\bot$-Rule. For example, if $C \sqcap \bot \in L(v)$, then $\bot \in L(v)$ can be produced by the $\sqcap$-Rule. Similarly, if $\exists r.\bot \in L(v)$, then the clash is produced by the $\exists$-Rule.

Definition 3. *A tableau $T = (V, L)$ defines an interpretation $\mathcal{I} = (\Delta^{\mathcal{I}}, \cdot^{\mathcal{I}})$ where:*

- $\Delta^{\mathcal{I}} = V,$
- $A^{\mathcal{I}} = \{x \in V \mid A \in L(x)\}$ *for each atomic concept $A \in N_C$,*
- $r^{\mathcal{I}} = \{\langle x, y \rangle \in V \times V \mid r \in L(x, y)\}$ *for each atomic role $r \in N_R$.*

Example 14. Consider the first tableau expansion from Example 13 (see the left of Fig. 2). This tableau defines an interpretation $\mathcal{I} = (\Delta^{\mathcal{I}}, \cdot^{\mathcal{I}})$ with $\Delta^{\mathcal{I}} = \{v_0, v_1\}$, $A^{\mathcal{I}} = \{v_1\}$, $B^{\mathcal{I}} = \emptyset$, and $r^{\mathcal{I}} = \{\langle v_0, v_1 \rangle\}$. Let us calculate the values of the other concepts appearing in the label of the tableau under this interpretation:

- $(\neg A)^{\mathcal{I}} = \{v_0\},$
- $(\exists r.A)^{\mathcal{I}} = \{v_0\},$
- $(\forall r.(\neg A))^{\mathcal{I}} = \{v_1\},$
- $(\forall r.(\neg A) \sqcup B)^{\mathcal{I}} = \{v_1\},$
- $((\exists r.A) \sqcap (\forall r.(\neg A) \sqcup B))^{\mathcal{I}} = \emptyset.$

As we can see, the interpretation $\mathcal{I}$ does not prove the satisfiability of the concept $C = (\exists r.A) \sqcap (\forall r.(\neg A) \sqcup B)$, for which the tableau is constructed, since $C^{\mathcal{I}} = \emptyset$.

Let us now consider the interpretation $\mathcal{J} = (\Delta^{\mathcal{J}}, \cdot^{\mathcal{J}})$ defined by the second tableau expansion from Example 13 (see the right-hand side of Fig. 2): $\Delta^{\mathcal{J}} = \{v_0, v_1\}$, $A^{\mathcal{J}} = \{v_1\}$, $B^{\mathcal{J}} = \{v_0\}$, and $r^{\mathcal{J}} = \{\langle v_0, v_1 \rangle\}$. It is easy to see that for this interpretation we have:

- $(\neg A)^{\mathcal{J}} = \{v_0\}$,
- $(\exists r.A)^{\mathcal{J}} = \{v_0\}$,
- $\forall r.(\neg A)^{\mathcal{J}} = \{v_1\}$,

- $(\forall r.(\neg A) \sqcup B)^{\mathcal{J}} = \{v_0, v_1\}$,
- $((\exists r.A) \sqcap (\forall r.(\neg A) \sqcup B))^{\mathcal{J}} = \{v_0\}$.

Since $C^{\mathcal{J}} = \{v_0\} \neq \emptyset$, the interpretation $\mathcal{J}$ proves that C is satisfiable.

The satisfiability of the concept C proved in Example 14 using the interpretation for the second tableau expansion is not a coincidence. As we show next, in general, if the tableau rules can be applied without obtaining a clash, then each concept appearing in the label of each tableau node is satisfiable in the corresponding model.

Remark 3. Note that each rule in Table 2, with the exception of the $\bot$-Rule, can only add concepts to the labels if they are sub-concepts of some existing concept in the labels (to which the rule applies). Hence, every concept appearing in the labels of tableau nodes is a sub-concept of the original concept C for which the tableau is constructed or $\bot$. In particular, each such concept is in NNF. Similarly, only roles appearing in C can be added to the labels of the tableau edges.

Lemma 6. *Let $T = (V, L)$ be a clash-free, fully expanded tableau and $\mathcal{I} = (\Delta^{\mathcal{I}}, \cdot^{\mathcal{I}})$ an interpretation defined by T. Then, for every $v \in V$ and $C \in L(v)$, we have $v \in C^{\mathcal{I}}$.*

Proof. By Remark 3, the concept C is in NNF. We prove the lemma by induction on the construction of C according to the grammar definition (6):

Case $C = A \in L(v)$: In this case $v \in C^{\mathcal{I}} = A^{\mathcal{I}} = \{x \mid A \in L(x)\}$ by definition of $\mathcal{I}$.

Case $C = \top$: Then, trivially $v \in V = \Delta^{\mathcal{I}} = \top^{\mathcal{I}} = C^{\mathcal{I}}$.

Case $C = \bot \in L(v)$: Then T has a clash, which is not possible according to the assumption of the lemma.

Case $C = \neg A \in L(v)$: Then $A \notin L(v)$ since otherwise $\{A, \neg A\} \subseteq L(v)$ and $\bot \in L(v)$ since the $\bot$-Rule is not applicable to T, which would again mean that T has a clash. Since $A \notin L(v)$, we have $v \notin A^{\mathcal{I}}$ by definition of $\mathcal{I}$. Hence, $v \in \Delta^{\mathcal{I}} \setminus A^{\mathcal{I}} = (\neg A)^{\mathcal{I}}$.

Case $C = D \sqcap E \in L(v)$: Since the $\sqcap$-Rule is not applicable to $D \sqcap E \in L(v)$, we have $D \in L(v)$ and $E \in L(v)$. Then, by induction hypothesis, $v \in D^{\mathcal{I}}$ and $v \in E^{\mathcal{I}}$. Hence, $v \in D^{\mathcal{I}} \cap E^{\mathcal{I}} = (D \sqcap E)^{\mathcal{I}} = C^{\mathcal{I}}$.

Case $C = D \sqcup E \in L(v)$: Since the $\sqcup$-Rule is not applicable to $D \sqcup E \in L(v)$, we have $D \in L(v)$ or $E \in L(v)$. Then, by induction hypothesis, $v \in D^{\mathcal{I}}$ or $v \in E^{\mathcal{I}}$. Hence, $v \in D^{\mathcal{I}} \cup E^{\mathcal{I}} = (D \sqcup E)^{\mathcal{I}} = C^{\mathcal{I}}$.

Case $C = \exists r.D$: Since the $\exists$-Rule is not applicable to $\exists r.D \in L(v)$, there exists some $w \in V$ such that $r \in L(v, w)$ and $D \in L(w)$. From $r \in L(v, w)$, by definition of $\mathcal{I}$, we obtain $\langle v, w \rangle \in r^{\mathcal{I}}$. From $D \in L(w)$, by induction hypothesis, we obtain $w \in D^{\mathcal{I}}$. Hence, from $\langle v, w \rangle \in r^{\mathcal{I}}$ and $w \in D^{\mathcal{I}}$ we obtain $v \in (\exists r.D)^{\mathcal{I}}$.

Case C = ∀r.D: In order to prove that $v \in C^{\mathcal{I}} = (\forall r.D)^{\mathcal{I}}$, take any $w \in \Delta^{\mathcal{I}} = V$ such that $\langle v, w \rangle \in r^{\mathcal{I}}$. We need to show that $w \in D^{\mathcal{I}}$. Since $\langle v, w \rangle \in r^{\mathcal{I}}$, by definition of $\mathcal{I}$, we have $r \in L(v, w)$. Since the ∀-Rule is not applicable to $x = v$, $y = w$ and $\forall r.D \in L(v) = L(x)$, we must have $D \in L(y) = L(w)$. Hence, by induction hypothesis, $w \in D^{\mathcal{I}}$, as required.   □

Corollary 1. *Let C be an ALC concept in NNF, and $T = (V, L)$ a clash-free, fully expanded tableau obtained by the tableau procedure for checking satisfiability for C. Then C is satisfiable.*

Proof. Due to the tableau initialization, and since the tableau rules in Table 2 never remove nodes or labels, we must have $C \in L(v)$ for some $v \in V$. Hence, by Lemma 6 $v \in C^{\mathcal{I}}$ for the interpretation $\mathcal{I}$ defined by T. Thus, $C^{\mathcal{I}} \neq \emptyset$ and C is satisfiable.   □

Corollary 1 means that if we have managed to apply all tableau rules without producing a clash for a concept C in NNF, then we have proved that C is satisfiable. Does converse of this property also hold? Specifically, if C is satisfiable, is it always possible to apply the tableau rules without producing a clash? We prove that it is indeed the case by showing that if $C^{\mathcal{I}} \neq \emptyset$ for some interpretation $\mathcal{I}$, then we can always construct a tableau that *mimics* this interpretation in a certain way.

Definition 4. *We say say that a tableau $T = (V, L)$ mimics an interpretation $\mathcal{I} = (\Delta^{\mathcal{I}}, \cdot^{\mathcal{I}})$ if there exists a mapping $\tau : V \to \Delta^{\mathcal{I}}$ such that:*

(1) for each $v \in V$ and each $C \in L(v)$, we have $\tau(v) \in C^{\mathcal{I}}$, and
(2) for each $\langle v, w \rangle \in V \times V$ and each $r \in L(v, w)$, we have $\langle \tau(v), \tau(w) \rangle \in r^{\mathcal{I}}$.

The mapping τ is called a mimic *of T in $\mathcal{I}$.*

For example, if $T = (V, L)$ is a clash-free fully expanded tableau, then T mimics the interpretation $\mathcal{I}$ defined by T (cf. Definition 3) since the identity mapping $\tau(v) = v \in V = \Delta^{\mathcal{I}}$ satisfies the requirements of Definition 4. Indeed, by Lemma 6, for every $C \in L(v)$, we have $\tau(v) = v \in C^{\mathcal{I}}$, and, by Definition 3, for every $\langle v, w \rangle \in V \times V$ and $r \in L(v, w)$, we have $\langle \tau(v), \tau(w) \rangle = \langle v, w \rangle \in r^{\mathcal{I}}$. However, a tableau T can also mimic other interpretations.

Example 15. Consider the interpretation $\mathcal{I} = (\Delta^{\mathcal{I}}, \cdot^{\mathcal{I}})$ with $\Delta^{\mathcal{I}} = \{a\}$, $A^{\mathcal{I}} = B^{\mathcal{I}} = \{a\}$ and $r^{\mathcal{I}} = \{\langle a, a \rangle\}$ and the tableau $T = (V, L)$ obtained after the second expansion in Example 13 (see the right-hand side of Fig. 2). Then the mapping $\tau : V \to \Delta^{\mathcal{I}}$ defined by $\tau(v_0) = \tau(v_1) = a$ is a mimic of T in $\mathcal{I}$. Indeed, it is easy to see that $A^{\mathcal{I}} = B^{\mathcal{I}} = (\exists r.A)^{\mathcal{I}} = (\forall r.(\neg A) \sqcup B)^{\mathcal{I}} = ((\exists r.A) \sqcap (\forall r.(\neg A) \sqcup B))^{\mathcal{I}} = \{a\}$, hence, $\tau(v) = a \in C^{\mathcal{I}}$ for every $v \in V$ and every $C \in L(v)$. Also, since $\langle \tau(v_0), \tau(v_1) \rangle = \langle a, a \rangle \in r^{\mathcal{I}}$, Condition (2) of Definition 4 holds for $r \in L(v_0, v_1)$.

Note that if a tableau $T = (V, L)$ contains a clash $\bot \in L(v)$ for some $v \in V$, then T cannot mimic any interpretation $\mathcal{I}$, since, otherwise $\tau(v) \in \bot^{\mathcal{I}} = \emptyset$. Note also that T can have a mimic even if it is not fully expanded. For example, if $C^{\mathcal{I}} \neq \emptyset$ for some concept C and interpretation $\mathcal{I} = (\Delta^{\mathcal{I}}, \cdot^{\mathcal{I}})$, then the initial tableau $T = (V, L)$ with $V = \{v_0\}$ and $L(v_0) = \{C\}$ mimics $\mathcal{I}$ since for each $a \in C^{\mathcal{I}} \neq \emptyset$ and $\tau \colon V \to \Delta^{\mathcal{I}}$ defined by $\tau(v_0) = a$, we have $\tau(v_0) \in C^{\mathcal{I}}$. We will next show that in such a case T can always be expanded so that it still mimics $\mathcal{I}$.

Lemma 7. *Let $T = (V, L)$ be a tableau that mimics an interpretation $\mathcal{I} = (\Delta^{\mathcal{I}}, \cdot^{\mathcal{I}})$ and R be some tableau rule from Table 2 that is applicable to T. Then R can be applied in such a way that the resulting tableau also mimics $\mathcal{I}$.*

Proof. Suppose that $\tau \colon V \to \Delta^{\mathcal{I}}$ is a mimic of T in $\mathcal{I}$. We show how to apply R and extend τ to a mimic of the expanded tableau by considering all possible cases for R:

Case $\sqcap$-Rule: If the $\sqcap$-Rule is applicable to T, then $D \sqcap E \in L(v)$ for some $v \in V$. Since τ is a mimic of T in $\mathcal{I}$, we have $\tau(v) \in (D \sqcap E)^{\mathcal{I}}$. The application of the $\sqcap$-Rule only adds D and E to $L(v)$. To show that T still mimics $\mathcal{I}$ after this rule application, it is sufficient to prove that $\tau(v) \in D^{\mathcal{I}}$ and $\tau(v) \in E^{\mathcal{I}}$. This, clearly, follows from $\tau(v) \in (D \sqcap E)^{\mathcal{I}} = D^{\mathcal{I}} \cap E^{\mathcal{I}}$.

Case $\sqcup$-Rule: If the $\sqcup$-Rule is applicable to T, then $D \sqcup E \in L(v)$ for some $v \in V$. Since τ is a mimic of T in $\mathcal{I}$, we have $\tau(v) \in (D \sqcup E)^{\mathcal{I}} = D^{\mathcal{I}} \cup E^{\mathcal{I}}$. Hence, $\tau(v) \in D^{\mathcal{I}}$ or $\tau(v) \in E^{\mathcal{I}}$. We show that in each of these two cases one can apply the $\sqcup$-Rule so that the resulting tableau still mimics $\mathcal{I}$. Indeed, if $\tau(v) \in D^{\mathcal{I}}$, we can apply the $\sqcup$-Rule by adding D to $L(v)$. Since $\tau(v) \in D^{\mathcal{I}}$, τ is still a mimic of T in $\mathcal{I}$ after this rule application. Similarly, if $\tau(v) \in E^{\mathcal{I}}$, we can apply the $\sqcup$-Rule by adding E to $L(v)$. Since $\tau(v) \in E^{\mathcal{I}}$, τ remains a mimic of T in $\mathcal{I}$.

Case $\exists$-Rule: If the $\exists$-Rule is applicable to T, then $\exists r.D \in L(v)$ for some $v \in V$. Since τ is a mimic of T in $\mathcal{I}$, we have $\tau(v) \in (\exists r.D)^{\mathcal{I}}$. The application of the $\exists$-Rule adds a new node w to V with $L(v, w) = \{r\}$ and $L(w) = \{D\}$. To show that T still mimics $\mathcal{I}$ after this rule application, we define $\tau(w)$ such that Conditions (1) and (2) of Definition 4 hold for the two added labels. Specifically, since $\tau(v) \in (\exists r.D)^{\mathcal{I}}$, there exists some $d \in D^{\mathcal{I}}$ such that $\langle \tau(v), d \rangle \in r^{\mathcal{I}}$. Define $\tau(w) := d$. Then, since $\tau(w) = d \in D^{\mathcal{I}}$, Condition (1) of Definition 4 holds for $D \in L(v)$. Since $\langle \tau(v), \tau(w) \rangle = \langle \tau(v), d \rangle \in r^{\mathcal{I}}$, Condition (2) of Definition 4 holds for $r \in L(v, w)$.

Case $\forall$-Rule: If the $\forall$-Rule is applicable to T, then $\forall r.D \in L(v)$ for some $v \in V$ and $r \in L(v, w)$ for some $w \in V$. Since τ is a mimic of T in $\mathcal{I}$, we have $\tau(v) \in (\forall r.D)^{\mathcal{I}}$ and $\langle \tau(v), \tau(w) \rangle \in r^{\mathcal{I}}$. The application of the $\forall$-Rule adds D to $L(w)$. To show that T still mimics $\mathcal{I}$ after the rule application, it is sufficient to prove that $\tau(w) \in D^{\mathcal{I}}$, which clearly follows from $\tau(v) \in (\forall r.D)^{\mathcal{I}}$ and $\langle \tau(v), \tau(w) \rangle \in r^{\mathcal{I}}$.

Case $\bot$-Rule: If the $\bot$-Rule is applicable to T, then $\{A, \neg A\} \subseteq L(v)$ for some $v \in V$. Since τ is a mimic of T in $\mathcal{I}$, we have $\tau(v) \in A^{\mathcal{I}}$ and $\tau(v) \in (\neg A)^{\mathcal{I}} = \Delta^{\mathcal{I}} \setminus A^{\mathcal{I}}$. Clearly, this is not possible, which means that this case cannot occur. $\qquad\square$

Algorithm 1. A tableau algorithm for checking satisfiability $\mathcal{ALC}$ concepts

---

CSat(C): Checking satisfiability of a concept C
input   : an $\mathcal{ALC}$ concept C
output : **yes** if $C^{\mathcal{I}} \neq \emptyset$ for some interpretation $\mathcal{I}$ and **no** otherwise

1 $C \leftarrow \mathbf{NNF}(C)$;
2 $V \leftarrow \{v_0\}$, $L \leftarrow \{v_0 \mapsto \{C\}\}$;
3 $T \leftarrow (V, L)$;
4 **while not FullyExpanded(T) do**
5 | $R \leftarrow \mathbf{ChooseApplicableRule}(T)$;
6 |__ $T \leftarrow \mathbf{ApplyRule}(T, R)$;
7 **if** $\bot \in \bigcup_{v \in V} L(v)$ **then**
8 | **return no**;
9 **else**
10 |__ **return yes**;

Algorithm 1 summarizes our tableau procedure for checking satisfiability of $\mathcal{ALC}$ concepts. After converting the input concept to NNF (Lines 1) and initializing the tableau (Line 2–3), the algorithm continuously applies the tableau rules from Table 2 until the tableau is fully expanded (Lines 4–6). If the resulting tableau contains a clash (Line 7) the algorithm returns *no*; if not, the algorithm returns *yes*.

Note that due to the ⊔-Rule, the result of applying a rule (Line 6) is not uniquely determined. Hence, Algorithm 1 is *non-deterministic*. We next show that this algorithm solves the concept satisfiability problem for $\mathcal{ALC}$, i.e., it is *correct*. Recall (see Appendix A.1) that a non-deterministic algorithm A *solves* a problem $P \colon X \to \{yes, no\}$ if, for each $x \in X$ such that $P(x) = no$, each run of A terminates with the result *no*, and for each $x \in X$ such that $P(x) = yes$, there exists *at least one run* for which the algorithm terminates and produces *yes*. Proving correctness of a (non-deterministic) algorithm is usually accomplished by proving several properties: (1) *Soundness*: if for an input $x \in X$, A returns *yes* then $P(x) = yes$, (2) *Completeness*: if $P(x) = yes$ then A returns *yes* for at least one run, and (3) *Termination*: A terminates for every input. Soundness of Algorithm 1 follows from Lemma 6 since the algorithm returns *yes* only if a clash-free fully expanded tableau is computed. Completeness follows from Lemma 7 since, for a satisfiable concept, one can always apply the rules such that a clash is avoided. It is thus remains to prove that Algorithm 1 always terminates.

Remark 4. It may seem that a run of Algorithm 1 is determined not only by the choice of a possible expansion of the non-deterministic ⊔-Rule (Line 6), but also by the choice of which next rule to apply in case there are several applicable rules (Line 5). However, since Lemma 7 holds for any applicable rule, the latter choice does not have any impact on the completeness of the algorithm. In other words, we may assume that the function **ChooseApplicableRule**(T)

is *deterministic*, i.e., it always returns the same value for the same input (unlike function **ApplyRule**(T, R) for which different values need to be considered in different runs). The choice of *which* next rule to apply is usually referred to as *don't care non-determinism* of the algorithm, whereas the choice of *how* a rule is applied (the ⊔-Rule in our case) is referred to as a *don't know non-determinism*.

Exercise 6. The function **ChooseApplicableRule**(T) of Algorithm 1 can be defined in many different ways. If several rules are applicable to a node, e.g., the ⊓-Rule and the ⊔-Rule, the function may determine which of these rules should be applied first by specifying a *rule precedence*. If the same rule is applicable to different nodes, the function, likewise, can choose to which node the rule should be applied first. Discuss, which of these choices are more likely to result in fewer and/or shorter runs of the tableau procedure?

In order to prove termination of Algorithm 1, we show that the size of the tableau T that is constructed for a concept C at each step of the algorithm is bounded by an exponential function in the size of C (the number of symbols in C). This implies that every run of Algorithm 1 terminates after at most exponentially many rule applications since each rule application increases the size of the tableau. By Remark 3, each node label can contain only concepts that are sub-concepts of C or $\bot$, and each edge label can contain only roles that appear in C. Therefore, the maximal size of the label for each node and edge for each pair of nodes is bounded by a linear function in the size of C. We next show that the number of different nodes of a tableau is at most exponential in the size of C.

Definition 5. *For each node $v \in V$ of a tableau $T = (V, L)$, we define its* level $\ell(v)$ *by induction on the rule application of the tableau procedure:*

- *For the node v_0 created during tableau initialization we set $\ell(v_0) = 0$;*
- *For a node w created by an application of the $\exists$-Rule to a node $v \in V$, we set $\ell(w) = \ell(v) + 1$.*

The following lemma gives a bound on the number of nodes at each level of a tableau:

Lemma 8. *Let $T = (V, L)$ be a (possibly not fully expanded) tableau obtained for a concept C of size n. Then for each $k \geq 0$, the number of nodes v with $\ell(v) = k$ is bounded by n^k.*

Proof. The proof is by induction over $k \geq 0$.

Case $k = 0$: There exists only one node $v = v_0 \in V$ with $\ell(v) = 0$ since all other nodes are constructed by the $\exists$-Rule.

Case $k > 0$: Take any node $w \in V$ with $\ell(w) = k$. Since $k > 0$, w can only be constructed by an application of the $\exists$-Rule to some node v with $\ell(v) = k - 1$. This rule has been applied to some concept $\exists r.D \in L(v)$ and, after this rule application, the $\exists$-Rule can no longer be applied to $\exists r.D \in L(v)$. Hence, each

w with $\ell(v) = k$ is uniquely associated with a pair $\langle \exists r.D, v \rangle$ where $\exists r.D$ is a sub-concept of the original concept C and $\ell(v) = k - 1$. Since the number of sub-concepts of C is bounded by n and, by the induction hypothesis, the number of nodes v with $\ell(v) = k - 1$ is bounded by n^{k-1}, the number of nodes w with $\ell(v) = k$ is bounded by $n \cdot n^{k-1} = n^k$. $\square$

Finally, we prove that the level of a tableau node cannot exceed the quantifier depth of the input concept.

Definition 6. *The quantifier depth of an $\mathcal{ALC}$ concept C is a number $qd(C)$ that is defined inductively over* (1) *as follows:*

- $qd(\top) = qd(\bot) = qd(A) = 0$ *for each $A \in N_C$,*
- $qd(C \sqcap D) = qd(C \sqcup D) = \max(qd(C), qd(D))$,
- $qd(\neg C) = qd(C)$,
- $qd(\exists r.C) = qd(\forall r.C) = qd(C) + 1$.

Example 16.

$$qd(\exists r.((\neg A) \sqcap \forall r.B)) = qd((\neg A) \sqcap \forall r.B) + 1$$
$$= \max(qd(\neg A), qd(\forall r.B)) + 1$$
$$= \max(qd(A), qd(B) + 1) + 1$$
$$= \max(0, 0 + 1) + 1 = 2.$$

Note that $qd(C)$ is not greater than the number of quantifier symbols ($\exists$ or $\forall$) in C, which is not greater than the length of C.

Lemma 9. *Let $T = (V, L)$ be a (possibly not fully expanded) tableau obtained for a concept C with $qd(C) = q$, and $v \in V$ a node with $\ell(v) = k$. Then for each $D \in L(v)$, we have $qd(D) \leq q - k$.*

Proof. We prove the lemma by induction on the size (i.e., on the construction) of T.

If T is created during the tableau initialization, then $D = C$, $v = v_0$, and $k = 0$. Hence $qd(D) = qd(C) = q = q - k$ as required.

Otherwise, T was created by applying one of the tableau expansion rules in Table 2. If $D \in L(v)$ was not added by this rule, we can apply the induction hypothesis to the (smaller) tableau before the rule application. Otherwise, we consider all possible cases of such a rule that can add $D \in L(v)$:

Case $\sqcap$-Rule: If D was added by the $\sqcap$-Rule, then, before this rule application, $D \sqcap E \in L(v)$ or $E \sqcap D \in L(v)$ for some E. By applying the induction hypothesis for this concept, we obtain $qd(D) \leq qd(D \sqcap E) \leq q - k$ or $qd(D) \leq qd(E \sqcap D) \leq q - k$.

Case $\sqcup$-Rule: If D was added by the $\sqcup$-Rule, then, before this rule application, $D \sqcup E \in L(v)$ or $E \sqcup D \in L(v)$ for some E. By applying the induction hypothesis for this concept, we obtain $qd(D) \leq qd(D \sqcup E) \leq q - k$ or $qd(D) \leq qd(E \sqcup D) \leq q - k$.

Case $\exists$-*Rule:* If D was added by the $\exists$-Rule, then the node v was also created by this rule and the rule was applied to some $w \in V$ with $\exists r.D \in L(w)$ for some role r. Then $\ell(v) = \ell(w) + 1$ and, by induction hypothesis, $qd(\exists r.D) \leq q - \ell(w) = q - (\ell(v) - 1) = q - k + 1$. Since $qd(\exists r.D) = qd(D) + 1$, we obtain $qd(D) = qd(\exists r.D) - 1 \leq q - k$.

Case $\forall$-*Rule:* If D was added by the $\forall$-Rule, then this rule was applied to some $w \in V$ with $\forall r.D \in L(w)$ and $r \in L(w, v)$. Since $r \in L(w, v)$ could be only added by the $\exists$-Rule, $\ell(v) = \ell(w) + 1$. By induction hypothesis for $\forall r.D \in L(w)$, we obtain $qd(\forall r.D) \leq q - \ell(w) = q - (\ell(v) - 1) = q - k + 1$. Since $qd(\forall r.D) = qd(D) + 1$, we obtain $qd(D) = qd(\forall r.D) - 1 \leq q - k$.

Case $\bot$-*Rule:* If $D = \bot$ was added by the $\bot$-Rule, then, before this rule application, we have $\{A, \neg A\} \subseteq L(v)$ for some atomic concept A. By induction hypothesis, $qd(A) \leq q - k$. Hence $qd(\bot) = 0 = qd(A) \leq q - k$. $\square$

Let $T = (V, L)$ be a tableau constructed during a run of the tableau procedure for a concept C with size n and $qd(C) = q \leq n$. Since every node $v \in V$ of a tableau $T = (V, L)$ always contains at least one concept D in the label, by Lemma 9 it follows that $0 \leq qd(D) \leq q - \ell(v)$. Hence $\ell(v) \leq q$ holds for every $v \in V$. Since, by Lemma 8 the number of nodes at level k is bounded by n^k, the total number of nodes in the tableau is bounded by $\sum_{0 \leq k \leq q} n^k \leq n^{q+1} = 2^{(\log_2 n) \cdot (q+1)} \leq 2^{n^2}$.

Essentially we have shown that the tableau procedure constructs a special kind of interpretation (represented by the tableau). The interpretations have a *tree shape*: each node except for the initial node (the *root* of the tree) is connected by an edge to exactly one *predecessor* node from which this node was created (by an application of the $\exists$-Rule). The depth of the tree is bounded by the quantifier depth of the concept C for which the tableau was constructed. The branching degree of the tree (the maximal number of successor nodes of each node) is bounded by the number of existential concepts occurring in C. Hence the size of the tree is bounded *exponentially* in the size of C.

Exercise 7. Show that the exponential upper bound on the size of the tableau cannot be improved. Specifically, for each $n \geq 0$ construct a concept C_n of polynomial size in n (i.e., the number of symbols in C_n is bounded by $p(n)$ for some polynomial function p) such that the fully expanded tableau for C_n contains at least 2^n nodes. Hint: the tableau rules should create a binary tree of depth n. Hence by Lemma 9, $qd(C_n) \geq n$. The label of each non-leaf node should contain two different concepts of the form $\exists R.D$.

The exponential bound on the tableau size implies the following complexity result:

Theorem 1. *Algorithm 1 solves the concept satisfiability problem in* $\mathcal{ALC}$ *in non-deterministic exponential time.*

Remark 5. It is possible to make some further improvements to Algorithm 1 to prove better complexity bounds. First note that to check satisfiability of a concept C, it is not necessary to keep the whole tableau in memory. Once all tableau expansion rules are applied to a node (and the node is checked for the presence of a clash), the node can be completely deleted. By processing nodes in a depth-first manner it is, therefore, possible to store at most linearly many nodes in memory at any given time (because the tableau has a linear depth). This gives a non-deterministic *polynomial space* algorithm solving this problem. A well-known result from complexity theory called Savitch's theorem then implies that there is a *deterministic* polynomial-space algorithm solving this problem, which is now an optimal complexity bound for checking concept satisfiability in $\mathcal{ALC}$ (see, e.g., [6, Sect. 5.1.1]).

3.2 TBox Reasoning

In this section, we extend the tableau procedure presented earlier to also take into account TBox axioms of the ontology. Given an $\mathcal{ALC}$ concept C and an $\mathcal{ALC}$ TBox $\mathcal{O}$, our goal now is to check the satisfiability of C w.r.t. $\mathcal{O}$, i.e., to check if there exists a model $\mathcal{I}$ of $\mathcal{O}$ such that $C^{\mathcal{I}} \neq \emptyset$.

As in the case of the previous procedure, before applying the tableau rules, we first need to convert the input into a suitable normal form. We say that a TBox axiom is in *normal form* (or is *normalized*) if it is a concept inclusion axiom of the form $\top \sqsubseteq D$ where D is a concept in NNF. Every TBox axiom can be converted into the normal form by applying the following simple rewriting rules:

$$
\begin{aligned}
C \equiv D &\Rightarrow C \sqsubseteq D, \quad D \sqsubseteq C, \\
C \sqsubseteq D &\Rightarrow \top \sqsubseteq \neg C \sqcup D && \text{if } C \neq \top, \\
\top \sqsubseteq D &\Rightarrow \top \sqsubseteq \text{NNF}(D) && \text{if } D \text{ is not in NNF.}
\end{aligned}
$$

Exercise 8. Similarly to Exercise 5, show that TBox normalization preserves concept satisfiability w.r.t. the TBox. That is, a concept C is satisfiable w.r.t. a TBox $\mathcal{O}$ if and only if C is satisfiable w.r.t. the normalization of $\mathcal{O}$. Hint: show that for each rewrite step $\alpha \Rightarrow \beta$ above and each interpretation $\mathcal{I}$ we have $\mathcal{I} \models \alpha$ if and only if $\mathcal{I} \models \beta$.

To take the resulting axioms into account in our tableau procedure, we need to add an additional tableau expansion rule shown in Table 3. We can now use a simple modification of Algorithm 1, where, in addition to a (normalized) concept C, the input also contains a (normalized) ontology $\mathcal{O}$ and, in addition to the rules in Table 2, a new rule from Table 3 can be chosen and applied at Steps 5 and 6.

Example 17. Consider $C = A$ and $\mathcal{O} = \{A \sqcap \forall r.B \sqsubseteq \exists r.A\}$. We check satisfiability of C w.r.t. $\mathcal{O}$ using the tableau procedure. The concept C is already in NNF. The axiom in $\mathcal{O}$ is normalized as follows:

Table 3. The additional tableau expansion rule for handling (normalized) TBox axioms

Rule	Conditions	Expansions
T-Rule	$\top \sqsubseteq D \in \mathcal{O},\ D \notin L(x)$	Set $L(x) := L(x) \cup \{D\}$

$$A \sqcap \forall r.B \sqsubseteq \exists r.A \quad \Rightarrow \quad \top \sqsubseteq \neg(A \sqcap \forall r.B) \sqcup \exists r.A,$$
$$\Rightarrow \quad \top \sqsubseteq ((\neg A) \sqcup \exists r.\neg B) \sqcup \exists r.A.$$

The tableau is initialized to $T = (V, L)$ with $V = \{v_0\}$ and $L(v_0) = \{A\}$, and expanded by the following rule applications:

1. T-Rule: $L(v_0) := L(v_0) \cup \{((\neg A) \sqcup \exists r.\neg B) \sqcup \exists r.A\}$,
2. $\sqcup$-Rule: $L(v_0) := L(v_0) \cup \{(\neg A) \sqcup \exists r.\neg B\}$,
3. $\sqcup$-Rule: $L(v_0) := L(v_0) \cup \{\exists r.\neg B\}$,
4. $\exists$-Rule: $L(v_0, v_1) := \{r\}, \quad L(v_1) := \{\neg B\}$,
5. T-Rule: $L(v_1) := L(v_1) \cup \{((\neg A) \sqcup \exists r.\neg B) \sqcup \exists r.A\}$,
6. $\sqcup$-Rule: $L(v_1) := L(v_1) \cup \{(\neg A) \sqcup \exists r.\neg B\}$,
7. $\sqcup$-Rule: $L(v_1) := L(v_1) \cup \{\neg A\}$.

$$
\begin{array}{ll}
v_0 \bullet & A, \quad ((\neg A) \sqcup \exists r.\neg B) \sqcup \exists r.A \\
& (\neg A) \sqcup \exists r.\neg B, \quad \exists r.\neg B \\
r \downarrow & \\
v_1 \bullet & \neg B, \quad ((\neg A) \sqcup \exists r.\neg B) \sqcup \exists r.A \\
& (\neg A) \sqcup \exists r.\neg B, \quad \neg A
\end{array}
$$

Fig. 3. A possible tableau expansion for $C = A$ and $\mathcal{O} = \{\top \sqsubseteq (\neg A \sqcup \exists r.\neg B) \sqcup \exists r.A\}$

Figure 3 shows the resulting tableau expansion. Note that the new T-Rule is applied for both nodes v_0 and v_1 (Steps 1 and 5). Without applying this rule, no other rule would be applicable. Note that at Step 3 we have applied the $\sqcup$-Rule by adding the second disjunct $\exists r.\neg B$ to $L(v_0)$ because adding the first disjunct $\neg A$ would result in a clash since $A \in L(v_0)$. The same disjunction also appears in $L(v_1)$, but since $A \notin L(v_1)$, we could apply the $\sqcup$-Rule by adding the first disjunct $\neg A$ to $L(v_1)$ (Step 7).

Since after Step 7 the tableau is fully expanded and does not contain a clash, we conclude that C is satisfiable w.r.t. $\mathcal{O}$.

Intuitively, the new T-Rule ensures that the interpretation $\mathcal{I} = (\Delta^{\mathcal{I}}, \cdot^{\mathcal{I}})$ defined by $T = (V, L)$ (see Definition 3) satisfies all axioms in the (normalized) ontology $\mathcal{O}$ once all tableau rules are applied. Indeed, Lemma 6 still holds for the extended tableau procedure, since the proof of the lemma is by induction on the construction of a concept $C \in L(v)$ and we did not add any new concept constructors. Now, since T is fully expanded, for every normalized axiom $\top \sqsubseteq D \in \mathcal{O}$ and every $v \in V$, we have $D \in L(v)$ due to the $\top$-Rule. Hence, by Lemma 6, $v \in D^{\mathcal{I}}$ for every $v \in V = \Delta^{\mathcal{I}}$. Consequently, $\mathcal{I} \models \top \sqsubseteq D$ for each $\top \sqsubseteq D \in \mathcal{O}$. This implies that the extended Algorithm 1 remains sound.

Completeness of the extension of Algorithm 1 can also easily be shown. If C is satisfiable w.r.t. $\mathcal{O}$, then there exists a model $\mathcal{I} \models \mathcal{O}$ such that $C^{\mathcal{I}} \neq \emptyset$. We extend the proof of Lemma 7 to show that in this case, one can apply the tableau rules in such a way that T always mimics $\mathcal{I}$, thus, avoiding the production of a clash. For this we just need to update the proof with the case for the newly added rule:

Case $\top$ – Rule: If the $\top$-Rule is applicable to T, then $D \notin L(v)$ for some $v \in V$ and some $\top \sqsubseteq D \in \mathcal{O}$. The application of the $\top$-Rule adds only D to $L(v)$. To show that T still mimics $\mathcal{I}$ after this rule application, it is sufficient to prove that $\tau(v) \in D^{\mathcal{I}}$. Since $\top \sqsubseteq D \in \mathcal{O}$ and $\mathcal{I} \models \mathcal{O}$ we have $\top^{\mathcal{I}} = \Delta^{\mathcal{I}} \subseteq D^{\mathcal{I}}$. Hence $\tau(v) \in \Delta^{\mathcal{I}} \subseteq D^{\mathcal{I}}$.

Table 4. An additional expansion rule to handle TBox axioms of the form $C \sqsubseteq D$

Rule	Conditions	Expansions
$\sqsubseteq$-Rule	$C \sqsubseteq D \in \mathcal{O}$, $C \in L(x)$, $D \notin L(x)$	Set $L(x) := L(x) \cup \{D\}$

Exercise 9. In order to understand why the TBox axioms require a transformation to the form $\top \sqsubseteq D$, suppose we generalize the normal form to also permit axioms $C \sqsubseteq D$ where both C and D are in NNF, and formulate a new $\sqsubseteq$-Rule to handle axioms of this form as given in Table 4. Does the modified tableau algorithm remain sound and complete? Which of the lemmas cannot be proved any longer?

What happens if we only allow normalized axioms of the form $\top \sqsubseteq D$ and $A \sqsubseteq D$ where A is an atomic concept and D is a concept in NNF. Is the tableau algorithm with the $\top$-Rule and the $\sqsubseteq$-Rule sound and complete for this case?

Exercise 10. Description logic $\mathcal{ALCH}$ is an extension of the description logic $\mathcal{ALC}$, in which ontologies can contain *role inclusion axioms* of the form $r \sqsubseteq s$, where r and s are roles. An interpretation $\mathcal{I}$ satisfies $r \sqsubseteq s$ if $r^{\mathcal{I}} \subseteq s^{\mathcal{I}}$.

Extend the tableau procedure by adding a new rule to handle role inclusion axioms. Prove that this procedure is sound and complete. Use this procedure to show that the concept $C = A \sqcap \neg \exists s.(A \sqcap B) \sqcap \forall r.B$ is unsatisfiable w.r.t. $\mathcal{O} = \{A \sqsubseteq \exists r.A, r \sqsubseteq s\}$.

We have shown that the tableau algorithm extended with the $\top$-Rule remains sound and complete. In order to show that it solves the concept satisfiability problem w.r.t. TBoxes, it remains to show that it terminates for every input. Unfortunately, the latter is not the case as shown in the next example. Intuitively, since the $\top$-Rule adds a concept to every node label, this new concept can, in particular, trigger an application of the $\exists$-Rule, which, in turn, creates new nodes, for which the $\top$-Rule is applicable again.

Example 18. Consider $C = A$ and $\mathcal{O} = \{A \sqsubseteq \exists r.A\}$. We check the satisfiability of C w.r.t. $\mathcal{O}$ using the extended tableau procedure. The concept C is already in NNF, so we just need to normalize the axiom in $\mathcal{O}$:

$$A \sqsubseteq \exists r.A \quad \Rightarrow \quad \top \sqsubseteq (\neg A) \sqcup \exists r.A.$$

The tableau is initialized to $T = (V, L)$ with $V = \{v_0\}$ and $L(v_0) = \{A\}$, and expanded as shown in Fig. 4. Notice that unlike in Example 17, if we were to apply the $\sqcup$-Rule for $(\neg A) \sqcup \exists r.A \in L(v_1)$ by choosing the first disjunct $\neg A$, we would trigger a clash since $L(v_1)$ also contains A (added by $\exists$-Rule). Hence, the creation of infinitely many tableau nodes cannot be avoided.

1. $\top$-Rule: $L(v_0) := L(v_0) \cup \{(\neg A) \sqcup \exists r.A\}$,
2. $\sqcup$-Rule: $L(v_0) := L(v_0) \cup \{\exists r.A\}$,
3. $\exists$-Rule: $L(v_0, v_1) := \{r\}$, $L(v_1) := \{A\}$,
4. $\top$-Rule: $L(v_1) := L(v_1) \cup \{(\neg A) \sqcup \exists r.A\}$,
5. $\sqcup$-Rule: $L(v_1) := L(v_1) \cup \{\exists r.A\}$,
6. $\exists$-Rule: $L(v_1, v_2) := \{r\}$, $L(v_2) := \{A\}$,
7. $\top$-Rule: ...

$v_0 \quad A, \quad (\neg A) \sqcup \exists r.A$
$\quad\quad \exists r.A$
r
$v_1 \quad A, \quad (\neg A) \sqcup \exists r.A$
$\quad\quad \exists r.A$
r
$v_2 \quad A, \quad (\neg A) \sqcup \exists r.A$
$\quad\quad \exists r.A$

Fig. 4. The only clash-free tableau expansion for $C = A$ and $\mathcal{O} = \{\top \sqsubseteq (\neg A) \sqcup \exists r.A\}$

As discussed in Remark 5, to check satisfiability of a concept (also, with respect to an ontology), it is not necessary to keep the whole tableau in memory. We just need to verify that a clash-free tableau *exists*. This idea can be developed further to regain termination of the tableau algorithm with TBoxes. Notice that in Example 18, all nodes contain identical concepts in the labels. This means that if a rule is applicable to one node then it can be applied in exactly the same way to any other node with the same content. Hence, if a clash is obtained in this node, it is also obtained in the other node. Consequently, it is not necessary to apply the tableau rules to all nodes in order to verify that there exists a clash-free tableau expansion. Some rule applications can be *blocked*.

Definition 7. *A* blocking condition *is a function that assigns to every tableau* $T = (V, L)$ *a nonempty subset* $W \subsetneq V$ *of active nodes such that for every node* $v_1 \in W$ *and every node* $v_2 \notin W$ *such that* $L(v_1, v_2) \neq \emptyset$, *there exists a node* $w \in W$ *with* $L(v_2) \subseteq L(w)$. *In this case we say that a node* v_2 *is (directly)* blocked *by node* w. *Each node in* $V \setminus W$ *is called a* blocked *node.*

Algorithm 2. A tableau algorithm for checking satisfiability of $\mathcal{ALC}$ concepts with respect to $\mathcal{ALC}$ ontologies

---

COSat$(C, \mathcal{O})$: Checking satisfiability of a concept C w.r.t. an ontology $\mathcal{O}$
input : an $\mathcal{ALC}$ concept C, an $\mathcal{ALC}$ ontology $\mathcal{O}$
output : yes if $C^{\mathcal{I}} \neq \emptyset$ for some model $\mathcal{I}$ of $\mathcal{O}$ and **no** otherwise

1 $C \leftarrow \mathbf{NNF}(C)$;
2 $\mathcal{O} \leftarrow \mathbf{Normalize}(\mathcal{O})$;
3 $V \leftarrow \{v_0\}$, $L \leftarrow \{v_0 \mapsto \{C\}\}$;
4 $T \leftarrow (V, L)$;
5 **while not** $\mathbf{FullyExpandedUpToBlocking}(T)$ **do**
6 $\quad\mid\quad R \leftarrow \mathbf{ChooseApplicableRule}(T)$;
7 $\quad\mid\quad T \leftarrow \mathbf{ApplyRule}(T, R)$;

8 **if** $\bot \in \bigcup_{v \in V} L(v)$ **then**
9 $\quad\mid\quad$ **return no**;
10 **else**
11 $\quad\mid\quad$ **return yes**;

Example 19. Let $T = (V, L)$ be the tableau obtained after Step 6 in Example 18, i.e., $V = \{v_0, v_1, v_2\}$, $L(v_0) = L(v_1) = \{A, (\neg A) \sqcup \exists r.A, \exists r.A\}$, $L(v_2) = \{A\}$, and $L(v_0, v_1) = L(v_1, v_2) = \{r\}$. Then a blocking condition for T can be defined by setting $W = \{v_0\}$ since for $\langle v_0, v_1 \rangle \in E$ we have $L(v_1) \subseteq L(v_0)$.

We leave it open, how exactly the set of active nodes of a tableau is determined, so that different blocking strategies can be used in different algorithms. We show next that one can restrict the tableau algorithm to apply rules only to active nodes.

Definition 8. *Let $T = (V, L)$ be a tableau with a subset $W \subseteq V$ of active nodes. We say that a tableau rule from Tables 2 or 3 is applicable to a node $v \in V$ if the conditions of this rule are satisfied for a mapping $x \mapsto v$. We say that a tableau T is* fully expanded up to blocking *if no tableau rule is applicable to any active node $w \in W$.*

Example 20. It is easy to see that the tableau T from Example 19 with $W = \{v_0\}$ is fully expanded up to blocking since no tableau rule (for the ontology $\mathcal{O}$ from Example 18) is applicable to v_0.

Algorithm 2 is a modification of Algorithm 1 for checking satisfiability of concepts w.r.t. ontologies. Apart from a new step for normalization of the input ontology (Line 1), the algorithm uses a blocking condition to verify if the tableau is fully expanded up to blocking according to Definition 8 (Line 5), and to select a rule applicable to an active node (Line 6). We assume that the initial node v_0 always remains active.

We next show that the updated algorithm remains sound and complete. The introduction of a blocking condition does not have any impact on completeness:

the proof of Lemma 7 (including the new case for the $\top$-Rule) remains as before. Soundness of the new algorithm is, however, nontrivial: if the tableau is fully expanded with blocking, it does not mean that it is fully expanded without blocking. To prove soundness, we modify Definition 3 by taking the blocking condition into account:

Definition 9. *A tableau $T = (V, L)$ and a subset of active nodes $W \subseteq V$ define an interpretation $\mathcal{I} = (\Delta^{\mathcal{I}}, \cdot^{\mathcal{I}})$ such that:*

- *$\Delta^{\mathcal{I}} = W$,*
- *$A^{\mathcal{I}} = \{x \in W \mid A \in L(x)\}$ for each atomic concept $A \in N_C$,*
- *$r^{\mathcal{I}} = \{\langle x, y \rangle \in W \times W \mid \exists z \in V : r \in L(x, z) \text{ and } z = y \text{ or } z \text{ is blocked by } y\}$*
  for each atomic role $r \in N_R$.

A few comments about Definition 9 are in order. First note that this definition coincides with Definition 3 when $W = V$. The definition of $r^{\mathcal{I}}$ can be explained as follows: if $r \in L(x, z)$ and both nodes x and z are active, then $\langle x, z \rangle \in r^{\mathcal{I}}$. This corresponds to the case '$z = y$' of the definition for $r^{\mathcal{I}}$. If x active but z is not, then, since $L(x, z) \neq \emptyset$, z should be blocked by some $y \in W$. In this case, $r^{\mathcal{I}}$ contains all such pairs $\langle x, y \rangle$. This corresponds to the case 'z is blocked by y' of the definition for $r^{\mathcal{I}}$. If x not active then the label $r \in L(x, z)$ is ignored.

$\mathcal{I} = (\Delta^{\mathcal{I}}, \cdot^{\mathcal{I}})$, where:
- $\Delta^{\mathcal{I}} = \{v_0, v_2\}$,
- $A^{\mathcal{I}} = \{v_0, v_2\}$, $B^{\mathcal{I}} = \{v_2\}$, $C^{\mathcal{I}} = \{v_0\}$,
- $r^{\mathcal{I}} = \{\langle v_0, v_0 \rangle, \langle v_0, v_2 \rangle, \langle v_2, v_2 \rangle, \langle v_2, v_0 \rangle\}$.

Fig. 5. A tableau with blocking (v_0 and v_2 are are active nodes, v_1 and v_3 are blocked nodes) and the interpretation defined by this tableau

Example 21. Consider the tableau $T = (V, L)$ illustrated on the left-hand side of Fig. 5:

- $V = \{v_0, v_1, v_2, v_3\}$,
- $L(v_0) = \{A, C\}$, $L(v_1) = \{C\}$, $L(v_2) = \{A, B\}$, $L(v_3) = \{A\}$,
- $L(v_0, v_1) = L(v_0, v_2) = L(v_2, v_3) = \{r\}$.

Suppose that the set of active nodes is $W = \{v_0, v_2\}$. Note that v_1 is blocked by v_0 since $L(v_1) = \{C\} \subseteq \{A, C\} = L(v_0)$, v_3 is blocked by v_2 since $L(v_3) = \{A\} \subseteq \{A, B\} = L(v_2)$, and v_3 is blocked by v_0 since $L(v_3) = \{A\} \subseteq \{A, C\} = L(v_0)$. Then T and W define an interpretation shown in the right of Fig. 5.

Let $T = (V, L)$ be a tableau obtained by applying the tableau expansion rules for a concept C and an ontology $\mathcal{O}$. Suppose that T is fully expanded up to blocking for $W \subseteq V$ and does not contain a clash. Let $\mathcal{I}$ be the interpretation defined by T and W according to Definition 9. Our goal is to show that $\mathcal{I} \models \mathcal{O}$ and $C^{\mathcal{I}} \neq \emptyset$, which implies that C is satisfiable w.r.t. $\mathcal{O}$. To do this, we prove an analog of Lemma 6 for our new interpretation $\mathcal{I}$.

Lemma 10. *Let $T = (V, L)$ be a clash-free, fully expanded tableau up to blocking for $W \subseteq V$ and $\mathcal{I} = (\Delta^{\mathcal{I}}, \cdot^{\mathcal{I}})$ an interpretation defined by T and W according to Definition 9. Then for every $v \in W$ and every $C \in L(v)$, we have $v \in C^{\mathcal{I}}$.*

Proof. Just like for Lemma 6, we prove this lemma by induction on the construction of C according to the grammar definition (6). The only changes compared to the previous proof are those cases where the definition of $r^{\mathcal{I}}$ was used:

Case $C = \exists r.D$: Since the $\exists$-Rule is not applicable to $\exists r.D \in L(v)$, there exists some $w \in V$ such that $r \in L(v, w)$ and $D \in L(w)$. Define an active node $w' \in W$ as follows. If $w \in W$, we set $w' = w$. Otherwise, w must be blocked by some $w' \in W$. Then, by the definition of $r^{\mathcal{I}}$, we have $\langle v, w' \rangle \in r^{\mathcal{I}}$. Note also that $D \in L(w) \subseteq L(w')$. Since $w' \in W$, by induction hypothesis $w' \in D^{\mathcal{I}}$. From $\langle v, w' \rangle \in r^{\mathcal{I}}$ and $w' \in D^{\mathcal{I}}$ we obtain $v \in (\exists r.D)^{\mathcal{I}}$.

Case $C = \forall r.D$: In order to prove that $v \in C^{\mathcal{I}} = (\forall r.D)^{\mathcal{I}}$, take any $w' \in \Delta^{\mathcal{I}} = W$ such that $\langle v, w' \rangle \in r^{\mathcal{I}}$. We prove that $w' \in D^{\mathcal{I}}$. Since $\langle v, w' \rangle \in r^{\mathcal{I}}$, by definition of $r^{\mathcal{I}}$, there exists some $w \in V$ such that $r \in L(v, w)$ and either $w = w'$ or w is blocked by w'. In both cases $L(w) \subseteq L(w')$. Since the $\forall$-Rule is not applicable to $\forall r.D \in L(v)$ and $r \in L(v, w)$, we must have $D \in L(w) \subseteq L(w')$. Since $w' \in W$, from $D \in L(w')$ by induction hypothesis, we obtain $w' \in D^{\mathcal{I}}$, as required.   □

Exercise 11. Identify the places where the properties of a blocking condition (Definition 7) have been used in the proof of Lemma 10. Can the blocking condition be relaxed in such a way that more nodes of a tableau can potentially be blocked, but the proof of Lemma 10 can be still repeated? For example, do we really need that *all concepts* of $L(v_2)$ are contained in $L(w)$?

Finally, we consider the question of termination of Algorithm 2. Clearly, the algorithm does not terminate for every blocking condition. For example, as shown in Example 18, the algorithm does not terminate if all nodes are active, i.e., $W = V$. Hence, we need to make some further assumptions about the blocking condition in order to show termination.

Definition 10. *The* eager blocking *condition (for a root node w_0) assigns to every tableau $T = (V, L)$ a minimal (w.r.t. set inclusion) set of active nodes $W \subseteq V$ containing w_0 that satisfies the condition of Definition 7.*

Intuitively, the eager blocking condition for $T = (V, L)$ can be implemented as follows. We first set $W = \{w_0\}$. Then, repeatedly for every $v_1 \in W$ and

$v_2 \in V \setminus W$ such that $L(v_1, v_2) \neq \emptyset$, check if there exists $w \in W$ such that $L(v_2) \subseteq L(w)$. If there is no such element, we add v_2 to W. This process continues until no further nodes can be added. Note that the resulting set W depends on the order in which the nodes v_2 are processed. In practice, the set W can dynamically be updated when applying tableau rules. The eager blocking condition is related to the notion of *anywhere blocking* [42].

Lemma 11. *Let C and $\mathcal{O}$ be inputs of Algorithm 2 with the combined size n (i.e., the total number of symbols). Then each run of Algorithm 2 terminates in at most doubly exponential time in the size of n provided an eager blocking condition is used.*

Proof. Without loss of generality, we may assume that C and $\mathcal{O}$ are normalized since this step can be performed in linear time. Let $T = (V, L)$ be a tableau obtained during the run of the algorithm. For each node $v \in V$ of the tableau, we define its level $\ell(n)$ as in Definition 5. We will prove that $\ell(v) \leq 2^n$ for each node $v \in V$.

Assume to the contrary that there exists $w \in V$ with $\ell(w) = 2^n + 1$. Then w must have been created by the $\exists$-Rule from some $v \in W$ with $\ell(v) = \ell(w) - 1 = 2^n$, where $W \subseteq V$ is the set of active nodes of the tableau at the moment of this rule application. Since $\ell(v) = 2^n$, there must exist nodes $v_0, v_1, \ldots, v_{2^n} = v$ such that $L(v_{i-1}, v_i) \neq \emptyset$ ($1 \leq i \leq 2^n$). It is easy to see that $v_i \in W$ for all i with $0 \leq i \leq 2^n$. Indeed, otherwise there exists a maximal such i such that $v_i \notin W$ ($0 \leq i \leq 2^n$). Since $v_{2^n} = v \in W$, then $i < 2^n$ and $v_0 \neq v_{i+1} \in W$. But then one can remove v_{i+1} from W without violating the conditions of Definition 7 since $L(w, v_{i+1}) = \emptyset$ for all $w \in W$. This contradicts our assumption that W is a minimal set of active nodes containing v_0.

Now consider the sets of concepts $S_i = L(v_i)$ in the labels of v_i ($0 \leq i \leq 2^n$). Since each node label can contain only concepts that appear either in C or in $\mathcal{O}$ (possibly as sub-concepts) and the number of such concepts is bounded by the total combined length n of the input, there can be at most 2^n different subsets among S_i ($0 \leq i \leq 2^n$). By the pigeonhole principle, there exist some indexes i and j ($0 \leq i < j \leq 2^n$) such that $S_i = S_j$. But then node $v_j \neq v_0$ can be removed from W without violating the conditions of Definition 7 since for each $v \in W$ with $L(v, v_j) \neq \emptyset$, there exists $w = v_i \in W$ such that $L(v_i) \subseteq L(w)$ because $L(w) = L(v_i) = L(v_j)$. This again contradicts our assumption that W is a minimal set of active nodes. The obtained contradiction, therefore, proves that $\ell(v) \leq 2^n$ for every $v \in V$.

Now, by Lemma 8, the total number of tableau nodes is bounded by $\sum_{0 \leq k \leq 2^n} n^k \leq n^{2^n + 1} = 2^{(\log_2 n) \cdot (2^n + 1)} \leq 2^{2^{n^2}}$. Since each node contains at most n concept labels and every application of a tableau rule introduces at least one of them, each run of Algorithm 2 terminates after at most double exponentially many steps. $\square$

Theorem 2. *Algorithm 2 solves the concept satisfiability problem with respect to ontologies expressed in $\mathcal{ALC}$ in non-deterministic doubly exponential time.*

As with Algorithm 1, the complexity bound provided by Algorithm 2 is not optimal and can be improved to *deterministic exponential* time (see, e.g., [6, Sect. 5.1.2]).

Note that the requirements about the blocking condition used in Lemma 11 can be relaxed. Indeed, in the proof of the lemma we only used that a node v_j is directly blocked by an *ancestor* node v_i, i.e., a node from which v_j was created by a sequence of $\exists$-Rule applications.

Exercise 12 (Advanced). Is it possible to improve the upper bound shown in the proof of Lemma 11 to *single exponential* time? Which additional assumptions about the blocking condition are necessary for this? Hint: for every tableau $T = (V, L)$ used in the computation, consider the set of all subsets of the labels of the nodes: $P = \{S \subseteq L(v) \mid v \in V\}$. How can this set change after a tableau rule application? How many times can this set change during the tableau run? What is the maximal possible number of consequent rule applications that do not change this set?

4   Axiom Pinpointing

In Sect. 2.3, we have discussed several ontology reasoning problems and how they can help in detecting modeling errors in ontologies. For example, inconsistency of an ontology indicates that the modeled domain cannot match any model of the ontology since the ontology does not have models. In Sect. 3 we have shown how to check ontologies for consistency and solve other reasoning problems using tableau procedures. Knowing that an ontology is inconsistent, however, does not tell much about what exactly *causes* the inconsistency let alone how to *repair* it.

Recall from Sect. 2.3, that all reasoning problems can be reduced to concept subsumption checking. For example, by Lemma 5, an ontology $\mathcal{O}$ is unsatisfiable if and only if $\mathcal{O} \models \top \sqsubseteq \bot$. *Axiom pinpointing* methods can help the user to identify the exact axioms that are responsible for this or any other entailment.

Definition 11. *A* justification *for an entailment $\mathcal{O} \models \alpha$ is a subset of axioms $J \subseteq \mathcal{O}$ such that $J \models \alpha$ and for every $J' \subsetneq J$, we have $J' \not\models \alpha$.*

In other words, a justification for an entailment $\mathcal{O} \models \alpha$ is a minimal set of axioms of the ontology that entails α. Note that since $\mathcal{O} \models \alpha$, at least one justification for the entailment exists. Indeed, either $J_0 = \mathcal{O}$ satisfies the condition of Definition 11, or there exits $J_1 \subsetneq J_0$ such that $J_1 \models \alpha$. Similarly, either J_1 is a justification or there exists some $J_2 \subsetneq J_1$ such that $J_2 \models \alpha$, etc. At some point this process stops since $\mathcal{O}$ contains only finitely many axioms and J_i $(i \geq 0)$ gets smaller with every step. Therefore, the last set J_k will be a justification for $\mathcal{O} \models \alpha$.

Note that we say *a* justification instead of *the* justification. Indeed, Definition 11 does not imply that a justification must be unique as the following example shows.

Example 22. Consider the following entailment:

$$\mathcal{O} = \{A \sqsubseteq B, B \sqsubseteq C, A \sqsubseteq C, A \sqcap B \sqsubseteq \bot\} \models \alpha = A \sqsubseteq C.$$

This entailment has three different justifications:

- $J_1 = \{A \sqsubseteq B, B \sqsubseteq C\}$,
- $J_2 = \{A \sqsubseteq C\}$,
- $J_3 = \{A \sqsubseteq B, A \sqcap B \sqsubseteq \bot\}$.

Indeed, it is easy to see that $J_i \models \alpha$ for $1 \leq i \leq 3$. For example, $J_3 \models A \sqsubseteq C$ because for every model $\mathcal{I} \models J_3$ we have $A^{\mathcal{I}} \subseteq A^{\mathcal{I}} \cap B^{\mathcal{I}} \subseteq \bot^{\mathcal{I}} = \emptyset \subseteq C^{\mathcal{I}}$. We can show that each J_i satisfies the remaining condition of Definition 11 by enumerating all proper subsets of J_i $(1 \leq i \leq 3)$:

- J_1 has only the proper subsets $M_0 = \emptyset$, $M_1 = \{A \sqsubseteq B\}$ and $M_2 = \{B \sqsubseteq C\}$,
- J_2 has only the proper subset $M_0 = \emptyset$,
- J_3 has only the proper subsets $M_0 = \emptyset$, $M_1 = \{A \sqsubseteq B\}$, and $M_3 = \{A \sqcap B \sqsubseteq \bot\}$.

We can show that none of these subsets M_i entails α by presenting the corresponding counter-models $\mathcal{I}_i = (\Delta^{\mathcal{I}_i}, \cdot^{\mathcal{I}_i})$ such that $\mathcal{I}_i \models M_i$ but $\mathcal{I}_i \not\models \alpha$ $(0 \leq i \leq 3)$:

- For $M_0 = \emptyset$ take $\mathcal{I}_0 = (\Delta^{\mathcal{I}_0}, \cdot^{\mathcal{I}_0})$ with $\Delta^{\mathcal{I}_0} = \{a\}$, $A^{\mathcal{I}_0} = \{a\}$ and $C^{\mathcal{I}_0} = \emptyset$. Clearly, $\mathcal{I}_0 \models M_0$ but $\mathcal{I}_0 \not\models A \sqsubseteq C$ since $A^{\mathcal{I}_0} = \{a\} \not\subseteq \emptyset = C^{\mathcal{I}_0}$.
- For $M_1 = \{A \sqsubseteq B\}$ take $\mathcal{I}_1 = (\Delta^{\mathcal{I}_1}, \cdot^{\mathcal{I}_1})$ with $\Delta^{\mathcal{I}_1} = \{a\}$, $A^{\mathcal{I}_1} = B^{\mathcal{I}_1} = \{a\}$, and $C^{\mathcal{I}_1} = \emptyset$. Clearly, $\mathcal{I}_1 \models M_1$ because $A^{\mathcal{I}_1} = \{a\} \subseteq \{a\} = B^{\mathcal{I}_1}$ but $\mathcal{I}_1 \not\models A \sqsubseteq C$ similarly as for $\mathcal{I}_0$.
- For $M_2 = \{B \sqsubseteq C\}$ take $\mathcal{I}_2 = (\Delta^{\mathcal{I}_2}, \cdot^{\mathcal{I}_2})$ with $\Delta^{\mathcal{I}_2} = \{a\}$, $A^{\mathcal{I}_2} = \{a\}$, and $B^{\mathcal{I}_2} = C^{\mathcal{I}_2} = \emptyset$. Clearly, $\mathcal{I}_2 \models M_2$ because $B^{\mathcal{I}_2} = \emptyset \subseteq \emptyset = C^{\mathcal{I}_2}$ but $\mathcal{I}_2 \not\models A \sqsubseteq C$ similarly as for $\mathcal{I}_0$ and $\mathcal{I}_1$.
- For $M_3 = \{A \sqcap B \sqsubseteq \bot\}$ take $\mathcal{I}_3 = \mathcal{I}_2$ from the previous case. Clearly, $\mathcal{I}_2 \models M_3$ because $(A \sqcap B)^{\mathcal{I}_2} = A^{\mathcal{I}_2} \cap B^{\mathcal{I}_2} = \{a\} \cap \emptyset = \emptyset \subset \bot^{\mathcal{I}_2} = \emptyset$ but $\mathcal{I}_2 \not\models A \sqsubseteq C$.

How many justifications may an entailment have? The following example shows that the number of justifications can be exponential in the size of the ontology.

Example 23. Consider the following ontology $\mathcal{O}$ and $\alpha = A_0 \sqsubseteq A_n$:

$$\mathcal{O} = \{A_{i-1} \sqsubseteq B \sqcap A_i, A_{i-1} \sqsubseteq C \sqcap A_i \mid 1 \leq i \leq n\},$$

where A_i $(0 \leq i \leq n)$, B, and C are atomic concepts. Note that, for each i with $1 \leq i \leq n$, we have $\mathcal{O} \models A_{i-1} \sqsubseteq A_i$ because $A_{i-1} \sqsubseteq B \sqcap A_i \models A_{i-1} \sqsubseteq A_i$ (or $A_{i-1} \sqsubseteq C \sqcap A_i \models A_{i-1} \sqsubseteq A_i$). Consequently, $\mathcal{O} \models \alpha = A_0 \sqsubseteq A_n$. However, there are 2^n minimal subsets $J \subseteq \mathcal{O}$ such that $J \models \alpha$. Indeed, since each axiom $A_{i-1} \sqsubseteq A_i$ follows from two different axioms, J must include one of them for each i $(1 \leq i \leq n)$. Hence, there are 2^n possible variants for each J.

Algorithm 3. Minimizing entailment

Minimize($\mathcal{O}, \alpha$): compute a justification for $\mathcal{O} \models \alpha$
input   : an ontology $\mathcal{O}$ and an axiom α such that $\mathcal{O} \models \alpha$
output : a minimal subset $J \subseteq \mathcal{O}$ such that $J \models \alpha$ (cf. Definition 11)

1 $J \leftarrow \mathcal{O}$;
2 **for** $\beta \in \mathcal{O}$ **do**
3   **if** $J \setminus \{\beta\} \models \alpha$ **then**
4     $\lfloor \; J \leftarrow J \setminus \{\beta\}$;

5 **return** J;

Specifically, let $S \subseteq \{i \mid 1 \leq i \leq n\}$ be any subset of indexes between 1 and n. There are in total 2^n such subsets. For each such subset S, define

$$J_S = \{A_{i-1} \sqsubseteq B \sqcap A_i \mid i \in S\} \cup \{A_{i-1} \sqsubseteq C \sqcap A_i \mid i \notin S\} \subseteq \mathcal{O}.$$

Clearly, $J_{S_1} \neq J_{S_2}$ for each $S_1 \neq S_2$. Furthermore, note that for each i with $1 \leq i \leq n$, we have $J_S \models A_{i-1} \sqsubseteq A_i$: if $i \in S$ then $J \ni A_{i-1} \sqsubseteq B \sqcap A_i \models A_{i-1} \sqsubseteq A_i$; if $i \notin S$ then $J \ni A_{i-1} \sqsubseteq C \sqcap A_i \models A_{i-1} \sqsubseteq A_i$. Hence $J_S \models \alpha$.

To show that each J_S is a justification for $\mathcal{O} \models \alpha$, it remains to show that $J' \not\models \alpha$ for every $J' \subsetneq J_S$. Indeed, if $J' \subsetneq J_S$ then for some k with $1 \leq k \leq n$, we have $A_{k-1} \sqsubseteq B \sqcap A_k \notin J_S$ and $A_{k-1} \sqsubseteq C \sqcap A_k \notin J_S$. Let $\mathcal{I} = (\Delta^{\mathcal{I}}, \cdot^{\mathcal{I}})$ be an interpretation with $\Delta^{\mathcal{I}} = \{a\}$, $A_i^{\mathcal{I}} = B^{\mathcal{I}} = C^{\mathcal{I}} = \{a\}$ for $0 \leq i < k$, and $A_i^{\mathcal{I}} = \emptyset$ for $k \leq i \leq n$. It is easy to see that $\mathcal{I} \models A_{i-1} \sqsubseteq B \sqcap A_i$ and $\mathcal{I} \models A_{i-1} \sqsubseteq C \sqcap A_i$ for each i with $1 \leq i < k$ or with $k < i \leq n$. Indeed, if $1 \leq i < k$, then $A_{i-1}^{\mathcal{I}} = \{a\} \subseteq \{a\} \cap \{a\} = (B \sqcap A_i)^{\mathcal{I}} = (C \sqcap A_i)^{\mathcal{I}}$. If $k < i \leq n$, then $A_{i-1}^{\mathcal{I}} = \emptyset \subseteq \{a\} \cap \emptyset = (B \sqcap A_i)^{\mathcal{I}} = (C \sqcap A_i)^{\mathcal{I}}$. Hence, $\mathcal{I} \models J'$. Since $A_0^{\mathcal{I}} = \{a\} \not\subseteq \emptyset = A_n^{\mathcal{I}}$, we have $\mathcal{I} \not\models \alpha$. Consequently, $J' \not\models \alpha$.

Assuming we have an algorithm for testing entailment of axioms, e.g., the tableau procedure described in Sect. 3, we are now concerned with the question of how to *compute* justifications in particular for the entailment of concept inclusions.

4.1   Computing One Justification

Computing one justification for $\mathcal{O} \models \alpha$ is relatively easy. Starting from $J = \mathcal{O}$, we repeatedly remove axioms from J if this does not break the entailment $J \models \alpha$. At a certain point, no axioms can be removed without breaking the entailment, which implies that J is justification for $\mathcal{O} \models \alpha$. Algorithm 3 summarizes this idea.

Example 24. The following table shows a run of Algorithm 3 on the input $\mathcal{O}$ and α from Example 22. Each row of the table shows the value of the variables J and β in the beginning of each for-loop iteration (Lines 2–4). The last column shows the result of the evaluation of the if-statement in Line 3. The last line shows the (returned) value of J after the last iteration of the loop.

J	β	$J \setminus \{\beta\} \models^? \alpha = A \sqsubseteq C$
$\{A \sqsubseteq B, B \sqsubseteq C, A \sqsubseteq C, A \sqcap B \sqsubseteq \bot\}$	$A \sqsubseteq B$	*yes*
$\{B \sqsubseteq C, A \sqsubseteq C, A \sqcap B \sqsubseteq \bot\}$	$B \sqsubseteq C$	*yes*
$\{A \sqsubseteq C, A \sqcap B \sqsubseteq \bot\}$	$A \sqsubseteq C$	*no*
$\{A \sqsubseteq C, A \sqcap B \sqsubseteq \bot\}$	$A \sqcap B \sqsubseteq \bot$	*yes*
$\{A \sqsubseteq C\}$	-	-

As we can see, the algorithm returns the justification $J_2 = \{A \sqsubseteq C\}$.

The correctness of Algorithm 3 relies on the fact that the entailment relation $\mathcal{O} \models \alpha$ between an ontology $\mathcal{O}$ and an axioms α is *monotonic* over axiom additions to $\mathcal{O}$:

Lemma 12. *Let J_1 and J_2 be two sets of $\mathcal{ALC}$ axioms such that $J_1 \subseteq J_2$, and α an $\mathcal{ALC}$ axiom. Then $J_1 \models \alpha$ implies $J_2 \models \alpha$.*

Proof. It is equivalent to show that $J_2 \not\models \alpha$ implies $J_1 \not\models \alpha$. If $J_2 \not\models \alpha$ then there exists a model $\mathcal{I} \models J_2$ such that $\mathcal{I} \not\models \alpha$. Since $J_1 \subseteq J_2$ and $\mathcal{I} \models J_2$, we have $\mathcal{I} \models J_1$. Since $\mathcal{I} \models J_1$ and $\mathcal{I} \not\models \alpha$, we obtain $J_1 \not\models \alpha$, as required.   □

Note that in the proof of Lemma 12 we did not rely on any specific constructors of $\mathcal{ALC}$. We have only used that the entailment relation $J \models \alpha$ is defined by means of interpretations, i.e., $J \models \alpha$ iff $\mathcal{I} \models \alpha$ for every $\mathcal{I} \models J$ and $\mathcal{I} \models J$ iff $\mathcal{I} \models \beta$ for every $\beta \in J$. Although most standard DLs (including those that underpin the OWL standard) have such a *classical semantics*, there are some *non-monotonic DLs* in which the entailment relation is defined in other ways, e.g., as a result of performing certain operations [11,14,18]. From now on we assume that we deal only with monotonic (classical) entailment relations. We show that Algorithm 3 is correct in such cases.

Theorem 3. *Let J be the output of Algorithm 3 for the input $\mathcal{O}$ and α such that $\mathcal{O} \models \alpha$. Then J is a justification for $\mathcal{O} \models \alpha$.*

Proof. Clearly, $J \models \alpha$ since we only assign subsets that entail α to the variable J (in Lines 1 and 4). If J is not a justification for $\mathcal{O} \models \alpha$, by Definition 11 there exists some $J' \subsetneq J$ such that $J' \models \alpha$. Since $J' \subsetneq J$, there exists some $\beta \in J \setminus J' \subseteq \mathcal{O}$. Let J'' be the value of the variable J of Algorithm 3 at the beginning of the for-loop (Line 2) when $\beta \in \mathcal{O}$ was processed. Since $\beta \notin J' \subseteq J \subseteq J''$, we have $J' \subseteq J \setminus \{\beta\} \subseteq J'' \setminus \{\beta\}$. Since $J' \models \alpha$, by Lemma 12, $J'' \setminus \{\beta\} \models \alpha$. Hence β must have been removed from J'' at Line 4, and consequently, $\beta \notin J$. This contradicts $\beta \in J \setminus J'$, which proves that there is no $J' \subsetneq J$ such that $J' \models \alpha$. Hence J is a justification for $\mathcal{O} \models \alpha$.   □

Finally, observe that a run of Algorithm 3 requires exactly n subsumption tests. Hence, the complexity of computing one justification is bounded by a linear function over the complexity of entailment checking. In particular, one justification for concept subsumptions in $\mathcal{ALC}$ can be computed in exponential time.

4.2 Computing All Justifications

A justification for $\mathcal{O} \models \alpha$ contains axioms that are responsible for *one* reason for the entailment. As we have seen in Examples 22 and 23, there can be several different justifications. To repair an unwanted entailment $\mathcal{O} \models \alpha$, it is, therefore, necessary to change an axiom in every justification of $\mathcal{O} \models \alpha$. How do we compute *all* justifications?

Note that the output of Algorithm 3 depends on the *order* in which the axioms in $\mathcal{O}$ are enumerated in the for-loop (Line 2). Different orders of the axioms can result in different removals and, consequently, different justifications.

Example 25. Consider the run of Algorithm 3 on the input $\mathcal{O}$ and α from Example 22, where the axioms in $\mathcal{O}$ are enumerated in the reverse order as in Example 24.

J	β	$J \setminus \{\beta\} \models^? \alpha = A \sqsubseteq C$
$\{A \sqcap B \sqsubseteq \bot, A \sqsubseteq C, B \sqsubseteq C, A \sqsubseteq B\}$	$A \sqcap B \sqsubseteq \bot$	*yes*
$\{A \sqsubseteq C, B \sqsubseteq C, A \sqsubseteq B\}$	$A \sqsubseteq C$	*yes*
$\{B \sqsubseteq C, A \sqsubseteq B\}$	$B \sqsubseteq C$	*no*
$\{B \sqsubseteq C, A \sqsubseteq B\}$	$A \sqsubseteq B$	*no*
$\{B \sqsubseteq C, A \sqsubseteq B\}$	-	-

In this case, the algorithm returns the justification $J_1 = \{A \sqsubseteq B, B \sqsubseteq C\}$.

Exercise 13. For which order of axioms in $\mathcal{O}$ does Algorithm 3 return the justification $J_3 = \{A \sqsubseteq B, A \sqcap B \sqsubseteq \bot\}$ from Example 22?

Exercise 14. Prove that for each justification J of an entailment $\mathcal{O} \models \alpha$ there exists some order of axioms in $\mathcal{O}$ for which Algorithm 3 with the input $\mathcal{O}$ and α returns J.

The property stated in Exercise 14 means that for computing all justifications of $\mathcal{O} \models \alpha$, it is sufficient to run Algorithm 3 for all possible orders of axioms in $\mathcal{O}$. Since the number of permutations of elements in an n-element set is $n!$,[4] the

---

[4] $n! = n \cdot (n-1) \cdot (n-2) \cdots 2 \cdot 1$, there are n possibilities to choose the first element, $n-1$ to choose the second element from the remaining ones, $n-2$ to choose the third one, etc.

algorithm terminates after exactly $n \cdot n!$ entailment tests; since $n \cdot n! \leq n^{n+1} = 2^{(log_2 n) \cdot (n+1)} \leq 2^{n^2}$, this value is bounded by an exponential function in n. As shown in Example 23, the number of justifications can be exponential in n, so the exponential behavior of an algorithm for computing *all* justifications cannot be avoided, in general. Unfortunately, the described algorithm is not very practical since it performs exponentially many subsumption tests for *all* inputs, even if, e.g., $\mathcal{O} \models \alpha$ has just one justification, which is $\mathcal{O}$ itself. This is because this algorithm is *not goal-directed*: the computation of each next justification does not depend on the justifications computed before.

How can we find a more goal-directed algorithm? Suppose that we have computed a justification J_1 using Algorithm 3. The next justification J_2 must be different from J_1, so J_2 should miss at least one axiom from J_1. Hence the next justification J_2 can be found by finding $\beta_1 \in J_1$ such that $\mathcal{O} \setminus \{\beta_1\} \models \alpha$ and calling Algorithm 3 for the input $\mathcal{O} \setminus \{\beta_1\}$ and α. The next justification J_3, similarly, should miss something from J_1 and something from J_2, so it can be found by finding some $\beta_1 \in J_1$ and $\beta_2 \in J_2$ such that $\mathcal{O} \setminus \{\beta_1, \beta_2\} \models \alpha$ and calling Algorithm 3 for the input $\mathcal{O} \setminus \{\beta_1, \beta_2\}$ and α. In general, when justifications J_i $(1 \leq i \leq k)$ are computed, the next justification can be found by calling Algorithm 3 for the input $\mathcal{O} \setminus \{\beta_i \mid 1 \leq i \leq k\}$ and α such that $\beta_i \in J_i$ $(1 \leq i \leq k)$ and $\mathcal{O} \setminus \{\beta_i \mid 1 \leq i \leq k\} \models \alpha$. Enumeration of subsets $\mathcal{O} \setminus \{\beta_i \mid 1 \leq i \leq k\}$ can be organized using a data structure called a hitting set tree.

Definition 12. *A hitting set tree (short: HS-tree) for the entailment $\mathcal{O} \models \alpha$ is a labeled tree $T = (V, E, L)$ with $V \neq \emptyset$ such that:*

1. *each non-leaf node $v \in V$ is labeled with a justification $L(v) = J$ for $\mathcal{O} \models \alpha$ and, for each $\beta \in J$, v has an outgoing edge $\langle v, w \rangle \in E$ with label $L(v, w) = \beta$*
2. *each leaf node $v \in V$ is labeled by a special symbol $L(v) = \bot$.*

For each $v \in V$ let $H(v)$ be the set of edge labels appearing on the path from v to the root node of H. Then the following properties should additionally hold:

3. *for each non-leaf node $v \in V$ we have $L(v) \cap H(v) = \emptyset$,*
4. *for each leaf node $v \in V$ we have $\mathcal{O} \setminus H(v) \not\models \alpha$.*

Figure 6 shows an example of two different HS-trees for the entailment from Example 22. Note that the justification J_2 labels two different nodes of the left tree. We next prove that every HS-tree must contain every justification at least once.

Lemma 13. *Let $T = (V, E, L)$ be an HS-tree for the entailment $\mathcal{O} \models \alpha$. Then, for each justification J for $\mathcal{O} \models \alpha$, there exists a node $v \in V$ such that $L(v) = J$.*

Proof. Let $v \in V$ be a node with a maximal (w.r.t. set inclusion) set $H(v)$ (see Definition 12) such that $H(v) \cap J = \emptyset$, i.e., for every other node $w \in V$ either $H(w) \subseteq H(v)$ or $H(w) \cap J \neq \emptyset$. We prove that $L(v) = J$.

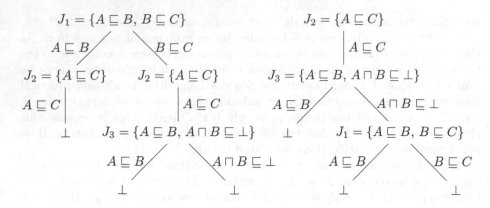

Fig. 6. Two HS-trees for $\mathcal{O} = \{A \sqsubseteq B, B \sqsubseteq C, A \sqsubseteq C, A \sqcap B \sqsubseteq \bot\} \models \alpha = A \sqsubseteq C$

Observe that since $H(v) \cap J = \emptyset$ and $J \subseteq \mathcal{O}$, we have $J \subseteq \mathcal{O} \setminus H(v)$. Since $J \models \alpha$, by monotonicity of entailment, we obtain $\mathcal{O} \setminus H(v) \models \alpha$. Therefore, by Condition 4 of Definition 12, v cannot be a leaf node. Hence, $L(n) = J'$ for some justification J' of $\mathcal{O} \models \alpha$. If $J = J'$ we have proved what is required. Otherwise, since J is a justification for $\mathcal{O} \models \alpha$ and $J' \models \alpha$, we have $J' \not\subseteq J$. Hence, there exists some $\beta \in J' \setminus J$. By Condition 1 of Definition 12, there exists $\langle v, w \rangle \in E$ with $L(v, w) = \beta$. Furthermore, by Condition 3 of Definition 12, $J' \cap H(v) = \emptyset$. Hence, $\beta \notin H(v)$ since $\beta \in J'$. Hence, $H(w) = H(v) \cup \{\beta\} \not\subseteq H(v)$ and, since $\beta \notin J$ and $H(v) \cap J = \emptyset$, we have $H(w) \cap J = \emptyset$. This contradicts our assumption that $H(v)$ is a maximal set such that $H(v) \cap J = \emptyset$. This contradiction proves that $L(n) = J$ is the only possible case. □

We next show that each HS-tree $T = (V, E, L)$ for an entailment $\mathcal{O} \models \alpha$ has at most exponentially many nodes in the number of axioms in $\mathcal{O}$. Take any $\langle v, w \rangle \in E$. Then v is not a leaf node. Hence, by Condition 1 of Definition 12, $L(v) = J$ for some justification J for $\mathcal{O} \models \alpha$ and $L(v, w) \in J$. By Condition 3, $J \cap H(v) = \emptyset$. Hence $L(v, w) \notin H(v)$. This implies that for each node $v \in V$ each axiom $\beta \in H(v)$ appears on the path from v to the root node exactly once. Hence the depth of H is bounded by the maximal number of axioms in $H(v)$, which is bounded by the number of axioms in $\mathcal{O}$. Similarly, since each non-leaf node v has exactly one successor for every $\beta \in L(v) \subseteq \mathcal{O}$, the branching factor of H is also bounded by the number of axioms in $\mathcal{O}$. This analysis gives us the following bound on the size of T:

Lemma 14. *Every HS-tree T for $\mathcal{O} \models \alpha$ has at most $\sum_{0 \leq k \leq n} n^k$ nodes where n is the number of axioms in $\mathcal{O}$.*

Exercise 15. Prove that T has at most $(n+1)!$ nodes. Hint: show that every path of T from the root to the leaf has a unique sequence of axioms on the labels of edges. Is this bound better than the bound from Lemma 14?

An HS-tree $T = (V, E, L)$ for an entailment $\mathcal{O} \models \alpha$ can be constructed as follows. We start by creating the root node $v_0 \in V$. Then we repeatedly assign labels of nodes and edges as follows. For each $v \in V$, if $L(v)$ was not yet assigned, we calculate $H(v)$. If $\mathcal{O} \setminus H(v) \not\models \alpha$, we label $L(v) = \bot$ according to Condition 4 of Definition 12. Otherwise, we compute a justification J for $\mathcal{O} \setminus H(v) \models \alpha$ using Algorithm 3 and set $L(v) = J$. Note that J satisfies Condition 3 of Definition 12 since $J \subseteq \mathcal{O} \setminus H(v)$. Next, for each $\beta \in J$, we create a successor node w of v and label $L(v, w) = \beta$. This ensures that Condition 1 of Definition 12 is satisfied for v. Since, by Lemma 14, H has a bounded number of nodes, this process eventually terminates. The described algorithm is known as Reiter's *Hitting Set Tree algorithm* (or short: *HST-algorithm*) [22, 46].

Exercise 16. Construct an HS-tree for the entailment $\mathcal{O} \models \alpha$ from Example 23 for the parameter $n = 2$ using the HST-algorithm.

Note that we call Algorithm 3 exactly once per node. Then Lemma 14 gives us the following bound on the number of entailment tests performed by the HST-algorithm:

Lemma 15. *An HS-tree for an entailment $\mathcal{O} \models \alpha$ can be constructed using at most $\sum_{1 \leq k \leq n+1} n^k$ entailment tests, where n is the number of axioms in $\mathcal{O}$.*

Note that unlike the algorithm sketched in Exercise 14, the input for each call of Algorithm 3 depends on the results returned by the previous calls.

Exercise 17. Suppose that an entailment $\mathcal{O} \models \alpha$ has a single justification $J = \mathcal{O}$. How many entailment tests will be performed by the HST-algorithm if $\mathcal{O}$ contains n axioms?

The HST-algorithm can further be optimized in several ways. First, it is not necessary to store the complete HS-tree in memory. For computing a justification J at each node v, it is sufficient to know just the set $H(v)$. For each successor w of v associated with some $\beta \in H(v)$, the set $H(w)$ can be computed as $H(w) = H(v) \cup \{\beta\}$. Hence, it is possible to compute all justifications by recursively processing and creating the sets $H(v)$ as shown in Algorithm 4. The algorithm saves all justifications in a set S, which is initially empty (Line 1). The justifications are computed by processing the sets $H(v)$; the sets that are not yet processed are stored in the queue Q, which initially contains $H(v_0) = \emptyset$ for the root node v_0 (Line 2). The elements of Q are then repeatedly processed in a loop (Lines 3–10) until Q becomes empty. First, we choose any $H \in Q$ (Line 4) and remove it from Q (Line 5). Then, we test whether $\mathcal{O} \setminus H \models \alpha$ (Line 6). If the entailment holds, this means that the corresponding node v of the HS-tree with $H(v) = H$ is not a leaf node. We then compute a justification J using Algorithm 3 and add it to S (Lines 7–8). Further, for each $\beta \in J$, we create the set $H(w) = H(v) \cup \{\beta\}$ for the corresponding successor node w of v and add $H(w)$ to Q for later processing (Lines 9–10). If the entailment $\mathcal{O} \setminus H \not\models \alpha$ does not hold, this means that we have reached a leaf of the HS-tree and no further children of this node should be created.

Algorithm 4. Computing all justifications by the Hitting Set Tree algorithm

---

ComputeJustificationsHST$(\mathcal{O}, \alpha)$: compute all justifications for $\mathcal{O} \models \alpha$
input   : an ontology $\mathcal{O}$ and an axiom α such that $\mathcal{O} \models \alpha$
output : the set of all minimal subsets $J \subseteq \mathcal{O}$ such that $J \models \alpha$

1 $S \leftarrow \emptyset$;
2 $Q \leftarrow \{\emptyset\}$;
3 **while** $Q \neq \emptyset$ **do**
4 | $H \leftarrow$ **choose** $H \in Q$;
5 | $Q \leftarrow Q \setminus \{H\}$;
6 | **if** $\mathcal{O} \setminus H \models \alpha$ **then**
7 | | $J \leftarrow \mathrm{Minimize}(\mathcal{O} \setminus H, \alpha)$;
8 | | $S \leftarrow S \cup \{J\}$;
9 | | **for** $\beta \in J$ **do**
10 | | | $Q \leftarrow Q \cup \{H \cup \{\beta\}\}$;

11 **return** S;

Exercise 18. Prove directly that Algorithm 4 returns *all* justifications for the given entailment $\mathcal{O} \models \alpha$. For this, show that the following invariant always holds in the main loop (Lines 3–10): if J is a justification for $\mathcal{O} \models \alpha$, then either $J \in S$ or there exists $H \in Q$ such that $J \subseteq \mathcal{O} \setminus H$.

Note that it is not specified in which *order* the elements should be taken from Q in Line 4 of Algorithm 4. Indeed, *correctness* of the algorithm does not depend on the order in which the sets $H \in Q$ are processed. However, *performance* of the algorithm may depend on this order. If the elements of Q are processed according to the *first-in-first-out (short: FIFO)* strategy, i.e., we take elements in the order in which they were inserted to the queue, this means that the nodes of the HS-tree are processed in a *breadth-first* way, so the queue may contain *exponentially many* unprocessed sets H at some point in time. If the elements of Q are processed according to the *last-in-first-out (short: LIFO)* strategy, i.e., we remove elements in the reversed order in which they were inserted, this means that the nodes of the HS-tree are processed in a *depth-first* way, so the queue always contains at most n^2 sets (at most n sets for nodes of some tree path plus at most $n - 1$ other successors for each of these nodes). Consequently, the LIFO strategy should be more memory efficient. This optimization is related to the optimization discussed in Remark 5, which allows for executing a tableau procedure in polynomial space.

A few further optimizations can be used to improve the *running time* of Algorithm 4 in certain cases. Note that some justifications can be computed multiple times as shown for the example on the left-hand side of Fig. 6. It is possible to *detect* such repetitions by checking if any justification $J \in S$ computed so far is a subset of $\mathcal{O} \setminus H$ for the currently processed set H. In this case, a (potentially expensive) call of Algorithm 3 in Line 7 of Algorithm 4 can be avoided by reusing J. Of course, testing $J \subseteq \mathcal{O} \setminus H$ for *all* $J \in S$ can also be expensive since S may contain exponentially many justifications. In practice, it makes sense to perform this test only for small J, for which the test is more likely to succeed. Another possible repetition is when some $H \in Q$, which was already processed before, is processed again. In this case, not only the previously computed justification $J \subseteq \mathcal{O} \setminus H$ can be reused, but it is also not necessary to create the successor sets $H \cup \{\beta\}$ for $\beta \in J$ since also those sets should have been created before. Of course, to check if a set H was processed before, we need to save all previously processed sets H, which is not done in the base version of Algorithm 4 since this information can increase the memory consumption of the algorithm. Hence, this is an example of an optimization that trades memory consumption for potentially improving the running time. Another optimization is to test if H is a superset of some set H' for which the test $\mathcal{O} \setminus H' \models \alpha$ was *negative*. Clearly, in this case, the test $\mathcal{O} \setminus H \models \alpha$ is negative as well by monotonicity of the entailment, so such H can immediately be disregarded by the algorithm without performing this test.

4.3 Computing All Repairs

The main idea of the HST-algorithm is to systematically compute two kinds of sets: (1) justifications J for the entailment $\mathcal{O} \models \alpha$ and (2) sets H that contain one element from each justification J on a branch. The name of the algorithm comes from the notion of a hitting set, which characterizes the latter sets.

Definition 13. *Let P be a set of sets of some elements. A set H is a* hitting set *for P if $H \cap S \neq \emptyset$ for each $S \in P$. A hitting set H for P is minimal if every $H' \subsetneq H$ is not a hitting set for P.*

Intuitively, a hitting set for P is a set H that contains at least one element from every set $S \in P$. An HS-tree is then a tree $T = (V, E, L)$ such that for each $v \in V$, $H(v)$ is a hitting set of the set of justifications on the path from v to the root of T. The leaf nodes v of T are labeled by hitting sets $H(v)$ such $\mathcal{O} \setminus H(v) \not\models \alpha$. Intuitively, the set $H(v)$ represents a set such that the removal of $H(v)$ from $\mathcal{O}$ breaks the entailment $\mathcal{O} \models \alpha$.

Definition 14. *A set R is a* repair *for the entailment $\mathcal{O} \models \alpha$ if $\mathcal{O} \setminus R \not\models \alpha$. A minimal repair for $\mathcal{O} \models \alpha$ is a repair R such that for every $R' \subsetneq R$ we have $\mathcal{O} \setminus R \models \alpha$.*

Notice some similarities between the notion of a minimal repair and the notion of a justification (cf. Definition 11): justifications are minimal subsets of $\mathcal{O}$ which entail the conclusion α, whereas minimal repairs are complements of the maximal subsets of $\mathcal{O}$ which do not entail α. Notice that each repair R for $\mathcal{O} \models \alpha$ should contain one axiom from every justification J for $\mathcal{O} \models \alpha$. Indeed, if $J \cap R = \emptyset$ for some justification J and repair R, then $J \subseteq \mathcal{O} \setminus R$, which violates the conditions $J \models \alpha$ and $J \subseteq \mathcal{O} \setminus R \not\models \alpha$ due to monotonicity of the entailment. This means that every justification must be a minimal hitting set of the set of all minimal repairs and, likewise, every minimal repair is a minimal hitting set of the set of all justifications. This property is known as the *hitting set duality* (between justifications and minimal repairs).

The HTS-algorithm can easily be extended to compute repairs in addition to justifications. Indeed, as mentioned above, if $v \in V$ is a leaf node, then $H(v)$ is a repair for $\mathcal{O} \models \alpha$ since $\mathcal{O} \setminus H(v) \not\models \alpha$ by Condition 4 of Definition 12.

Example 26. The leaf nodes of the HS-tree on the left-hand side of Fig. 6 correspond to the repairs:

$$R_1 = \{A \sqsubseteq B, A \sqsubseteq C\},$$
$$R_2 = \{B \sqsubseteq C, A \sqsubseteq C, A \sqsubseteq B\},$$
$$R_3 = \{B \sqsubseteq C, A \sqsubseteq C, A \sqcap B \sqsubseteq \bot\}.$$

Notice that the repair R_2 is not minimal since $R_1 \subsetneq R_2$.

A natural question is whether *all* repairs for the entailment will be computed by the described extension of the HST-algorithm. This is not true for arbitrary repairs: indeed, the whole ontology $\mathcal{O}$ from Example 22 is clearly a repair for the entailment $\mathcal{O} \models \alpha = A \sqsubseteq C$, since $\mathcal{O} \setminus \mathcal{O} = \emptyset \not\models \alpha$ because $A \sqsubseteq C$ is not a tautology. However, as shown in Example 26, the repair $\mathcal{O}$ was not computed. It turns out, however, that the extended HST-algorithm computes all *minimal* repairs.

Exercise 19. Prove an analogy of Lemma 13 showing that if R is a minimal repair for $\mathcal{O} \models \alpha$ then each HS-tree $T = (V, E, L)$ for $\mathcal{O} \models \alpha$ contains a leaf node $v \in V$ such that $H(v) = R$. Hint: take a node $v \in V$ with a *maximal* $H(v)$ such that $H(v) \subseteq R$ and prove that $H(v) = R$ for this node. Use the property that $J \cap R \neq \emptyset$ for every justification J for $\mathcal{O} \models \alpha$.

To compute repairs using Algorithm 4, it is sufficient to add an else-block for the test in Line 6, in which the set H for which this test fails, is added to the set of repairs.

4.4   Computing Justifications and Repairs Using SAT Solvers

In the previous section, we have discussed the *hitting set duality* property between justifications and repairs: every justification is a hitting set for the set of all minimal repairs and every minimal repair is a hitting set of the set of all justifications. Hitting set duality takes a prominent place in the HST-algorithm, but we can use this property as the basis of a *direct* algorithm for computing justifications and minimal repairs.

Suppose that we have already computed some set S of justifications and some set P of minimal repairs for the entailment $\mathcal{O} \models \alpha$. How can we find a new justification or a minimal repair? As mentioned, each new justification

---

Algorithm 5. Maximizing non-entailment

Maximize$(\mathcal{O}, M, \alpha)$: compute a maximal subset $N \subseteq \mathcal{O}$ such that
$$M \subseteq N \text{ and } N \not\models \alpha$$
input : an ontology $\mathcal{O}$, a subset $M \subseteq \mathcal{O}$, and an axiom α such that
$M \not\models \alpha$

output : $N \subseteq \mathcal{O}$ such that $M \subseteq N \not\models \alpha$ but $N' \models \alpha$ for every N' with
$N \subsetneq N' \subseteq \mathcal{O}$

1 $N \leftarrow M$;
2 **for** $\beta \in \mathcal{O} \setminus M$ **do**
3   **if** $N \cup \{\beta\} \not\models \alpha$ **then**
4          $N \leftarrow N \cup \{\beta\}$;

5 **return** N;

---

must be a hitting set for P, i.e., it should contain one axiom from every repair $R \in P$. Furthermore, it should be different from any of the previously computed justifications, i.e., it should miss one axiom from every $J \in S$. Suppose we have found a subset $M \subseteq \mathcal{O}$ satisfying these two requirements:

$$\forall R \in P : M \cap R \neq \emptyset, \tag{7}$$

$$\forall J \in S : J \setminus M \neq \emptyset. \tag{8}$$

If $M \models \alpha$, then, using Algorithm 3, we can extract a *minimal* subset $J' \subseteq M$ such that $J' \models \alpha$. Note that J' still misses at least one axiom from each $J \in S$ since (8) is preserved under removal of axioms from M. Therefore, J' is a *new justification* for $\mathcal{O} \models \alpha$. If $M \not\models \alpha$, then, similarly, by adding axioms $\beta \in \mathcal{O}$ to M preserving $M \not\models \alpha$, we can find a *maximal* superset N of M ($M \subseteq N \subseteq \mathcal{O}$) such that $N \not\models \alpha$: see Algorithm 5. Note that (7) is preserved under additions of elements to M, hence, $R' = \mathcal{O} \setminus N$ is a *new minimal repair* for $\mathcal{O} \models \alpha$. Thus, using any set M satisfying (7) and (8) we can find either a new justification or a new minimal repair.

How to find a set M satisfying Conditions (7) and (8)? These conditions require solving a rather complex combinatorial problem. Propositional (SAT) solvers, offer a convenient and effective way of solving such problems. In the following, we describe a propositional encoding of Conditions (7) and (8). The interested reader can find some background information on Propositional Logic and SAT in Appendix A.2.

To formulate the propositional encoding, we assign to each axiom $\beta \in \mathcal{O}$ a fresh propositional variable p_β. Then, every interpretation $\mathcal{I}$ determines a set $M = M(\mathcal{I}) = \{\beta \in \mathcal{O} \mid p_\beta^{\mathcal{I}} = 1\}$ of axioms whose corresponding propositional variable is true. We construct a propositional formula F such that $F^{\mathcal{I}} = 1$ if and only if $M(\mathcal{I})$ satisfies (7) and (8) for the given sets S of justifications and P of minimal repairs. Thus, to find a subset M satisfying (7) and (8), it is sufficient to find a model $\mathcal{I}$ of F and compute $M(\mathcal{I})$. We define F as follows:

$$F = F(S, P) = \bigwedge_{J \in S} \bigvee_{\beta \in J} \neg p_\beta \wedge \bigwedge_{R \in P} \bigvee_{\beta \in R} p_\beta. \tag{9}$$

Example 27. Let $\mathcal{O}$ be the ontology from Example 22. We assign propositional variables to axioms from $\mathcal{O}$ as follows:

- $A \sqsubseteq B \rightsquigarrow p_1$,   - $A \sqsubseteq C \quad \rightsquigarrow p_3$,
- $B \sqsubseteq C \rightsquigarrow p_2$,   - $A \sqcap B \sqsubseteq \bot \rightsquigarrow p_4$.

Let S be the set of justifications J_1 and J_2 from Example 22 and P a set containing only repair R_1 from Example 26. Then according to (9) we have:

$$F = F(S, P) = (\neg p_1 \vee \neg p_2) \wedge (\neg p_3) \wedge (p_1 \vee p_3).$$

F has a model $\mathcal{I}$ with $p_1^\mathcal{I} = 1$ and $p_2^\mathcal{I} = p_3^\mathcal{I} = p_4^\mathcal{I} = 0$, which gives $M(\mathcal{I}) = \{A \sqsubseteq B\}$.

Once the set M determined by a model $\mathcal{I}$ of F is found, we can extract either a new justification J or a new repair R from M by minimizing entailment using Algorithm 3 or maximizing non-entailment using Algorithm 5. After that, we can update F according to (9) and compute a new model of F, if there exist any.

Example 28. Continuing Example 27, observe that $M(\mathcal{I}) = \{A \sqsubseteq B\} \not\models \alpha = A \sqsubseteq C$. By running Algorithm 5 for $\mathcal{O}$, $M = M(\mathcal{I})$ and α we compute N as follows:

N	β	$M \cup \{\beta\} \models^? \alpha = A \sqsubseteq C$
$\{A \sqsubseteq B\}$	$B \sqsubseteq C$	yes
$\{A \sqsubseteq B\}$	$A \sqsubseteq C$	yes
$\{A \sqsubseteq B\}$	$A \sqcap B \sqsubseteq \bot$	yes
$\{A \sqsubseteq B\}$	-	-

Hence, $R = \mathcal{O} \setminus N = \{B \sqsubseteq C, A \sqsubseteq C, A \sqcap B \sqsubseteq \bot\}$ is a new minimal repair for $\mathcal{O} \models \alpha$ (repair R_3 from Example 26). After we add this repair to P and re-compute F according to (9), we obtain a formula with an additional conjunct:

$$F = F(S, P) = (\neg p_1 \vee \neg p_2) \wedge (\neg p_3) \wedge (p_1 \vee p_3) \wedge (p_2 \vee p_3 \vee p_4).$$

The interpretation $\mathcal{I}$ from Example 27 is no longer a model for F, but we can find a new model $\mathcal{I}$ of F with $p_1^\mathcal{I} = p_4^\mathcal{I} = 1$ and $p_2^\mathcal{I} = p_3^\mathcal{I} = 0$. For this model, we have $M(\mathcal{I}) = \{A \sqsubseteq B, A \sqcap B \sqsubseteq \bot\} \models \alpha$. By running Algorithm 3 for M and α, we obtain a new justification $J = \{A \sqsubseteq B, A \sqcap B \sqsubseteq \bot\}$ (justification J_3 from Example 22). The new justification, when added to S, gives us another conjunct for F:

Algorithm 6. Computing all justifications using a SAT solver

ComputeJustificationsSAT$(\mathcal{O}, \alpha)$: compute all justifications for
$$\mathcal{O} \models \alpha$$
input   : ontology $\mathcal{O}$ and axiom α such that $\mathcal{O} \models \alpha$
output : the set of all minimal subsets $J \subseteq \mathcal{O}$ such that $J \models \alpha$

1 $S \leftarrow \emptyset$;
2 $F \leftarrow \top$;
3 **while** $\exists \mathcal{I} : F^{\mathcal{I}} = 1$ **do**
4 $\quad$ $\mathcal{I} \leftarrow$ **choose** $\mathcal{I} : F^{\mathcal{I}} = 1$;
5 $\quad$ $M \leftarrow \{\beta \mid p_\beta^{\mathcal{I}} = 1\}$;
6 $\quad$ **if** $M \models \alpha$ **then**
7 $\quad\quad$ $J \leftarrow \mathrm{Minimize}(M, \alpha)$;
8 $\quad\quad$ $S \leftarrow S \cup \{J\}$;
9 $\quad\quad$ $F \leftarrow F \wedge \bigvee\{\neg p_\beta \mid \beta \in J\}$;
10 $\quad$ **else**
11 $\quad\quad$ $N \leftarrow \mathrm{Maximize}(\mathcal{O}, M, \alpha)$;
12 $\quad\quad$ $F \leftarrow F \wedge \bigvee\{p_\beta \mid \beta \in \mathcal{O} \setminus N\}$;

13 **return** S;

$$F = F(S, P) = (\neg p_1 \vee \neg p_2) \wedge (\neg p_3) \wedge \underline{(\neg p_1 \vee \neg p_4)} \wedge (p_1 \vee p_3) \wedge (p_2 \vee p_3 \vee p_4).$$

This formula F is now unsatisfiable.

Note that if F is unsatisfiable, then S already contains all justifications for $\mathcal{O} \models \alpha$ and P contains all minimal repairs. Indeed, if S does not contain some justification J for $\mathcal{O} \models \alpha$ then $M = J$ clearly satisfies (7) and (8), hence, the interpretation $\mathcal{I} = \mathcal{I}(M)$ defined by $p_\alpha^{\mathcal{I}} = 1$ if and only if $\alpha \in M$, is a model of F. Similarly, if P does not contain some minimal repair R for $\mathcal{O} \models \alpha$, then $M = \mathcal{O} \setminus R$ satisfies (7) and (8), hence, the interpretation $\mathcal{I} = \mathcal{I}(M)$ is likewise a model of F. To conclude, either F is satisfiable and from its model we can compute a new justification or a minimal repair and extend F with the corresponding conjunct or F is unsatisfiable, in which case we have computed all justifications and minimal repairs.

Algorithm 6 summarizes the described procedure for computing all justifications using a SAT solver. We start by creating an empty set S of justifications (Line 1) and a formula F that is always true (Line 2). Then, in a loop (Lines 3–12), as long as F is satisfiable (which is checked using a SAT solver), we take any model $\mathcal{I}$ of F (Line 4), extract the corresponding set $M = M(\mathcal{I})$ that it defines (Line 5), and check the entailment $M \models \alpha$. If the entailment holds, using Algorithm 3 we compute a justification for $M \models \alpha$ (Line 7), which, by monotonicity of entailment, is also a justification for $\mathcal{O} \models \alpha$. This justification is then added to S (Line 8) and F is extended with a new conjunct for this justification according to (9) (Line 9). If the entailment does not hold, we compute a maximal superset N of M such that $N \not\models \alpha$ using Algorithm 5 (Line 11) and extend F with the corresponding conjunct for the new repair $R = \mathcal{O} \setminus N$ according to (9) (Line 12). As soon as F becomes unsatisfiable, we return the set S of computed justifications (Line 13).

Example 29. Consider the entailment $\mathcal{O} \models \alpha$ from Example 22 and propositional encoding of axioms in $\mathcal{O}$ from Example 27. The following table shows a run of Algorithm 6 for the inputs $\mathcal{O}$ and α. Every row in this table corresponds to one iteration of the while-loop (Lines 3–12). The first column gives the value of the interpretation $\mathcal{I}$ for F computed in this iteration. The second column shows the value of M computed for this interpretation and whether the entailment $M \models \alpha$ holds. The third column shows the result of minimizing the entailment or maximizing the non-entailment using Algorithms 3 and 5. The last column shows the conjunct that is added to F for the corresponding justification or repair.

$p_1^{\mathcal{I}}$	$p_2^{\mathcal{I}}$	$p_3^{\mathcal{I}}$	$p_4^{\mathcal{I}}$	$M \models^? \alpha$	$\min(M) \models \alpha / \max(M) \not\models \alpha$	C
0	0	0	0	$\emptyset \not\models \alpha$	$\{A \sqsubseteq B\} \not\models \alpha$	$p_2 \vee p_3 \vee p_4$
0	1	0	0	$\{B \sqsubseteq C\} \not\models \alpha$	$\{B \sqsubseteq C, A \sqcap B \sqsubseteq \bot\} \not\models \alpha$	$p_1 \vee p_3$
1	1	0	0	$\{A \sqsubseteq B, B \sqsubseteq C\} \models \alpha$	$\{A \sqsubseteq B, B \sqsubseteq C\} \models \alpha$	$\neg p_1 \vee \neg p_2$
0	0	1	1	$\{A \sqsubseteq C, A \sqcap B \sqsubseteq \bot\} \models \alpha$	$\{A \sqsubseteq C\} \models \alpha$	$\neg p_3$
1	0	0	1	$\{A \sqsubseteq B, A \sqcap B \sqsubseteq \bot\} \models \alpha$	$\{A \sqsubseteq B, A \sqcap B \sqsubseteq \bot\} \models \alpha$	$\neg p_1 \vee \neg p_4$

Algorithm 6 can be easily turned into an algorithm for computing repairs (in addition or instead of justifications), by saving the repairs $\mathcal{O} \setminus N$ for N computed in Line 11.

Let us briefly discuss similarities and differences between Algorithm 4 and Algorithm 6. Both algorithms work by systematically exploring subsets of $\mathcal{O}$ and minimizing entailments from such subset to compute justifications. Algorithm 4 constructs such subsets $(\mathcal{O} \setminus H)$ manually by removing one axiom appearing in the previously computed justification (if there is any) in all possible ways. Algorithm 6 enumerates such subsets M with a help of a SAT solver. The main difference is that Algorithm 4 may encounter the same subsets many times (on different branches), whereas the propositional encoding used in Algorithm 6 ensures that such subsets never repeat. The following example shows a situation where Algorithm 4 performs exponentially many iterations of the while-loop, whereas Algorithm 6 has only quadratically many iterations.

Example 30. Consider axioms $\beta_i = A \sqsubseteq B \sqcap D_i$, $\gamma_i = B \sqsubseteq C \sqcap D_i$ $(1 \leq i \leq n)$, and the ontology $\mathcal{O} = \{\beta_i, \gamma_i \mid 1 \leq i \leq n\}$, where A, B, C, and D_i $(1 \leq i \leq n)$ are atomic concepts. Clearly $\mathcal{O} \models \alpha = A \sqsubseteq C$. Furthermore, there are exactly n^2 justifications for $\mathcal{O} \models \alpha$: $J_{ij} = \{\beta_i, \gamma_j\}$ $(1 \leq i, j \leq n)$ and exactly 2 minimal repairs: $R_1 = \{\beta_i \mid 1 \leq i \leq n\}$, $R_2 = \{\gamma_j \mid 1 \leq j \leq n\}$. Hence Algorithm 6 will perform exactly $n^2 + 2 + 1$ calls to a SAT solver with a formula F of the size at most $c \cdot (n^2 \cdot 2 + 2 \cdot n)$ for some constant c.[5] On the other hand, each HS-tree $T = (V, E, L)$ for $\mathcal{O} \models \alpha$ has at least 2^n nodes. Indeed, every non-leaf node $v \in V$ must be labeled by some $L(v) = J_{ij}$ with $1 \leq i, j \leq n$, which contains two axioms. Hence every non-leaf node of $v \in V$ must have two successor nodes (see Condition 1 of Definition 12). For every leaf node $v \in V$, the value $H(v)$ must be a repair for $\mathcal{O} \models \alpha$, so $H(v)$ must be a super-set of either R_1 or R_2. Hence $H(v)$ contains at least n elements, which means that the path from v to the root of T has at least n edges. Therefore, T is a binary tree whose leafs have the level n or higher. Hence T has at least 2^n nodes.

Of course, an iteration of Algorithm 4 cannot be directly compared to an iteration of Algorithm 6. Both iterations use at most one call to Algorithm 3, but Algorithm 6 may also require a call to Algorithm 5, as well as checking satisfiability of F. The latter requires solving an NP-compete problem, for which no polynomial algorithm is known so far. In order to check satisfiability of F, a SAT solver usually tries several (in worst-case exponentially many) propositional interpretations until a model of F is found. As each such interpretation $\mathcal{I}$ corresponds to a subset $M(\mathcal{I}) \subseteq \mathcal{O}$, this process can be compared to the enumeration of subsets in Algorithm 4. However, a SAT solver usually implements a number of sophisticated optimizations, which make the search for models very efficient in practice, whereas the subset enumeration strategy used Algorithm 4 is rather simplistic. Hence Algorithm 6 is likely to win in speed. On the other hand, Algorithm 6 requires saving all justifications (and minimal repairs) in the propositional formula F, which might result in a formula of exponential size, if the number of such justifications or repairs is exponential. In this regard, Algorithm 4 could be more memory efficient since saving (all) justifications is optional (see the discussion at the end of Sect. 4.2). Hence both algorithms have their own advantages and disadvantages.

5   Summary and Outlook

In this course, we have looked in-depth into the most common algorithms for reasoning and explanation in Description Logics. We have seen that the development of such algorithms is a complicated process already for the relatively simple DL $\mathcal{ALC}$. To show correctness of algorithms, one usually needs to prove several theoretical properties, such as soundness, completeness and termination. The algorithmic complexity analysis is helpful to understand the worst-case behavior of algorithms and to compare different algorithms across several dimensions such as (non-deterministic) time and space complexity. Identifying the exact computational complexity for various DLs and reasoning problems has, therefore, been one of the central research topics in DLs. The DL Complexity Navigator[6] provides an interactive overview of many of these results.

---

[5] The conjuncts for J_{ij} in F consist of two negated propositional variables, the conjunct for R_1 and R_2 in F consist of n propositional variables.

[6] http://www.cs.man.ac.uk/~ezolin/dl/.

Proving correctness and complexity results often requires understanding of model-theoretic properties of the languages. As we have seen in Sect. 3, for reasoning with $\mathcal{ALC}$ ontologies, it is sufficient to restrict the search to a special kind of *tree model* represented by tableaux. This so-called *tree model property* was argued to be one of the main reasons for decidability and the relatively low complexity of *Modal Logics*, the siblings of Description Logics [66]. For pure $\mathcal{ALC}$ concept satisfiability, i.e., without background ontologies, it is sufficient to consider tree models of a bounded depth (Sect. 3.1). With additional background ontologies, the tree models are no longer finite and special *blocking techniques* are required to ensure termination of tableau algorithms (Sect. 3.2). When moving to very expressive DLs, such as $\mathcal{SROIQ}$ [29] (the language underpinning the OWL 2 Direct Semantics), eventually the tree model property is lost and proving termination of tableau procedures, while still ensuring soundness and completeness, becomes increasingly difficult. It is not very surprising that when increasing the *expressivity* of languages, i.e., when adding new ways to construct concepts and axioms, the complexity of the reasoning problems increases as well. For example, the time complexity of all standard reasoning problems in $\mathcal{SROIQ}$ becomes *non-deterministic doubly exponential* [31], whereas it is "only" *deterministic exponential* for $\mathcal{ALC}$ (see the remark after Theorem 2).

The theoretical analysis of algorithms does not always give an accurate prediction about their practical performance. Often a situation that triggers the worst-case behavior of an algorithm represents some corner case, which rarely appears in practice. When it comes to practical efficiency, some other properties of algorithms become more important. For example, despite a relatively high algorithmic complexity (see Theorems 1 and 2), tableau algorithms remain among the fastest DL reasoning algorithms to date. This phenomenon can be explained by a range of optimization techniques that have been developed for tableau algorithms in the past two decades.

All state-of-the-art tableau reasoners, e.g., FaCT++ [62], HermiT [42], Konclude [58], MoRE [48], and Pellet [56], apply a significant range of optimizations. The optimizations can be categorized into those for *preprocessing, consistency checking*, and for *higher level reasoning tasks*. Examples of higher level reasoning tasks are *classification*, where one computes all subsumption relationships between atomic concepts or *materialization*, where one extends the ontology, for each individual (pair of individuals), with assertions to capture the atomic concepts (roles) of which the individual (the pair of individuals) is an instance.

Most reasoning systems preprocess the input ontologies. The simplest form of preprocessing is the presented conversion into negation normal form (Definition 2 in Sect. 3), which is not used for improving performance, but rather to allow for using fewer tableau rules. Other standard preprocessing optimizations include *lexical normalization* and *simplification*, which aim at identifying syntactic equivalences, contradictions and tautologies [4, Sect. 9.5]. A well-known and very important optimization for improving performance is *absorption*, which aims at rewriting general concept inclusion axioms to avoid non-determinism in the tableau algorithm. For example, here we suggested to convert an axiom of the form $A \sqcap B \sqsubseteq C$ into $\top \sqsubseteq \neg A \sqcup \neg B \sqcup C$ to allow for handling them with the $\top$-Rule. This introduces, however, a non-deterministic decision for each axiom and each node. Instead, practical tableau systems use a variant of the $\sqsubseteq$-Rule introduced in Table 4 restricted to atomic concepts on the left-hand side, i.e., for an axiom of the form $A \sqsubseteq C$ in the ontology, a node with A in its label, but C not in its label, the node's label is extended with C. With this rule, one can transform $A \sqcap B \sqsubseteq C$ into $A \sqsubseteq \neg B \sqcup C$, which already reduces the amount of non-determinism. Binary absorption [30] further allows for a conjunction of (two) atomic concepts on the left-hand side of a general concept inclusion, i.e., one can completely avoid the non-deterministic decisions for our example axiom $A \sqcap B \sqsubseteq C$. Further absorption techniques include *role absorption* [61], *nominal absorption* [55], and partial absorption [57].

As outlined in Sect. 3, consistency checking is the core reasoning task of a tableau-based reasoner. Since these checks typically occur very often, many optimizations are known including *model merging* techniques [24], *lazy unfolding*, *semantic branching*, *boolean constraint propagation*, *dependency directed backtracking* and *backjumping*, and *caching*. We refer interested readers to the DL Handbook [4, Sect. 9] for a more detailed descriptions of the latter optimizations. The HermiT reasoner further tries to reduce non-determinism by combining hypertableau [8] and hyper-resolution [47] techniques. In order to reduce the size of the tableau, modern DL reasoners several blocking strategies such as *anywhere blocking* [42] or *core blocking* [19].

Higher level reasoning tasks are usually reduced to a multitude of consistency checks such that they benefit from the optimizations of this task as much as possible. Many OWL reasoners, solve the classification problem using an Enhanced Traversal (ET) classification algorithm [5] similar to the one used in early description logic reasoners. To construct a concept hierarchy, the algorithm starts with the empty hierarchy and then iteratively inserts each concept from the ontology into the hierarchy. Each insertion step typically requires one or more subsumption tests—checks whether a subsumption relationship holds between two concepts—in order to determine the proper position of a class in the hierarchy constructed thus far. A more recent alternative to the ET algorithm is the *known/possible set classification* approach [20].

Despite the wide range of implemented optimization techniques, the reasoning performance might not be sufficient for some applications. The OWL 2 standard addresses this by introducing so-called OWL profiles [41], which are fragments of OWL 2 that restrict the allowed constructors in order to allow for tractable reasoning procedures. For example, the OWL 2 EL profile (based on the Description Logic $\mathcal{EL}$, a fragment of $\mathcal{ALC}$) allows for a one-pass classification of ontologies, i.e., repetitive subsumption tests are not needed. Some reasoners, e.g., Konclude, combine tableau procedures with tractable algorithms for handling those parts of an ontologies that are in the OWL 2 EL profile. Similarly, MoRe combines the (hyper-)tableau reasoner HermiT with the specialized OWL 2 EL reasoner ELK [34].

Many optimizations try to avoid unnecessary operations by making algorithms more *goal-directed* and thus reducing the *search space*. We have seen several examples of such optimizations in Sect. 4 when considering algorithms for computing justifications and repairs. Such optimizations typically do not reduce the worst case complexity of algorithms but they can significantly improve their behavior in typical cases. For example, Algorithm 3 for computing one justification, in practice, does not start with the whole ontology $J = \mathcal{O}$ (Line 3), but with a subset $J \subseteq \mathcal{O}$ such that $J \models \alpha$. If a small subset J like this is found, the number of subsequent entailment tests performed by the algorithm can significantly be reduced. The initial subset J can be found, for example, by starting with $J = \emptyset$ and repeatedly adding to J axioms from $\mathcal{O}$ until $J \models \alpha$. This part of the algorithm, called the *expansion phase*, requires additional entailment tests. To find a J that is as small as possible, one usually tries to first add axioms that are most *likely* to cause the entailment $J \models \alpha$, e.g., the axioms $\beta \in \mathcal{O}$ that contain symbols from α or from the previously added axioms in J. The initial subset $J \subseteq \mathcal{O}$ such that $\mathcal{J} \models \alpha$ can also be found using algorithms for computing modules of ontologies. A *(logical) module of $\mathcal{O}$ for a set of symbols Σ* is a subset $M \subseteq \mathcal{O}$ such that for every axiom β formulated using only symbols in Σ, if $\mathcal{O} \models \beta$ then $M \models \beta$. In our case we are interested in Σ consisting of all symbols in α. Some types of modules, e.g., *locality-based modules* can be computed in polynomial time without performing any subsumption tests [16]. It is also possible to reduce the number of entailment tests when minimizing the entailment $\mathcal{J} \models \alpha$ by removing several axioms at a time instead of one axiom like in Algorithm 3. Further details of optimization techniques for computing justifications can be found in the PhD thesis of Horridge [26].

Another way to optimize an algorithm is to use an existing "off-the-shelf" tool that is already optimized for solving a certain class of problems. One of the most popular examples of such tools are SAT solvers. In Sect. 4.4 we have shown how SAT solvers can be used for computing justifications and repairs (see Algorithm 6). This algorithm or variations thereof are implemented in several tools such as EL+SAT [53,67], EL2MUS [1], and SATPin [39]. The SAT solvers used in these tools are not only only used to find new candidate subsets M for justifications or complements of repairs, but also to check the entailments $\mathcal{O} \models \alpha$. This has been possible by using different, *consequence-based* algorithms for reasoning with ontologies. In contrast to tableau algorithms, consequence-based algorithms do not construct (representations of) models, but instead derive logical consequence of axioms using a number of dedicated inference rules. Thus, to prove the entailment $\mathcal{O} \models \alpha$, it is sufficient to show how the axiom α can be *derived* using these rules and the axioms in $\mathcal{O}$. Each *inference step* $\alpha_1, \ldots, \alpha_n \vdash \alpha$ used in this derivation can be encoded as a propositional formula $p_{\alpha_1} \wedge \cdots \wedge p_{\alpha_n} \rightarrow p_\alpha$, thus reducing DL entailment to propositional (Horn) entailment. Consequence-based procedures have been first formulated for the simple DL $\mathcal{EL}$ to show that entailment in this language can be solved in polynomial time [13]. The above mentioned tools for computing justifications are targeted to this language. Since then, consequence-based procedures have also been extended to more expressive (even non-polynomial) DLs [3,7,15,32,54].

One of the benefits of consequence-based algorithms is that they can be used to provide better explanations for the obtained reasoning results. Justification for entailments $\mathcal{O} \models \alpha$ tell *which* axioms of the ontology are responsible for the entailment, but not *how* the entailed axiom was obtained from them. This limitation has been mainly due to the *black-box* nature of the tableau-based reasoning algorithms: since tableau algorithms are based on constructing models, they cannot provide information supporting *positive* entailment tests $\mathcal{O} \models \alpha$ since in such cases no counter-model for the entailment $\mathcal{O} \models \alpha$ exists (see Lemma 2). In contrast, consequence-based algorithms can provide explanations for entailment in the form of derivations (or proofs). In practice, computing derivations for a given subsumption has not been an easy task because if in addition to computing all consequences, we also save all inference steps by which they were produced, the amount of memory required to store all this information can double. A *goal-directed* procedure for generation of inferences [33] can be used to mitigate this problem. The (black-box) algorithms for computing justifications have also been extended to provide some inference steps that derive (simple) *intermediate conclusions*, which can improve understanding of explanations [27,28].

Ontologies and the reasoning techniques described in this course are successfully employed in many domains, e.g., to reason over the environment of (autonomous) cars [17,68], in information integration tasks [35,38], or, most prominently, in medicine, life sciences, and bio-informatics [21,25,63]. The standardization efforts of the World Wide Web Consortium (W3C) for the DL-based Web Ontology Language OWL have certainly helped in promoting the use of logic-based knowledge representation and reasoning. While modern search engines have picked up the ideas of using structured or formal knowledge, this is often not in the form OWL (or DL) ontologies. For example, Google's knowledge graph and Facebook's Social Graph are based on proprietary formats. The same holds for Wikidata, although Semantic Web standards are also supported (e.g., a SPARQL [23] query interface and data dumps in the Resource Description Format (RDF) [52] are available). We attribute this to several reasons: While large companies such as Google recognized the importance of structured knowledge, they rather use their proprietary formats, possibly for business reasons. A contributing challenge is also that even the tractable fragments of DLs do not offer the performance required at Web scale. Furthermore, the knowledge in the Web is inherently inconsistent, which is challenging for logic-based approaches. DLs and OWL also lack features that are important for some applications. For example, Wikidata captures when a fact was true, e.g., the former German chancellor Helmut Kohl was married to Hannelore Kohl from 1960 to 2001. This is difficult to model using DLs since roles can only relate two elements, but research to address these issues is on-going [37,40]. Summing up, it is widely accepted today that structuring and formalizing knowledge is important and that significant advances were made in the last years; nevertheless, research is still needed in several directions.

A   Appendix

For the convenience of interested readers, in this appendix we recap some background material used in this course, such as the basic notions for describing the (theoretical) complexity of algorithms, and the propositional satisfiability problem.

A.1   Computational Complexity

A *decision problem* (for an input set X) is simply a mapping $P\colon X \to \{yes, no\}$. Note that X can be an arbitrary set of objects. For example, for the concept subsumption problem, X consists of all possible *pairs* $\langle \mathcal{O}, C \sqsubseteq D \rangle$ where the first component is an ontology $\mathcal{O}$ and the second component is a concept subsumption $C \sqsubseteq D$. An algorithm A *solves* (or *decides*) a decision problem P for X, if A accepts each value $x \in X$ as input, terminates for all these values, and returns the (correct) result $A(x) = P(x)$.

There are several dimensions according to which one can measure the *computational complexity* of problems and algorithms. We say that an algorithm A has an *(upper) time complexity* $f(n)$ if for each input $x \in X$ with the size (e.g., the number of symbols) n, the algorithm A terminates after at most $f(n)$ steps. A problem P for X is *solvable in time* $f(n)$ if there exists an algorithm A that solves P and has the time complexity $f(n)$. We say that a problem P is solvable in *polynomial time* if there exists a polynomial function $f(n)$ such that P is solvable in time $f(n)$. A problem P is solvable in *exponential time* (*doubly exponential time*, ...) if there exists a polynomial function $f(n)$ such that P is solvable in time $2^{f(n)}$ ($2^{2^{f(n)}}$, ...). Analogously to the algorithmic time complexity, one can define the algorithmic *space complexity*: a problem P for X is solvable in space $f(n)$ if there exists an algorithm A that solves P such that for each input $x \in X$ with the size n, the algorithm A uses at most $f(n)$ units of memory at every step of the computation.

Another dimension of the computational complexity is based on the notion of a non-deterministic computation. An algorithm A is said to be *non-deterministic* if the result of some operations that it can perform is not uniquely determined. Thus, the algorithm can produce different results for different *runs* even with the same input. A non-deterministic algorithm A *solves* a problem P for X if, for each $x \in X$ such that $P(x) = no$, each run of A terminates with the result *no*, and for each $x \in X$ such that $P(x) = yes$, there exists *at least one run* for which the algorithm terminates and produces *yes*. The intuition is that, if one has an unlimited number of identical computers, then one can solve the problem P by starting the algorithm A *in parallel* on all of these computers; if $P(x) = yes$, one of them is guaranteed to return *yes* (provided the results of all non-deterministic instructions are chosen at random).

The time and space complexity measures are also extended to non-deterministic algorithms. For example, a non-deterministic algorithm A has the *(upper) time complexity* $f(n)$ if, for every input $x \in X$ of the size n, *every run* of A terminates after at most $f(n)$ steps. We say that a problem P for X is *solvable in non-deterministic time* $f(n)$ if there exists a non-deterministic algorithm A that solves P and has the time complexity $f(n)$. Thus, a problem P is solvable in *non-deterministic polynomial (exponential, doubly exponential, ...) time* if P is solvable in non-deterministic time $f(n)$, where $f(n)$ is a polynomial (exponential, doubly exponential) function. The non-deterministic space complexity is defined similarly.

A common way to solve a problem is to reduce it to another problem, for which a solution is known. A decision problem $P_1 \colon X \to \{yes, no\}$ *is (many-one) reducible* to a decision problem $P_2 \colon Y \to \{yes, no\}$ if there exists an algorithm $R \colon X \to Y$ (that takes an input from X and produces an output from Y) such that for every $x \in X$, we have $P_1(x) = P_2(R(x))$. In this case the algorithm R is called *a reduction* from P_1 to P_2. Depending on the time or space complexity of the algorithm R (i.e., the maximal number of steps or memory units consumed for inputs of size n), the complexity bounds of the problems are also transferred by the reduction. Usually one is interested in *polynomial reductions*, where the number of steps for computing each $R(x)$ is bounded by a polynomial function in the size of x. In this case, if the complexity of P_2 is polynomial, exponential, or doubly exponential (for deterministic or non-deterministic, time or space complexity), then P_1 has the same complexity as P_2.

A.2 Propositional Logic and SAT

The *vocabulary* of Propositional Logic consists of a countably infinite set P of *propositional variables, Boolean constants*: $\top$ (Verum), $\bot$ (Falsum), and *Boolean operators*: $\land$ (conjunction), $\lor$ (disjunction), $\neg$ (negation) and $\to$ (implication). *Propositional formulas* are constructed from these symbols according to the grammar:

$$F, G ::= p \mid \top \mid \bot \mid F \land G \mid F \lor G \mid \neg F \mid F \to G, \tag{10}$$

where $p \in P$. A *propositional interpretation* $\mathcal{I}$ assigns to each propositional variable $p \in P$ a *truth value* $p^{\mathcal{I}} \in \{1, 0\}$ (1 means '*true*', 0 means '*false*') and is extended to other propositional formulas by induction over the grammar definition (10) as follows:

- $\top^{\mathcal{I}} = 1$ and $\bot^{\mathcal{I}} = 0$ for each $\mathcal{I}$,
- $(F \land G)^{\mathcal{I}} = 1$ if and only if $F^{\mathcal{I}} = 1$ and $G^{\mathcal{I}} = 1$,
- $(F \lor G)^{\mathcal{I}} = 1$ if and only if $F^{\mathcal{I}} = 1$ or $G^{\mathcal{I}} = 1$,
- $(\neg F)^{\mathcal{I}} = 1$ if and only if $F^{\mathcal{I}} = 0$,
- $(F \to G)^{\mathcal{I}} = 1$ if and only if $F^{\mathcal{I}} = 0$ or $G^{\mathcal{I}} = 1$.

If $F^{\mathcal{I}} = 1$ then we say that $\mathcal{I}$ is a *model* of F (or F is *satisfied* in $\mathcal{I}$). We say that F *is satisfiable* if F has at least one model; otherwise F is *unsatisfiable*. A *propositional satisfiability problem* (short: *SAT*) is the following decision problem:

- Given: a propositional formula F,

– Return: *yes* if F is satisfiable and *no* otherwise.

SAT is a classical example of a *non-deterministic polynomial* (short: *NP*) problem: it can be solved using an algorithm that non-deterministically choses a propositional interpretation $\mathcal{I}$, computes (in polynomial time) the value $F^{\mathcal{I}}$ and returns *yes* if $F^{\mathcal{I}} = 1$ and *no* if $F^{\mathcal{I}} = 0$. It can be shown that each problem solvable by a non-deterministic polynomial algorithm has a polynomial reduction to SAT, which means that SAT is actually an *NP-complete* problem. Currently, the most efficient algorithms for solving SAT are based on (extensions of) the *Davis-Putnam-Logemann-Loveland* (short: *DPLL*) procedure, which systematically explores interpretations in a goal-directed way. A program that implements an algorithm for solving SAT is called a *SAT-solver*. Usually a SAT-solver not only decides satisfiability of a given propositional formula F, but can also output a *model* of F in case F is satisfiable.

References

1. Arif, M.F., Mencía, C., Marques-Silva, J.: Efficient MUS enumeration of horn formulae with applications to axiom pinpointing. CoRR abs/1505.04365 (2015)
2. Baader, F.: Description logics. In: Tessaris, S., et al. (eds.) Reasoning Web 2009. LNCS, vol. 5689, pp. 1–39. Springer, Heidelberg (2009). https://doi.org/10.1007/978-3-642-03754-2_1
3. Baader, F., Brandt, S., Lutz, C.: Pushing the $\mathcal{EL}$ envelope. In: Proceedings of the 19th International Joint Conference on Artificial Intelligence (IJCAI 2005), pp. 364–369 (2005)
4. Baader, F., Calvanese, D., McGuinness, D., Nardi, D., Patel-Schneider, P. (eds.): The Description Logic Handbook: Theory, Implementation, and Applications, 2nd edn. Cambridge University Press, Cambridge (2007)
5. Baader, F., Franconi, E., Hollunder, B., Nebel, B., Profitlich, H.J.: An empirical analysis of optimization techniques for terminological representation systems. Appl. Intell. 4(2), 109–132 (1994)
6. Baader, F., Horrocks, I., Lutz, C., Sattler, U.: An Introduction to Description Logic. Cambridge University Press, Cambridge (2017)
7. Bate, A., Motik, B., Grau, B.C., Cucala, D.T., Simancik, F., Horrocks, I.: Consequence-based reasoning for description logics with disjunctions and number restrictions. J. Artif. Intell. Res. **63**, 625–690 (2018)
8. Baumgartner, P., Furbach, U., Niemelä, I.: Hyper tableaux. In: Alferes, J.J., Pereira, L.M., Orlowska, E. (eds.) JELIA 1996. LNCS, vol. 1126, pp. 1–17. Springer, Heidelberg (1996). https://doi.org/10.1007/3-540-61630-6_1
9. Bienvenu, M., Bourgaux, C.: Inconsistency-tolerant querying of description logic knowledge bases. In: Pan, J.Z., et al. (eds.) Reasoning Web 2016. LNCS, vol. 9885, pp. 156–202. Springer, Cham (2017). https://doi.org/10.1007/978-3-319-49493-7_5
10. Bienvenu, M., Ortiz, M.: Ontology-mediated query answering with data-tractable description logics. In: Faber, W., Paschke, A. (eds.) Reasoning Web 2015. LNCS, vol. 9203, pp. 218–307. Springer, Cham (2015). https://doi.org/10.1007/978-3-319-21768-0_9
11. Bonatti, P.A., Faella, M., Petrova, I.M., Sauro, L.: A new semantics for overriding in description logics. Artif. Intell. **222**, 1–48 (2015)

12. Botoeva, E., Konev, B., Lutz, C., Ryzhikov, V., Wolter, F., Zakharyaschev, M.: Inseparability and conservative extensions of description logic ontologies: a survey. In: Pan, J.Z., et al. (eds.) Reasoning Web 2016. LNCS, vol. 9885, pp. 27–89. Springer, Cham (2017). https://doi.org/10.1007/978-3-319-49493-7_2
13. Brandt, S.: Polynomial time reasoning in a description logic with existential restrictions, GCI axioms, and - what else? In: de Mántaras, R.L., Saitta, L. (eds.) Proceedings of the 16th European Conference on Artificial Intelligence (ECAI 2004), pp. 298–302. IOS Press (2004)
14. Casini, G., Straccia, U.: Defeasible inheritance-based description logics. J. Artif. Intell. Res. **48**, 415–473 (2013)
15. Cucala, D.T., Grau, B.C., Horrocks, I.: Consequence-based reasoning for description logics with disjunction, inverse roles, number restrictions, and nominals. In: Lang, J. (ed.) Proceedings of the 27th International Joint Conference on Artificial Intelligence (IJCAI 2018), pp. 1970–1976. ijcai.org (2018)
16. Cuenca Grau, B., Horrocks, I., Kazakov, Y., Sattler, U.: Modular reuse of ontologies: theory and practice. J. Artif. Intell. Res. **31**, 273–318 (2008)
17. Feld, M., Müller, C.: The automotive ontology: managing knowledge inside the vehicle and sharing it between cars. In: Proceedings of the 3rd International Conference on Automotive User Interfaces and Interactive Vehicular Applications, AutomotiveUI 2011, pp. 79–86. ACM, New York (2011). http://doi.acm.org/10.1145/2381416.2381429
18. Giordano, L., Gliozzi, V., Olivetti, N., Pozzato, G.L.: A non-monotonic description logic for reasoning about typicality. Artif. Intell. **195**, 165–202 (2013)
19. Glimm, B., Horrocks, I., Motik, B.: Optimized description logic reasoning via core blocking. In: Giesl, J., Hähnle, R. (eds.) IJCAR 2010. LNCS (LNAI), vol. 6173, pp. 457–471. Springer, Heidelberg (2010). https://doi.org/10.1007/978-3-642-14203-1_39
20. Glimm, B., Horrocks, I., Motik, B., Shearer, R., Stoilos, G.: A novel approach to ontology classification. J. Web Semant. **14**, 84–101 (2012)
21. Golbreich, C., Zhang, S., Bodenreider, O.: The foundational model of anatomy in OWL: experience and perspectives. J. Web Semant. **4**(3), 181–195 (2006)
22. Greiner, R., Smith, B.A., Wilkerson, R.W.: A correction to the algorithm in Reiter's theory of diagnosis. In: Readings in Model-Based Diagnosis, pp. 49–53. Morgan Kaufmann Publishers Inc. (1992)
23. Group, T.W.W. (ed.): SPARQL 1.1 Overview. W3C Recommendation, 21 March 2013. http://www.w3.org/TR/sparql11-overview/
24. Haarslev, V., Möller, R., Turhan, A.-Y.: Exploiting pseudo models for TBox and ABox reasoning in expressive description logics. In: Goré, R., Leitsch, A., Nipkow, T. (eds.) IJCAR 2001. LNCS, vol. 2083, pp. 61–75. Springer, Heidelberg (2001). https://doi.org/10.1007/3-540-45744-5_6
25. Hoehndorf, R., Dumontier, M., Gkoutos, G.V.: Evaluation of research in biomedical ontologies. Briefings Bioinform. **14**(6), 696–712 (2012)
26. Horridge, M.: Justification based explanation in ontologies. Ph.D. thesis, University of Manchester, UK (2011)
27. Horridge, M., Parsia, B., Sattler, U.: Laconic and precise justifications in OWL. In: Sheth, A., et al. (eds.) ISWC 2008. LNCS, vol. 5318, pp. 323–338. Springer, Heidelberg (2008). https://doi.org/10.1007/978-3-540-88564-1_21
28. Horridge, M., Parsia, B., Sattler, U.: Justification oriented proofs in OWL. In: Patel-Schneider, P.F., et al. (eds.) ISWC 2010. LNCS, vol. 6496, pp. 354–369. Springer, Heidelberg (2010). https://doi.org/10.1007/978-3-642-17746-0_23

29. Horrocks, I., Kutz, O., Sattler, U.: The even more irresistible $\mathcal{SROIQ}$. In: Doherty, P., Mylopoulos, J., Welty, C.A. (eds.) Proceedings 10th International Conference on Principles of Knowledge Representation and Reasoning (KR 2006), pp. 57–67. AAAI Press (2006)

30. Hudek, A.K., Weddell, G.E.: Binary absorption in tableaux-based reasoning for description logics. In: Proceedings of the 19th International Workshop on Description Logics (DL 2006), vol. 189. CEUR (2006)

31. Kazakov, Y.: $\mathcal{RIQ}$ and $\mathcal{SROIQ}$ are harder than $\mathcal{SHOIQ}$. In: Brewka, G., Lang, J. (eds.) Proceedings of the 11th International Conference on Principles of Knowledge Representation and Reasoning (KR 2008), pp. 274–284. AAAI Press (2008)

32. Kazakov, Y.: Consequence-driven reasoning for Horn $\mathcal{SHIQ}$ ontologies. In: Proceedings of the 21st International Joint Conference on Artificial Intelligence (IJCAI 2009), pp. 2040–2045. IJCAI (2009)

33. Kazakov, Y., Klinov, P.: Goal-directed tracing of inferences in EL ontologies. In: Mika, P., et al. (eds.) ISWC 2014. LNCS, vol. 8797, pp. 196–211. Springer, Cham (2014). https://doi.org/10.1007/978-3-319-11915-1_13

34. Kazakov, Y., Krötzsch, M., Simančík, F.: ELK: a reasoner for OWL EL ontologies. System description, University of Oxford (2012)

35. Kharlamov, E., et al.: Ontology based data access in statoil. Web Semant. Sci. Serv. Agents World Wide Web **44**, 3–36 (2017)

36. Kontchakov, R., Zakharyaschev, M.: An introduction to description logics and query rewriting. In: Koubarakis, M., et al. (eds.) Reasoning Web 2014. LNCS, vol. 8714, pp. 195–244. Springer, Cham (2014). https://doi.org/10.1007/978-3-319-10587-1_5

37. Krötzsch, M., Marx, M., Ozaki, A., Thost, V.: Attributed description logics: reasoning on knowledge graphs. In: Lang, J. (ed.) Proceedings of the Twenty-Seventh International Joint Conference on Artificial Intelligence, IJCAI 2018, Stockholm, Sweden, 13–19 July 2018. pp. 5309–5313. ijcai.org (2018). https://doi.org/10.24963/ijcai.2018/743

38. Maier, A., Schnurr, H.-P., Sure, Y.: Ontology-based information integration in the automotive industry. In: Fensel, D., Sycara, K., Mylopoulos, J. (eds.) ISWC 2003. LNCS, vol. 2870, pp. 897–912. Springer, Heidelberg (2003). https://doi.org/10.1007/978-3-540-39718-2_57

39. Manthey, N., Peñaloza, R., Rudolph, S.: Efficient axiom pinpointing in $\mathcal{EL}$ using SAT technology. In: Lenzerini, M., Peñaloza, R. (eds.) Proceedings of the 29th International Workshop on Description Logics (DL 2016). CEUR Workshop Proceedings, vol. 1577. CEUR-WS.org (2016). http://ceur-ws.org/Vol-1577/paper_33.pdf

40. Motik, B.: Representing and querying validity time in RDF and OWL: a logic-based approach. J. Web Semant. **12**, 3–21 (2012). https://doi.org/10.1016/j.websem.2011.11.004

41. Motik, B., Cuenca Grau, B., Horrocks, I., Wu, Z., Fokoue, A., Lutz, C. (eds.): OWL 2 Web Ontology Language: Profiles. W3C Recommendation, 27 October 2009. http://www.w3.org/TR/owl2-profiles/

42. Motik, B., Shearer, R., Horrocks, I.: Hypertableau reasoning for description logics. J. Artif. Intell. Res. **36**, 165–228 (2009)

43. Ortiz, M., Šimkus, M.: Reasoning and query answering in description logics. In: Eiter, T., Krennwallner, T. (eds.) Reasoning Web 2012. LNCS, vol. 7487, pp. 1–53. Springer, Heidelberg (2012). https://doi.org/10.1007/978-3-642-33158-9_1

44. OWL Working Group, W.: OWL 2 Web Ontology Language: Document Overview. W3C Recommendation, 27 October 2009. http://www.w3.org/TR/owl2-overview/

45. Peñaloza, R.: Explaining axiom pinpointing. In: Lutz, C., Sattler, U., Tinelli, C., Turhan, A.-Y., Wolter, F. (eds.) Description Logic, Theory Combination, and All That. LNCS, vol. 11560, pp. 475–496. Springer, Cham (2019). https://doi.org/10. 1007/978-3-030-22102-7_22

46. Reiter, R.: A theory of diagnosis from first principles. Artif. Intell. **32**(1), 57–95 (1987)

47. Robinson, J.A.: Automatic deduction with hyper-resolution. Int. J. Comput. Math. **1**(3), 227–234 (1965)

48. Armas Romero, A., Cuenca Grau, B., Horrocks, I.: MORe: modular combination of OWL reasoners for ontology classification. In: Cudré-Mauroux, P., et al. (eds.) ISWC 2012. LNCS, vol. 7649, pp. 1–16. Springer, Heidelberg (2012). https://doi. org/10.1007/978-3-642-35176-1_1

49. Rudolph, S.: Foundations of description logics. In: Polleres, A., et al. (eds.) Reasoning Web 2011. LNCS, vol. 6848, pp. 76–136. Springer, Heidelberg (2011). https:// doi.org/10.1007/978-3-642-23032-5_2

50. Sattler, U.: Reasoning in description logics: basics, extensions, and relatives. In: Antoniou, G., et al. (eds.) Reasoning Web 2007. LNCS, vol. 4636, pp. 154–182. Springer, Heidelberg (2007). https://doi.org/10.1007/978-3-540-74615-7_2

51. Schmidt-Schauß, M., Smolka, G.: Attributive concept descriptions with complements. J. Artif. Intell. **48**, 1–26 (1991)

52. Schreiber, G., Raimond, Y. (eds.): RDF 1.1 Primer. W3C Working Group Note, 24 June 2014. http://www.w3.org/TR/rdf11-primer/

53. Sebastiani, R., Vescovi, M.: Axiom pinpointing in lightweight description logics via horn-SAT encoding and conflict analysis. In: Schmidt, R.A. (ed.) CADE 2009. LNCS (LNAI), vol. 5663, pp. 84–99. Springer, Heidelberg (2009). https://doi.org/ 10.1007/978-3-642-02959-2_6

54. Simančík, F., Kazakov, Y., Horrocks, I.: Consequence-based reasoning beyond horn ontologies. In: Proceedings of the 22nd International Joint Conference on Artificial Intelligence (IJCAI 2011), pp. 1093–1098. AAAI Press/IJCAI (2011)

55. Sirin, E.: From wine to water: optimizing description logic reasoning for nominals. In: Proceedings of the 10th International Conference on Principles of Knowledge Representation and Reasoning (KR 2006), pp. 90–99. AAAI Press (2006)

56. Sirin, E., Parsia, B., Grau, B.C., Kalyanpur, A., Katz, Y.: Pellet: a practical OWL-DL reasoner. J. Web Semant. **5**(2), 51–53 (2007)

57. Steigmiller, A., Glimm, B., Liebig, T.: Optimised absorption for expressive description logics. In: Proceedings of the 27th International Workshop on Description Logics (DL 2014). CEUR Workshop Proceedings, vol. 1193. CEUR-WS.org (2014). https://www.uni-ulm.de/fileadmin/website_uni_ulm/iui.inst.090/ Publikationen/2014/StGL14b.pdf

58. Steigmiller, A., Liebig, T., Glimm, B.: Konclude: system description. J. Web Semant. **27–28**, 78–85 (2014)

59. Straccia, U.: All about fuzzy description logics and applications. In: Faber, W., Paschke, A. (eds.) Reasoning Web 2015. LNCS, vol. 9203, pp. 1–31. Springer, Cham (2015). https://doi.org/10.1007/978-3-319-21768-0_1

60. Tobies, S.: Complexity results and practical algorithms for logics in knowledge representation. Ph.D. thesis, RWTH Aachen, Germany (2001)

61. Tsarkov, D., Horrocks, I.: Efficient reasoning with range and domain constraints. In: Proceedings of the 17th International Workshop on Description Logics (DL 2004), vol. 104. CEUR (2004)

62. Tsarkov, D., Horrocks, I.: FaCT++ description logic reasoner: system description. In: Furbach, U., Shankar, N. (eds.) IJCAR 2006. LNCS (LNAI), vol. 4130, pp. 292–297. Springer, Heidelberg (2006). https://doi.org/10.1007/11814771_26

63. Tudorache, T., Nyulas, C.I., Noy, N.F., Musen, M.A.: Using semantic web in ICD-11: three years down the road. In: Alani, H., et al. (eds.) ISWC 2013. LNCS, vol. 8219, pp. 195–211. Springer, Heidelberg (2013). https://doi.org/10.1007/978-3-642-41338-4_13

64. Turhan, A.-Y.: Reasoning and explanation in $\mathcal{EL}$ and in expressive description logics. In: Aßmann, U., Bartho, A., Wende, C. (eds.) Reasoning Web 2010. LNCS, vol. 6325, pp. 1–27. Springer, Heidelberg (2010). https://doi.org/10.1007/978-3-642-15543-7_1

65. Turhan, A.-Y.: Introductions to description logics – a guided tour. In: Rudolph, S., Gottlob, G., Horrocks, I., van Harmelen, F. (eds.) Reasoning Web 2013. LNCS, vol. 8067, pp. 150–161. Springer, Heidelberg (2013). https://doi.org/10.1007/978-3-642-39784-4_3

66. Vardi, M.Y.: Why is modal logic so robustly decidable? In: Immerman, N., Kolaitis, P.G. (eds.) Descriptive Complexity and Finite Models, Proceedings of a DIMACS Workshop 1996, Princeton, New Jersey, USA, 14–17 January 1996. DIMACS Series in Discrete Mathematics and Theoretical Computer Science, vol. 31, pp. 149–183. DIMACS/AMS (1996)

67. Vescovi, M.: Exploiting SAT and SMT techniques for automated reasoning and ontology manipulation in description logics. Ph.D. thesis, University of Trento, Italy (2011). http://eprints-phd.biblio.unitn.it/477/

68. Zhao, L., Ichise, R., Mita, S., Sasaki, Y.: Core ontologies for safe autonomous driving. In: Villata, S., Pan, J.Z., Dragoni, M. (eds.) Proceedings of the ISWC 2015 Posters & Demonstrations Track co-located with the 14th International Semantic Web Conference (ISWC-2015), Bethlehem, PA, USA, 11 October 2015. CEUR Workshop Proceedings, vol. 1486. CEUR-WS.org (2015). http://ceur-ws.org/Vol-1486/paper_9.pdf

Explanation-Friendly Query Answering Under Uncertainty

Maria Vanina Martinez[1][✉] and Gerardo I. Simari[2]

[1] Department of Computer Science, Institute for Computer Science
(UBA–CONICET), Universidad de Buenos Aires (UBA),
C1428EGA Ciudad Autonoma de Buenos Aires, Argentina
mvmartinez@dc.uba.ar
[2] Department of Computer Science and Engineering,
Institute for Computer Science and Engineering (UNS–CONICET),
Universidad Nacional del Sur (UNS),
San Andres 800, 8000 Bahia Blanca, Argentina
gis@cs.uns.edu.ar

Abstract. Many tasks often regarded as requiring some form of intelligence to perform can be seen as instances of query answering over a semantically rich knowledge base. In this context, two of the main problems that arise are: (i) uncertainty, including both inherent uncertainty (such as events involving the weather) and uncertainty arising from lack of sufficient knowledge; and (ii) inconsistency, which involves dealing with conflicting knowledge. These unavoidable characteristics of real world knowledge often yield complex models of reasoning; assuming these models are mostly used by humans as decision-support systems, meaningful explainability of their results is a critical feature. These lecture notes are divided into two parts, one for each of these basic issues. In Part 1, we present basic probabilistic graphical models and discuss how they can be incorporated into powerful ontological languages; in Part 2, we discuss both classical inconsistency-tolerant semantics for ontological query answering based on the concept of repair and other semantics that aim towards more flexible yet principled ways to handle inconsistency. Finally, in both parts we ponder the issue of deriving different kinds of explanations that can be attached to query results.

1   Introduction

In this article, we address query answering under two different, though related, approaches to uncertainty: probabilistic reasoning, and inconsistency-tolerant reasoning—as we will see, incompleteness is another dimension to uncertainty that can be addressed by leveraging the power of ontology languages, which are at the core of the material that we aim to cover. We focus on Datalog+/− [17], a family of ontological languages that was born from the database theory community extending the well-known formalism of Datalog. This family is closely related to Description Logics (DLs); cf. Fig. 1 for a mapping of some of the basic

© Springer Nature Switzerland AG 2019
M. Krötzsch and D. Stepanova (Eds.): Reasoning Web 2019, LNCS 11810, pp. 65–103, 2019.
https://doi.org/10.1007/978-3-030-31423-1_2

constructs in description logics to Datalog+/− formulas—note that this is meant only to illustrate the general relationship between the two formalisms, and that there are constructs on either side that cannot be expressed in the other, such as number restrictions and disjunctions in Datalog+/− and predicates of arity greater than two in DLs.

We first put these notes into context by briefly presenting some historical details and basic aspects of explanations in AI. Then, in Sect. 2 we provide a brief introduction to Datalog+/−, the family of ontology languages that we use in the rest of the text. Sections 3 and 4 then describe the two main parts: probabilistic and inconsistency-tolerant reasoning, respectively; in each case, we conclude the section by exploring current capabilities and next steps that can be taken towards making these formalisms explainable. Finally, in Sect. 5 we provide a summary and discuss a roadmap for future work in these directions.

Context: A *Brief* Discussion about Explanations in AI

In order to put this material into context, we would like to briefly discuss the history surrounding one of the main topics of these notes. The meaning of *explanation*, and the related notions of *explainability* and *interpretability*, has been studied for quite some time in philosophy and related disciplines in the social sciences (cf. the recent work of [40] for a survey of these aspects). Essentially, this topic is of interest to these disciplines because explanations are usually meant to be *consumed by humans*—for instance, a (human) user would like to know why a certain weather forecast is likely to be true or, more importantly, why they are being denied a loan at the bank. In computer science, explanations were a core aspect of the *expert systems* that were developed over four decades ago [46,53]; ever since those foundational works, logic-based formalisms have often highlighted explainability as one of the strong points of developing AI in such a manner, contrasting with the fact that machine learning (ML) methods may in some cases perform very well but are incapable of offering users a satisfactory explanation. Structured argumentation is a good example,[1] in which *dialectical trees* are produced as part of the reasoning mechanism and can be examined by a user in order to gain insights into how conclusions are reached [25,26]; the work of [24] also explores how belief revision operators can be designed using argumentation-based comparisons of alternatives, which can also be offered as explanations. As a response to this—and the success of many ML-based approaches on concrete problems—in recent years, there has been a strong resurgence of research into how AI (mostly ML) tools can be made to be explainable; the term "XAI" (for explainable artificial intelligence) was thus born. This recent explosion in popularity has already led to interesting developments; in the context of reasoning under uncertainty (of particular interest here), the notion of *balanced* explanation—giving reasons both *why* and *why not* a given answer may be correct—is especially useful [30]. We refer the interested reader to [1,40,44] for some recent surveys developed from different points of view.

---

[1] Note, however, that the human aspect is not necessarily present, since the argumentation process could be carried out between software agents.

Description Logic Assertion	Datalog+/– Rule
CONCEPT INCLUSION: *Restaurant* $\sqsubseteq$ *Business*	$restaurant(X) \rightarrow business(X)$
CONCEPT PRODUCT: *Food* $\times$ *Food* $\sqsubseteq$ *TwoCourseMeal*	$food(X), food(Y) \rightarrow twoCourseMeal(X, Y)$
INVERSE ROLE INCLUSION: *InPromotionIn⁻* $\sqsubseteq$ *Serves*	$inPromotionIn(F, R) \rightarrow serves(R, F)$
ROLE TRANSITIVITY: trans(*LocatedIn*)	$locatedIn(X, Y), locatedIn(Y, Z) \rightarrow locatedIn(X, Z)$
PARTICIPATION: *Restaurant* $\sqsubseteq \exists Serves.Food$	$restaurant(R) \rightarrow \exists F \; serves(R, F) \wedge food(F)$
DISJOINTNESS: *City* $\sqcap$ *Country* $\sqsubseteq \perp$	$city(X), country(X) \rightarrow \perp$
FUNCTIONALITY: funct(*LocatedIn*)	$locatedIn(X, Y), locatedIn(X, Z) \rightarrow Y = Z$

Fig. 1. Translation of several different types of description logic axioms into Datalog+/–.

From this brief analysis we can conclude that there are many aspects that need to be further studied in order to arrive at adequate solutions to the problem of deriving explanations. On the one hand, logic-based models have a strong foundation that allows them to be better poised to offer explanations, but not much research has gone in to designing explanations that can be of use to actual users. On the other hand, ML-based solutions typically can be made to perform quite well on certain tasks, but there inner workings are more obscure. In these notes, we will thus focus on taking some first steps towards explaining the results given by two approaches to reasoning under uncertainty—we cannot hope to solve such a formidable family of problems completely just yet.

2   The Datalog+/– Family of Ontology Languages

We now present the basics of Datalog+/– [17]—relational databases, (Boolean) conjunctive queries, tuple- and equality-generating dependencies and negative constraints, the chase, and ontologies. The material presented in this section is mainly based on [47], which in turn contains some material originally appearing in [48].

2.1   Preliminary Concepts and Notations

Let us consider (i) an infinite universe of *(data) constants* Δ, which constitute the "normal" domain of a database), (ii) an infinite set of *(labelled) nulls* Δ_N (used as "fresh" Skolem terms, which are placeholders for unknown values, and can thus be seen as a special kind of variable), and (iii) an infinite set of *variables* $\mathcal{V}$ (used in queries, dependencies, and constraints). Different constants represent different values (this is generally known as the *unique name assumption*), while different nulls may represent the same value. We assume a lexicographic order on $\Delta \cup \Delta_N$, with every symbol in Δ_N following all symbols in Δ. We denote with **X** sequences of variables $X_1, \ldots, X_k$ with $k \geqslant 0$.

We will assume a *relational schema* $\mathcal{R}$, which is a finite set of *predicate symbols* (or simply *predicates*), each with an associated arity. As usual, a *term t* is a constant, null, or variable. An *atomic formula* (or *atom*) a has the form $p(t_1, \ldots, t_n)$, where p is an n-ary predicate, and $t_1, \ldots, t_n$ are terms. A term or atom is *ground* if it contains no nulls and no variables. An *instance I* for a relational schema $\mathcal{R}$ is a (possibly infinite) set of atoms with predicates from $\mathcal{R}$ and arguments from $\Delta \cup \Delta_N$. A *database* is a finite instance that contains only constants (i.e., its arguments are from Δ).

Homomorphisms. Central to the semantics of Datalog$+/-$ is the notion of *homomorphism* between relational structures. Let $A = \langle X, \sigma^A \rangle$ and $B = \langle Y, \sigma^B \rangle$ be two relational structures, where $dom(A) = X$ and $dom(B) = Y$ are the domains of A and B, and σ^A and σ^B are their signatures (which are composed of relations and functions), respectively. A *homomorphism* from A to B is a function $h : dom(A) \rightarrow dom(B)$ that "preserves structure" in the following sense:

– For each n-ary function $f^A \in \sigma^A$ and elements $x_1, \ldots, x_n \in dom(A)$, we have:

$$h\big(f^A(x_1, \ldots, x_n)\big) = f^B\big(h(x_1), \ldots, h(x_n)\big),$$

   and
– for each n-ary relation $R^A \in \sigma^A$ and elements $x_1, \ldots, x_n \in dom(A)$, we have:

$$\text{if } (x_1, \ldots, x_n) \in R^A, \text{ then } \big(h(x_1), \ldots, h(x_n)\big) \in R^B.$$

In the above statements, the superscripts used in function and relation symbols is simply a clarification of the structure in which they are being applied. Since we do not have function symbols, the first condition will not be necessary here (it is satisfied vacuously).[2]

For the purposes of Datalog$+/-$, we need to extend the concept of homomorphism to contemplate nulls. We then define homomorphisms from a set of atoms A_1 to a set of atoms A_2 as mappings $h \colon \Delta \cup \Delta_N \cup \mathcal{V} \rightarrow \Delta \cup \Delta_N \cup \mathcal{V}$ such that:

---

[2] As an aside, and using concepts that will be defined shortly, the fundamental result linking homomorphisms to conjunctive query answering over relational databases can be informally stated as follows: let Q be a BCQ, and J be a database instance; then, $J \models Q$ if and only if there exists a homomorphism from the *canonical database instance* I^Q (essentially, an instance built using the predicates and variables from Q) to J [5,19].

1. $c \in \Delta$ implies $h(c) = c$,
2. $c \in \Delta_N$ implies $h(c) \in \Delta \cup \Delta_N$,
2. $r(t_1, \ldots, t_n) \in A_1$ implies $h(r(t_1, \ldots, t_n)) = r(h(t_1), \ldots, h(t_n))) \in A_2$.

Similarly, one can extend h to a conjunction of atoms. Conjunctions of atoms are often identified with the *sets* of their atoms.

2.2 Syntax and Semantics of Datalog+/−

Given a relational schema $\mathcal{R}$, a Datalog+/− program consists of a finite set of tuple-generating dependencies (TGDs), negative constraints (NCs), and equality-generating dependencies (EGDs).

TGDs. A *tuple-generating dependency* (TGD) σ is a first-order (FO) rule that allows existentially quantified conjunctions of atoms in rule heads:

$$\sigma \; : \; \forall \mathbf{X} \forall \mathbf{Y} \; \Phi(\mathbf{X}, \mathbf{Y}) \rightarrow \exists \mathbf{Z} \; \Psi(\mathbf{X}, \mathbf{Z}) \text{ with } \mathbf{X}, \mathbf{Y}, \mathbf{Z} \subseteq \mathcal{V},$$

where $\Phi(\mathbf{X}, \mathbf{Y})$ and $\Psi(\mathbf{X}, \mathbf{Z})$ are conjunctions of atoms. Formulas Φ and Ψ are often referred to as the *body* and *head* of σ, respectively. By analyzing the general form of TGDs, one can see that variables in $\mathbf{X}$ and $\mathbf{Y}$ refer to objects that are already known, while those in $\mathbf{Z}$ correspond to the result of so-called *value invention*. For instance, in the TGD $person(X) \rightarrow \exists Y \; person(Y) \wedge father(Y, X)$, variable Y refers to a new object that is a person who is the father of X.

Since TGDs with multiple atoms in the head can be converted into sets of TGDs with only single atom in the head [14], from now on we assume that all sets of TGDs have only a single atom in their head. An instance I for $\mathcal{R}$ *satisfies* σ, denoted $I \models \sigma$, if whenever there exists a homomorphism h that maps the atoms of $\Phi(\mathbf{X}, \mathbf{Y})$ to atoms of I, there exists an extension h' of h that maps $\Psi(\mathbf{X}, \mathbf{Z})$ to atoms of I.

NCs. A *negative constraint* (NC) ν is a first-order rule that allows to express negation:

$$\nu \; : \; \forall \mathbf{X} \; \Phi(\mathbf{X}) \rightarrow \bot \text{ with } \mathbf{X} \subseteq \mathcal{V},$$

where $\Phi(\mathbf{X})$ a conjunction of atoms; formula Φ is usually referred to as the *body* of ν. An instance I for $\mathcal{R}$ *satisfies* ν, denoted $I \models \nu$, if for each homomorphism h, $h(\Phi(\mathbf{X}, \mathbf{Y})) \not\subseteq I$ holds.

EGDs. An *equality-generating dependency* (EGD) μ is a first-order rule of the form:

$$\mu \; : \; \forall \mathbf{X} \Phi(\mathbf{X}) \rightarrow X_i = X_j \text{ with } X_i, X_j \in \mathbf{X} \subseteq \mathcal{V},$$

where $\Phi(\mathbf{X})$ is conjunction of atoms; as above, formula Φ is usually referred to as the *body* of μ. An instance I for $\mathcal{R}$ *satisfies* μ, denoted $I \models \mu$, if whenever there is a homomorphism h such that $h(\Phi(\mathbf{X}, \mathbf{Y})) \subseteq I$, it holds that $h(X_i) = h(X_j)$.

In the following, we will sometimes omit the universal quantification in front of TGDs, NCs, and EGDs, and assume that all variables appearing in their bodies are universally quantified. We will sometimes use the words constraints and dependencies to refer to NCs and EGDs.

Programs and Ontologies. A *Datalog+/− program* Σ is a finite set $\Sigma_T \cup \Sigma_{NC} \cup \Sigma_E$ of TGDs, NCs, and EGDs. The schema of Σ, denoted $\mathcal{R}(\Sigma)$, is the set of predicates occurring in Σ. A *Datalog+/− ontology* $KB = (D, \Sigma)$ consists of a finite database D and a Datalog+/− program Σ. The following example illustrates a simple Datalog+/− ontology, used in the sequel as a running example.

Example 1. Consider the ontology $KB = (D, \Sigma)$, where D and $\Sigma = \Sigma_T \cup \Sigma_E$ are defined as follows:

$\Sigma_T = \{\ r_1 : restaurant(R) \rightarrow business(R),$

$\quad\quad r_2 : restaurant(R) \rightarrow \exists F\ food(F) \wedge serves(R, F),$

$\quad\quad r_3 : restaurant(R) \rightarrow \exists C\ cuisine(C) \wedge restaurantCuisine(R, C),$

$\quad\quad r_4 : business(B) \rightarrow \exists C\ city(C) \wedge locatedIn(B, C),$

$\quad\quad r_5 : city(C) \rightarrow \exists D\ country(D) \wedge locatedIn(C, D)\},$

$\Sigma_E = \{\ locatedIn(X, Y), locatedIn(X, Z) \rightarrow Y = Z\},$

$D = \{\ food(bifeDeChorizo),\quad food(soupAlOignon),$

$\quad\quad foodType(meat),\quad foodType(soup),$

$\quad\quad cuisine(argentine),\quad cuisine(french),$

$\quad\quad restaurant(laCabrera),\quad restaurant(laTartine),$

$\quad\quad city(buenosAires),\quad city(paris),$

$\quad\quad country(argentina),\quad country(france),$

$\quad\quad locatedIn(laCabrera, buenosAires),$

$\quad\quad serves(laCabrera, bifeDeChorizo),\quad serves(laTartine, soupeAlOignon)\}.$

This ontology models a very simple knowledge base for restaurants—it could be used, for instance, as the underlying model in an online recommendation and reviewing system (e.g., in the style of TripAdvisor or Yelp).   ∎

Models. The conjunction of the first-order sentences associated with the rules of a Datalog+/− program Σ is denoted Σ_P. A *model* of Σ is an instance for $\mathcal{R}(\Sigma)$ that satisfies Σ_p. For a database D for $\mathcal{R}$, and a set of TGDs Σ on $\mathcal{R}$, the set of *models* of D and Σ, denoted $mods(D, \Sigma)$, is the set of all (possibly infinite) instances I such that:

1. $D \subseteq I$, and
2. every $\sigma \in \Sigma$ is satisfied in I (i.e., $I \models \Sigma$).

The ontology is *consistent* if the set $mods(D, \Sigma)$ is not empty.

The semantics of Σ on an input database D, denoted $\Sigma(D)$, is a model I of D and Σ such that for every model I' of D and Σ there exists a homomorphism h such that $h(I) \subseteq I'$; such an instance is called *universal model* of Σ w.r.t. D. Intuitively, a universal model contains no more and no less information than what the given program requires.

In general, there exists more than one universal model of Σ w.r.t. D, but the universal models are (by definition) the same up to homomorphic equivalence, i.e., for each pair of universal models M_1 and M_2, there exist homomorphisms

h_1 and h_2 such that $h_1(M_1) \subseteq M_2$ and $h_2(M_2) \subseteq M_1$. Thus, $\Sigma(D)$ is unique up to homomorphic equivalence.

2.3 Conjunctive Query Answering

We now introduce conjunctive query answering for Datalog+/−. A *conjunctive query* (CQ) over $\mathcal{R}$ has the form:

$$q(\mathbf{X}) = \exists \mathbf{Y}\, \Phi(\mathbf{X}, \mathbf{Y}),$$

where $\Phi(\mathbf{X}, \mathbf{Y})$ is a conjunction of atoms (consisting also possibly of equalities, but not inequalities) involving variables in $\mathbf{X}$ and $\mathbf{Y}$, and possibly constants, but without nulls, and q is a predicate not occurring in $\mathcal{R}$. A *Boolean CQ* (BCQ) over $\mathcal{R}$ is a CQ of the form $q()$, often written as the set of all its atoms, without quantifiers. As mentioned above for the basic components of the language, formulas q and Φ are sometimes referred to as the *head* and *body* of the query, respectively.

The set of *answers* to a CQ $q(\mathbf{X}) = \exists \mathbf{Y}\, \Phi(\mathbf{X}, \mathbf{Y})$ over an instance I, denoted $q(I)$, is the set of all tuples t over Δ, for which there exists a homomorphism $h: \mathbf{X} \cup \mathbf{Y} \to \Delta \cup \Delta_N$ such that $h(\Phi(\mathbf{X}, \mathbf{Y})) \subseteq I$ and $h(\mathbf{X}) = t$. The *answer* to a BCQ $q()$ over a database instance I is *Yes*, denoted $D \models q$, if $q(I) \neq \emptyset$.

Formally, *query answering* under TGDs, i.e., the evaluation of CQs and BCQs on databases under a set of TGDs is defined as follows. The set of *answers* to a CQ q over a database D and a set of TGDs Σ, denoted $ans(q, D, \Sigma)$, is the set of all tuples t such that $t \in q(I)$ for all $I \in mods(D, \Sigma)$. The *answer* to a BCQ q over D and Σ is *Yes*, denoted $D \cup \Sigma \models q$, if $ans(q, D, \Sigma) \neq \emptyset$. Note that for query answering, homomorphically equivalent instances are indistinguishable, i.e., given two instances I and I' that are the same up to homomorphic equivalence, $q(I)$ and $q(I')$ coincide. Therefore, queries can be evaluated on any universal model.

The decision problem of *CQ answering* is defined as follows: given a database D, a set Σ of TGDs, a CQ q, and a tuple of constants t, decide whether $t \in ans(q, D, \Sigma)$.

For query answering of BCQs in Datalog+/− with TGDs, adding negative constraints is computationally easy, as for each constraint $\forall \mathbf{X} \Phi(\mathbf{X}) \to \bot$ one only has to check that the BCQ $\exists \mathbf{X}\, \Phi(\mathbf{X})$ evaluates to false in D under Σ; if one of these checks fails, then the answer to the original BCQ q is true, otherwise the constraints can simply be ignored when answering the BCQ q.

Adding EGDs over databases with TGDs along with negative constraints does not increase the complexity of BCQ query answering as long as they are *non-conflicting* [17]. Intuitively, this ensures that, if the chase (described next) fails (due to strong violations of EGDs), then it already fails on the database, and if it does not fail, then whenever "new" atoms are created in the chase by the application of the EGD chase rule, atoms that are logically equivalent to the new ones are guaranteed to be generated also in the absence of the EGDs, guaranteeing that EGDs do not influence the chase with respect to query answering.

Therefore, from now on, we assume that all the fragments of Datalog+/− have non-conflicting rules.

There are two main ways of processing rules to answer queries: *forward chaining* (the *chase*) and *backward chaining*, which uses the rules to rewrite the query in different ways with the aim of producing a query that directly maps to the facts. The key operation is the unification between part of a current goal (a conjunctive query or a fact) and part of a rule. Here, we will only cover the chase procedure, which is described next.

The TGD Chase. Query answering under general TGDs is undecidable [9] and the chase is used as a procedure to do query answering for Datalog+/−. Given a program Σ with only TGDs (see [17] for further details and for an extended chase with also EGDs), $\Sigma(D)$ can be defined as the least fixpoint of a monotonic operator (modulo homomorphic equivalence). This can be achieved by exploiting the *chase* procedure, originally introduced for checking implication of dependencies, and for checking query containment [28]. Roughly speaking, it executes the rules of Σ starting from D in a forward chaining manner by inferring new atoms, and inventing new null values whenever an existential quantifier needs to be satisfied. By "chase", we refer both to the procedure and to its output.

Let D be a database and σ a TGD of the form $\Phi(\mathbf{X}, \mathbf{Y}) \rightarrow \exists \mathbf{Z}\, \Psi(\mathbf{X}, \mathbf{Z})$. Then, σ is *applicable* to D if there exists a homomorphism h that maps the atoms of $\Phi(\mathbf{X}, \mathbf{Y})$ to atoms of D. Let σ be applicable to D, and h_1 be a homomorphism that extends h as follows: for each $Z_j \in \mathbf{Z}$, $h_1(Z_j) = z_j$, where z_j is a "fresh" null, i.e., $z_j \in \Delta_N$, z_j does not occur in D, and z_j lexicographically follows all other nulls already introduced. The *application of σ* on D adds to D the atom $h_1(\Psi(\mathbf{X}, \mathbf{Z}))$ if not already in D. The chase rule described above is also called *oblivious*.

The chase algorithm for a database D and a set of TGDs Σ consists of an exhaustive application of the TGD chase rule in a breadth-first (level-saturating) fashion, which outputs a (possibly infinite) chase for D and Σ.

Formally, the *chase of level up to 0* of D relative to Σ, denoted $chase^0(D, \Sigma)$, is defined as D, assigning to every atom in D the *(derivation) level* 0. For every $k \geqslant 1$, the *chase of level up to k* of D relative to Σ, denoted $chase^k(D, \Sigma)$, is constructed as follows: let $I_1, \ldots, I_n$ be all possible images of bodies of TGDs in Σ relative to some homomorphism such that (i) $I_1, \ldots, I_n \subseteq chase^{k-1}(D, \Sigma)$ and (ii) the highest level of an atom in every I_i is $k - 1$; then, perform every corresponding TGD application on $chase^{k-1}(D, \Sigma)$, choosing the applied TGDs and homomorphisms in a (fixed) linear and lexicographic order, respectively, and assigning to every new atom the *(derivation) level* k. The *chase* of D relative to Σ, denoted $chase(D, \Sigma)$, is defined as the limit of $chase^k(D, \Sigma)$ for $k \to \infty$. This, possibly infinite chase, is a *universal model* of D and Σ, i.e., there is a homomorphism from $chase(D, \Sigma)$ onto every $B \in mods(D, \Sigma)$ [17]—Fig. 2 provides an illustration. Thus, BCQs q over D and Σ can be evaluated on the chase for D and Σ, i.e., $D \cup \Sigma \models q$ is equivalent to $chase(D, \Sigma) \models q$. We will assume

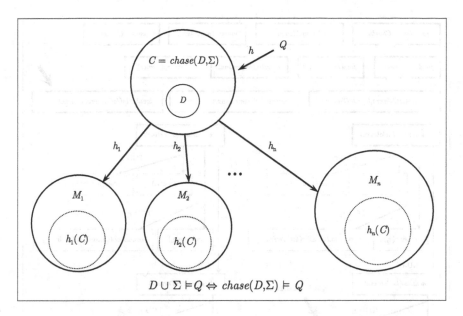

Fig. 2. The chase procedure yields a data structure—also commonly referred to as the chase—that allows to answer queries to a Datalog+/− ontology; it is a universal model, which means that it homomorphically maps to all possible models of the ontology.

that the nulls introduced in the chase are named via Skolemization—this has the advantage of making the chase unique; Δ_N is therefore the set of all possible nulls that may be introduced in the chase.

Example 2. Figure 3 shows the application of the chase procedure over the Datalog+/− ontology from Example 1. As an example, the TGD r_1 is applicable in D, since there is a mapping from atoms *restaurant(laCabrera)* and *restaurant(laTartine)* to the body of the rule. The application of r_1 generates atoms *business(laCabrera)* and *business(laTartine)*.

Consider the following BCQ:

$$q() = \exists X \; restaurant(laTartine) \wedge locatedIn(laTartine, X),$$

asking if there exists a location for restaurant *laTartine*. The answer is *Yes*; in the chase, we can see that after applying TGDs r_1 and r_4, we obtain the atom *locatedIn(laTartine, z_6)*, where z_6 is a null—we would also obtain the same answer, if we ask for restaurant *laCabrera*, because atom *locatedIn(laCabrera, z_5)* is also produced.

Now, consider the CQ:

$$q'(X, Y) = restaurant(X) \wedge locatedIn(X, Y);$$

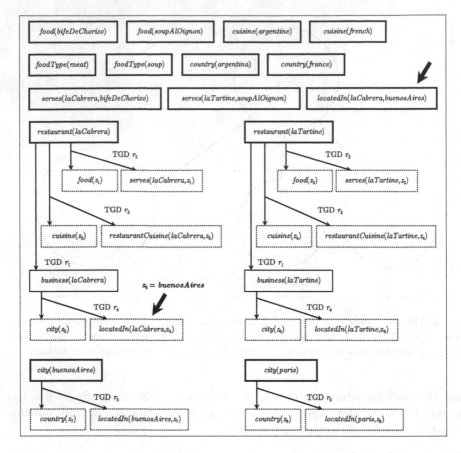

Fig. 3. The chase for the ontology in Example 1. The atoms in boxes with thicker border are part of the database, while those with dotted lines correspond to atoms with null values (denoted with z_i). The arrows point to the mapping of z_5 to the constant *"buenosAires"* during the chase procedure.

in this case, we only obtain one answer, namely (*laCabrera, buenosAires*), since null z_5 eventually maps to *buenosAires*. In the case of *laTartine*, we do not obtain an answer corresponding to the city where it is located: we know there exists a city, but we cannot say which one it is; in every model of KB, z_6 may take a different value from the domain. This can be seen in the chase (a universal model of KB), since z_6 does not unify with a constant after "chasing" D with Σ. ∎

Computational Complexity. The following complexity measures, partially proposed by Vardi [49], are commonly adopted in the literature:

- The *combined complexity* of CQ answering is calculated by considering all the components—the database, the set of dependencies, and the query—as part of the input.
- The *bounded-arity combined complexity* (or *ba-combined complexity*) is calculated by assuming that the arity of the underlying schema is bounded by an integer constant. In the context of DLs, the combined complexity is equivalent to the *ba*-combined complexity, since the arity of the underlying schema is at most two. In practical applications, the schema is usually small, and it can safely be assumed to be fixed—therefore, in this case the arity is also fixed.
- The *fixed-program combined complexity* (or *fp-combined complexity*) is calculated by considering the set of constraints to be fixed.
- The *data complexity* is calculated by taking only the database as input.

Some key facts about complexity and decidability of query answering with TGDs: (i) under general TGDs, the problem is undecidable [9], even when the query and set of dependencies are fixed [14]; (ii) the two problems of CQ and BCQ evaluation under TGDs are LOGSPACE-equivalent [13]; and (iii) the query output tuple (QOT) problem (as a decision version of CQ evaluation that asks if a tuple belongs to the output) and BCQ evaluation are AC_0-reducible to each other. Given the last two points, we focus only on BCQ evaluation, and any complexity results carry over to the other problems.

2.4 Datalog+/− Fragments: In Search of Decidability and Tractability

We now briefly discuss different restrictions that are designed to ensure decidability and tractability of conjunctive query answering with TGDs. While the addition of existential quantifiers in the heads of rules accounts for the "+" in Datalog+/−, these restrictions account for the "−".

Generally, restrictions can be classified into either *abstract* (semantic) or *concrete* (syntactic) properties. Three abstract properties are considered in [6]: (i) the chase is finite, yielding *finite expansion sets* (fes); (ii) the chase may not halt but the facts generated have a tree-like structure, yielding *bounded tree-width sets* (bts); and (iii) a backward chaining mechanism halts in finite time, yielding *finite unification sets* (fus). Other abstract fragments are: (iv) *parsimonious sets* (ps) [33], where the main property for this class is that the chase can be precociously terminated, and (v) *weakly-chase-sticky* TGDs [39] that considers information about the finiteness of predicate positions (positions are infinite if there is an instance D for which an unlimited number of different values appear in that position during the chase).

The main conditions on TGDs that guarantee the decidability of CQ answering are: (i) *guardedness* [13,15], (ii) *stickiness* [16], and (iii) *acyclicity*—each of these classes has a "weak" counterpart: *weak guardedness* [14], *weak stickiness* [16], and *weak acyclicity* [22,23]. Finally, other classes that fall outside this main classification are *Full TGDs* (those that do not have existentially quantified variables), *Tame TGDs* [27] (a combination of the guardedness and sticky-join properties), and *Shy TGDs* [33] (the *shyness* property holds if during the chase procedure nulls do not meet each other to join but only to propagate—nulls thus propagate from a single atom).

We refer the reader to [48] and [47] for a more complete discussion of known classes, a summary of the currently known containment relations between classes, and summaries of known complexity results.

3   Query Answering over Probabilistic Knowledge Bases

We begin by addressing query answering under probabilistic uncertainty. In Sect. 3.1 we provide a very brief overview of some well-known probabilistic graphical models; Sect. 3.2 is devoted to presenting the basics of the probabilistic Datalog+/− model, and finally Sect. 3.3 outlines paths towards deriving explanations to query answers over this framework.

3.1   Brief Overview of Basic Probabilistic Graphical Models

In the spirit of making this document relatively self-contained, we now provide a quick introduction to a few basic probabilistic graphical models. Such models are essentially ways to specify joint probability distributions over a set of random variables, based on graph structures—they will come into play when defining the semantics of Probabilistic Datalog+/− (cf. Sect. 3.2).

For each of the models, we assume we have a (finite) set of random variables $X = \{X_1, \ldots, X_n\}$. Each random variable X_i may take on *values* from a finite *domain* $Dom(X_i)$. A *value* for $X = \{X_1, \ldots, X_n\}$ is a mapping $x \colon X \to \bigcup_{i=1}^{n} Dom(X_i)$ such that $x(X_i) \in Dom(X_i)$; the *domain* of X, denoted $Dom(X)$, is the set of all values for X. We are generally interested in modeling the joint probability distribution over all values in $x \in Dom(X)$, which we denote $\Pr(x)$. We thus have that $0 \leqslant \Pr(x) \leqslant 1$ for all $x \in Dom(X)$, and $\sum_{x \in X} \Pr_x = 1$.

Bayesian Networks. A Bayesian Network (BN, for short) is comprised of: (i) A directed acyclic graph in which each node corresponds to a single random variable in X (and vice versa). If there is an edge from X_i to X_j, we say that X_i is a *parent* of X_j—this represents a direct dependence between the two variables. (ii) A conditional probability distribution $\Pr(X_i | Parents(X_i))$ for each node X_i, also sometimes called a *node probability table*.

One of the advantages of the BN model is that the graph structure encodes the probabilistic dependence between variables. For instance, each variable is independent of its non-descendents if we are given values for its parents. Therefore, the probability for any value $x = (x_1, \ldots, x_n) \in Dom(X)$ can be computed as follows:

$$\Pr(x_1, \ldots, x_n) = \prod_{i=1}^{n} \Pr\left(x_i \mid par\text{-}val(X_i)\right),$$

where $par\text{-}val(X_i)$ denotes the values of the variables in $Parents(X_i)$. Moreover, the absence of an edge between nodes represents conditional independence between the corresponding variables—the details of how such independence is characterized are non-trivial, and we refer the interested reader to the vast amount of material on BNs (cf. [41] for one of the earliest sources). Knowing the details behind variable (in)dependence is of great value if one is interested in tractable algorithms for computing probabilities, since the probability of the conjunction of independent variables is simply the product of their probabilities.

The most common problems (or probabilistic queries) associated with BNs are the following:

- PE (*Probability of evidence*, also known as *inference*): Compute the probability of a group of variables having a specific value. This problem is #P-complete.
- MAP (*Maximum A posteriori Probability*): Given evidence e over variables $E \subset X$, and variables $Y \subseteq X - E$, compute the value of y of Y that maximizes $\Pr(Y = y|E)$. This problem is NP^{PP}-complete in its decision version.
- MPE (*Most Probable Explanation*): Given evidence, find the assignment of values to the rest of the variables that has the highest probability. This problem is NP-complete in its decision version.

Even though all of these problems are computationally intractable in general, there exist special cases for which they can be solved in polynomial time, either exactly or approximately.

Markov Random Fields. A Markov Random Field (MRF) [41] (sometimes also referred to as Markov Network) is a probabilistic model that is similar to a Bayesian network (BN) in that it includes a graph $G = (V, E)$ in which each node corresponds to a variable, but, differently from a BN, the graph is undirected; in an MRF, two variables are connected by an edge in G iff they are conditionally dependent. Furthermore, the model contains a *potential function* ϕ_i for each (maximal) clique in the graph; potential functions are non-negative real-valued functions of the values of the variables in each clique (called the *state* of the clique). Here, we assume the *log-linear* representation of MRFs, which involves defining a set of *features* of such states; a feature is a real-valued function of the state of a clique (we only consider binary features in this work). Given a value $x \in Dom(X)$ and a feature f_j for clique j, the probability distribution represented by an MRF can be computed as follows:

$$P(X = x) = \frac{1}{Z} \exp \left(\sum_j w_j \cdot f_j(x) \right),$$

where j ranges over the set of cliques in the graph G, and $w_j = \log \phi_j(x_{\{j\}})$ (here, $x_{\{j\}}$ is the state of the j-th clique in x). The term Z is a normalization constant to ensure that the values given by the equation above are in $[0, 1]$ and sum to 1; it is given by:

$$Z = \sum_{x \in Dom(X)} \exp\left(\sum_j w_j \cdot f_j(x)\right).$$

Probabilistic inference in MRFs is intractable (#P-complete); however, approximate inference mechanisms, such as Markov Chain Monte Carlo (discussed briefly below), have been developed and successfully applied to problems in practice.

Markov Logic. Markov Logic Networks (MLNs) [45], also sometimes referred to as Markov Logic, combine first-order logic with Markov Random Fields. The main idea behind MLNs is to provide a way to soften the constraints imposed by a set of classical logic formulas. Instead of considering possible worlds that violate some formulas to be impossible, we wish to make them less probable. An MLN is a finite set L of pairs (F_i, w_i), where F_i is a formula in first-order logic, and w_i is a real number. Such a set L, along with a finite set of constants $C = \{c_1, \ldots, c_m\}$, defines a Markov network $M_{L,C}$ that contains: (i) one binary node corresponding to each element of the Herbrand base of the formulas in L (i.e., all possible ground instances of the atoms), where the node's value is 1 iff the atom is true; and (ii) one feature for every possible ground instance of a formula in L. The value of the feature is 1 iff the ground formula is true, and the weight of the feature is the weight corresponding to the formula in L. From this characterization and the description above of the graph corresponding to an MN, it follows that $M_{L,C}$ has an edge between any two nodes corresponding to ground atoms that appear together in at least one formula in L. Furthermore, the probability of $x \in Dom(X)$ in $M_{L,C}$ and thus in the MLN is defined by $P(X = x) = \frac{1}{Z} \exp(\sum_j w_j \cdot n_j(x))$, where $n_j(x)$ is the number of ground instances of F_j made true by x, and Z is defined analogously as above. This formula can be used in a generalized manner to compute the probability of any setting of a subset of random variables $X' \subseteq X$, as we show below.

Example 3. Consider the following simple MLN:

ψ_1: $(p(X) \Rightarrow q(X), 0.5)$,
ψ_2: $(p(X) \Rightarrow r(X), 2)$,
ψ_3: $(s(X) \Rightarrow r(X), 4)$.

Suppose we have s single constant a; grounding the formulas above relative to set of constants $\{a\}$, we obtain the set of ground atoms

$$\{p(a), q(a), r(a), s(a)\}.$$

Similarly, if we had two constants, a and b, we would get:

$$\{p(a), q(a), r(a), s(a), p(b), q(b), r(b), s(b)\}.$$

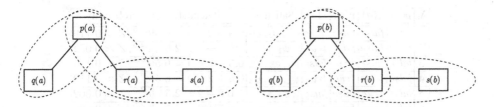

Fig. 4. The graphical representation of the MRF for the MLN from Example 3 (instantiated with set of constants $\{a, b\}$). There is one Boolean random variable for each ground atom in the grounding of the formulas ψ_1, ψ_2, and ψ_3 with respect to the set of constants. The dotted lines show the different cliques in the graph.

The graphical representation of the MRFs corresponding to these groundings are shown in Fig. 4.

Consider the former (with respect to a single constant). This MRF represents a probability distribution over the possible Boolean values for each node. Given that there are four ground atoms, there are $2^4 = 16$ possible settings of the variables in the MRF; Fig. 5 shows all such possible settings, along with other information used to compute probabilities. The normalizing factor Z is the sum of the probabilities of all worlds, which is computed as shown above by summing the exponentiated sum of weights times the number of ground formulas satisfied (equivalent to summing e to the power of each number in the "potential" column in Fig. 5), yielding $Z \approx 5593.0623$. Similarly, the probability that a formula, such as $p(a) \wedge q(a) \wedge \neg s(a)$, holds is the sum of the probabilities that all the satisfying worlds hold, which in this case corresponds to the worlds 13 and 15 (cf. Fig. 5); the resulting probability is $\frac{e^{4.5} + e^{0.5}}{5593.0623} \approx 0.0265$. ∎

Markov Chains. Lastly, we wish to briefly mention a somewhat different model that is geared towards *dynamically evolving* systems. A *Markov Chain* (MC, for short), is a stochastic process $\{X_n\}$, with $n \in \mathbb{N} \cup \{0\}$—essentially, an MC is a sequence of *states*, or values of variables. The *Markov property* holds if it is always the case that given $n \in \mathbb{N} \cup \{0\}$ and states $x_0, x_1, ..., x_n, x_{n+1}$, we have:

$$\Pr(X_{n+1} = x_{n+1} \mid X_n = x_n, ..., X_0 = x_0) = \Pr(X_{n+1} = x_{n+1} \mid X_n = x_n).$$

That is, the distribution of conditional probability of future states only depends on the *current* state; this property is also sometimes referred to as *memoryless*.

MCs can be represented as sequences of graphs where the edges of graph n are labeled with the probability of going from one state at moment n to other states at moment $n + 1$:

$$\Pr(X_{n+1} = x \mid X_n = x_n).$$

λ_i	$p(a)$	$q(a)$	$r(a)$	$s(a)$	Satisfies	Potential	Probability
1	false	false	false	false	ψ_1,ψ_2,ψ_3	$0.5+2+4=6.5$	$e^{6.5}/Z \approx 0.119$
2	false	false	false	true	ψ_1,ψ_2	$0.5+2=2.5$	$e^{2.5}/Z \approx 0.002$
3	false	false	true	false	ψ_1,ψ_2,ψ_3	$0.5+2+4=6.5$	$e^{6.5}/Z \approx 0.119$
4	false	false	true	true	ψ_1,ψ_2,ψ_3	$0.5+2+4=6.5$	$e^{6.5}/Z \approx 0.119$
5	false	true	false	false	ψ_1,ψ_2	$0.5+2=2.5$	$e^{2.5}/Z \approx 0.002$
6	false	true	false	true	ψ_1,ψ_2	$0.5+2=2.5$	$e^{2.5}/Z \approx 0.002$
7	false	true	true	false	ψ_1,ψ_2,ψ_3	$0.5+2+4=6.5$	$e^{6.5}/Z \approx 0.119$
8	false	true	true	true	ψ_1,ψ_2,ψ_3	$0.5+2+4=6.5$	$e^{6.5}/Z \approx 0.119$
9	true	false	false	false		0	$e^{0}/Z \approx 0$
10	true	false	false	true		0	$e^{0}/Z \approx 0$
11	true	false	true	false	ψ_2,ψ_3	$2+4=6$	$e^{6}/Z \approx 0.072$
12	true	false	true	true	ψ_2,ψ_3	$2+4=6$	$e^{6}/Z \approx 0.072$
13	true	true	false	false	ψ_1,ψ_3	$0.5+4=4.5$	$e^{4.5}/Z \approx 0.016$
14	true	true	false	true	ψ_1	0.5	$e^{0.5}/Z \approx 0$
15	true	true	true	false	ψ_1,ψ_2,ψ_3	$0.5+2+4=6.5$	$e^{6.5}/Z \approx 0.119$
16	true	true	true	true	ψ_1,ψ_2,ψ_3	$0.5+2+4=6.5$	$e^{6.5}/Z \approx 0.119$

Fig. 5. Details of how to compute potentials for each possible setting of the random variables (worlds) of the MRF for the MLN from Example 3 (grounded with a single constant).

The same information can be represented via a *transition matrix* M where

$$M[i,j] = \Pr(X_{n+1} = x_j \,|\, X_n = x_i).$$

Taking the power of this matrix with itself iteratively, we can answer queries regarding the probability that the system will be in a certain state after several time steps. One of the most important classes of MCs are the ones for which a *stationary distribution* exists—one that is invariant over time—since they represent stable stochastic processes.

One of the most important applications of MCs is as the basis of the *Markov Chain Monte Carlo* (MCMC) family of algorithms, which are random walk-based traversals of the state space that can be used to sample from an unknown probability distribution, thus arriving at approximations of the distribution itself or of queries of interest.

Brief Comparison Among Models. As a quick comparison of the strengths and weaknesses of the four models that we introduced above, we can point out a few salient aspects:

- Bayesian Networks are useful when identified dependencies are acyclic, and information is available regarding conditional dependencies (i.e., the structure of the graph, plus the probability tables).
- Markov Random Fields are more flexible in that they allow cycles and probabilities are derived from weights. The disadvantage associated with the latter is that the relationship between weights and probabilities is not always clear.

- Markov Logic Networks are essentially *first order templates* for MRFs; their main strength with respect to them is that a model can be derived given a set of constants, which can change depending on the situation in which it is intended for.
- One of the main applications of Markov Chains is as the basis of Markov Chain Monte Carlo (MCMC) methods that can be used to approximate unknown distributions, or distributions that are specified by other models like Bayesian Networks or Markov Random Fields that are intractable to compute exactly, or dynamical systems for which a closed form solution may not be possible.

As we will see in the next section, these and other probabilistic models can be leveraged as part of extended logic-based languages in order to deal with uncertainty in a principled manner.

Learning Models from Data. There are many, many different approaches and algorithms available for automatically or semi-automatically deriving these and other probabilistic models from available data—even a cursory treatment is outside the scope of this work. We refer the reader to the vast literature on these topics that has been developing for many years; good starting points can be found at [8] and [50].

3.2   Probabilistic Datalog+/−

In this section, considering the basic setup from Sects. 2 and 3.1, we introduce the syntax and the semantics of probabilistic Datalog+/−.

Syntax
As in Sect. 2, we assume an infinite universe of (data) constants Δ, an infinite set of labeled nulls Δ_N, and an infinite set of variables $\mathcal{V}$. Furthermore, as in Sect. 3.1, we assume a finite set of random variables X. Informally, a probabilistic Datalog+/− ontology consists of a finite set of probabilistic atoms, probabilistic TGDs, probabilistic negative constraints, and probabilistic separable EGDs, along with a probabilistic model that yields a full joint distribution over the values in $\times_{i=1}^{|X|} dom(X_i)$.

We first define probabilistic annotations and more specifically worlds. Intuitively, a probabilistic annotation λ is an assignment of values x_i to random variables X_i, representing the event in the probabilistic model where the X_i's have the value x_i. In particular, a world assigns a value to each random variable. In general, we use *true* and *false* to refer to the values 1 and 0 of Boolean random variables, respectively; furthermore, to simplify notation, we use X and $\neg X$ to denote $X = true$ and $X = false$, respectively.

Definition 1. A *(probabilistic) annotation* λ is a (finite) set of expressions $X_i = x_i$, where $X_i \in X$, $x_i \in Dom(X_i)$, and the X_i's are pairwise distinct. If $|\lambda| = |X|$, then λ is a *world*. We denote by *worlds(M)* the set of all worlds of a probabilistic model M.

We next attach probabilistic annotations λ to classical Datalog+/− formulas F to produce annotated formulas $F: \lambda$. Intuitively, F holds whenever the event associated with λ occurs. Note that whenever a random variable's value is left unspecified in a probabilistic annotation, the variable is unconstrained; in particular, a formula annotated with an empty probabilistic annotation means that the formula holds in every world. As we discuss in detail in Sect. 3.2, this kind of annotation works much in the same way as in many other probabilistic formalisms where possible worlds are induced via probabilistic distributions coming from outside the model; typical examples are the independent choice logic [42], P-LOG [7], and background variables in Bayesian networks (where probabilities come from exogenous events) [41].

Definition 2. If a is an atom, σ_T is a TGD, σ_{NC} is a negative constraint, σ_E is an EGD, and λ is a probabilistic annotation, then: (i) $a: \lambda$ is a *probabilistic atom*; (ii) $\sigma_T: \lambda$ is a *probabilistic TGD*; (iii) $\sigma_{NC}: \lambda$ is a *probabilistic (negative) constraint*; and (iv) $\sigma_E: \lambda$ is a *probabilistic EGD*. We also refer to probabilistic atoms, TGDs, (negative) constraints, and EGDs as *annotated formulas*. Annotated formulas of the form $F: \{\}$ are abbreviated as F.

We are now ready to define the notion of a probabilistic Datalog+/− ontology.

Definition 3. A *probabilistic Datalog+/− ontology* is a pair $\Phi = (O, M)$, where O is a finite set of probabilistic atoms, TGDs, (negative) constraints, and EGDs, and M is a probabilistic model.

Loosely vs. Tightly Coupled Ontologies. There are two ways in which probabilistic annotations can be combined with formulas: in *loosely coupled* ontologies, annotations cannot refer to elements in the ontology (the scope of variables reaches only the ontology or the annotation). As can be seen in the results in [29], this is an advantage from the complexity point of view, since the cost of computing probabilities in the probabilistic model does not grow with the database; however, it also represents a limitation in expressive power of the formalism. On the other hand, in *tightly coupled* probabilistic Datalog+/− variables in annotations can be shared with those in Datalog+/− formulas; an early version of this idea can be found in [34], which develops the same concept for $\mathcal{EL}^{++}$ ontologies.

The annotation of Datalog+/− formulas offers a clear modeling advantage by separating the two tasks of ontological modeling and of modeling the uncertainty around the axioms in the ontology. More precisely, in our formalism, it is possible to express the fact that the probabilistic nature of an ontological axiom is determined by elements that are *outside of the domain* modeled by the ontology. The probabilistic distribution of events (and *existence* of certain objects, for instance as part of a heuristic process) is a separate concern relative to the knowledge encoded in the "classical part" of the ontology.

In the rest of this section, we will resort to the following as a running example.

$$\alpha_1: \ a(x_1) \qquad\qquad\qquad : \{p(a), s(a)\}$$
$$\alpha_2: \ b(x_2) \qquad\qquad\qquad : \{p(a), s(a), \neg r(a)\}$$
$$\alpha_3: \ d(x_3) \qquad\qquad\qquad : \{p(a), q(a), \neg s(a)\}$$

$$\sigma_1: \ a(X) \rightarrow c(X) \qquad\quad : \{r(a)\}$$
$$\sigma_2: \ b(X) \rightarrow d(X) \qquad\quad : \{q(a)\}$$
$$\sigma_3: \ a(X) \rightarrow \exists Y \, p(X, Y)$$

$$v_1: \ a(X) \wedge b(Y) \rightarrow X = Y$$
$$v_2: \ b(X) \wedge c(X) \rightarrow \bot$$

Fig. 6. The probabilistic Datalog+/− ontology from Example 4.

Example 4. Let the Datalog+/− ontology $O = (D, \Sigma_T \cup \Sigma_E \cup \Sigma_{NC})$ be given by the database D, set of TGDs Σ_T, set of EGDs Σ_E, and set of (negative) constraints Σ_{NC}:

$$D = \{a(x_1), \, b(x_2), \, d(x_3)\};$$
$$\Sigma_T = \{\sigma_1: a(X) \rightarrow c(X), \ \sigma_2: b(X) \rightarrow d(X), \ \sigma_3: a(X) \rightarrow \exists Y p(X, Y)\};$$
$$\Sigma_E = \{v_1: a(X) \wedge b(Y) \rightarrow X = Y\};$$
$$\Sigma_{NC} = \{v_2: b(X) \wedge c(X) \rightarrow \bot\}.$$

Furthermore, consider the MLN M from Example 3. The annotated formulas in Fig. 6 are the result of annotating the formulas in the Datalog+/− ontology with expressions assigning *true* or *false* to a subset of the random variables that arise from the atoms described above: $\{p(a), q(a), r(a), s(a)\}$. Recall that annotated formulas $F: \{\}$ are abbreviated as F; they hold irrespective of the setting of the random variables. ∎

As described next, worlds *induce* certain subontologies of a probabilistic Datalog+/− ontology, according to whether or not they satisfy the annotation of each formula.

Definition 4. *Let $\Phi = (O, M)$ be a probabilistic Datalog+/−− ontology and λ be a world. Then, the (non-probabilistic) Datalog+/−− ontology induced from Φ by λ, denoted O_λ, is the set of all F_i such that $\lambda_i \subseteq \lambda$ for some $F_i: \lambda_i \in O$; any such F_i is relevant in λ.*

In the sequel, we consider only probabilistic Datalog+/− ontologies $\Phi = (O, M)$ in which the EGDs in every induced ontology O_λ are separable from the TGDs in O_λ.

The notion of *decomposition* of a probabilistic ontology provides a convenient way of referring to its constituent subontologies with respect to the worlds.

Definition 5. *Let $\Phi = (O, M)$ be a probabilistic Datalog+/−− ontology. Then, the decomposition (or decomposed form) of Φ, denoted $decomp(\Phi)$, is defined as follows:*

$$decomp(\Phi) = ([O_{\lambda_1}, \ldots, O_{\lambda_n}], M),$$

where $worlds(M) = \{\lambda_1, \ldots, \lambda_n\}$. To simplify notation, we assume that the worlds are ordered according to a lexicographical order of the values of the variables, and therefore the i-th ontology in a decomposition corresponds to the i-th world in this ordering.

Example 5. Consider the probabilistic Datalog+/− ontology $\Phi = (O, M)$, with $O = (D, \Sigma)$, from Example 4. There are 16 worlds, so the decomposition of Φ has the form

$$decomp(\Phi) = ([O_{\lambda_1}, \ldots, O_{\lambda_{16}}], M).$$

For example, the world λ_{16} is determined by $\{p(a), q(a), r(a), s(a)\}$, while λ_{14} is determined by $\{p(a), q(a), \neg r(a), s(a)\}$. It is then easy to see that $O_{\lambda_{16}} = (\{\alpha_1\}, \Sigma)$ and $O_{\lambda_{14}} = (\{\alpha_1, \alpha_2\}, \Sigma - \{\sigma_1\})$. ∎

We now define the *canonical composition*, the inverse of the decomposition.

Definition 6. *Let $\Psi = ([O_{\lambda_1}, \ldots, O_{\lambda_n}], M)$ be a probabilistic Datalog+/−− ontology in decomposed form. Then, the canonical composition of Ψ, denoted $decomp^{-1}(\Psi)$, is the probabilistic Datalog+/−− ontology $\Phi = (\bigcup_{i=1}^{n}\{F: \lambda_i \mid F \in O_{\lambda_i}\}, M)$.*

Example 6. Let $\Phi_{decomp} = ([O_{\lambda_1}, \ldots, O_{\lambda_{16}}], M)$ be the probabilistic ontology in decomposed form from Example 5. Although it is easy to verify that $decomp^{-1}(\Phi_{decomp})$ yields an ontology that is equivalent to Φ (from Example 4), this ontology actually contains several instances of the same formula, each with different annotations. For example, the atom α_1 appears four times, with the annotations $\{p(a), q(a), \neg r(a), \neg s(a)\}$, $\{p(a), q(a), \neg r(a), s(a)\}$, $\{p(a), q(a), r(a), \neg s(a)\}$, and $\{p(a), q(a), r(a), s(a)\}$, respectively. ∎

Semantics

Towards the semantics of probabilistic Datalog+/− ontologies, we first define classical interpretations and the satisfaction of annotated formulas in such interpretations. The former consist of a database and a world in the probabilistic model, while the latter is done by interpreting $F: \lambda$ as $F \Leftarrow \hat{\lambda}$ (or equivalently $F \vee \neg\hat{\lambda}$), where $\hat{\lambda} = \bigwedge_{X_i = x_i \in \lambda} X_i = x_i$.

Definition 7. *A classical interpretation $\mathcal{I} = (D, v)$ consists of a database D and a value $v \in D(X)$. We say that $\mathcal{I}$ satisfies an annotated formula $F: \lambda$, denoted $\mathcal{I} \models F: \lambda$, iff $D \models F$ whenever $v(X_i) = x_i$ for all $X_i = x_i \in \lambda$.*

We next define probabilistic interpretations Pr as finite probability distributions over classical interpretations, and the probability of formulas and their satisfaction in such Pr, as usual. Here, formulas are either annotated formulas (including classical Datalog+/− formulas as a special case) or events in the probabilistic model (i.e., Boolean combinations of expressions $X_i = x_i$, where $X_i \in X$ and $x_i \in Dom(X_i)$). Furthermore, we define the satisfaction of probabilistic Datalog+/− ontologies in Pr, where (i) all annotated formulas in the ontology are satisfied by Pr, and (ii) the probabilities that Pr assigns to worlds coincide with those of the probabilistic model.

Definition 8. A *probabilistic interpretation* Pr is a probability distribution Pr over the set of all classical interpretations such that only a finite number of classical interpretations are mapped to a non-zero value. The *probability* of a formula ϕ, denoted $Pr(\phi)$, is the sum of all $Pr(\mathcal{I})$ such that $\mathcal{I} \models \phi$. We say that Pr satisfies (or is a *model* of) ϕ iff $Pr(\phi) = 1$. Furthermore, Pr is a model of a probabilistic Datalog+/− ontology $\Phi = (O, M)$ iff: (i) $Pr \models \phi$ for all $\phi \in O$, and (ii) $Pr(\hat{\lambda}) = Pr_M(\hat{\lambda})$ for all $\lambda \in worlds(M)$, where $\hat{\lambda}$ is defined as above, and Pr_M is the probability in the model M.

Here, we are especially interested in computing the probabilities associated with ground atoms in a probabilistic Datalog+/− ontology, as defined next. Intuitively, the probability of a ground atom is defined as the infimum of the probabilities of that ground atom under all probabilistic interpretations that satisfy the probabilistic ontology.

Definition 9. Let Φ be a probabilistic Datalog+/− ontology, and a be a ground atom constructed from predicates and constants in Φ. The *probability* of a in Φ, denoted $Pr^{\Phi}(a)$, is the infimum of $Pr(a)$ subject to all probabilistic interpretations Pr such that $Pr \models \Phi$.

Based on this notion, we can define several different kinds of probabilistic queries, as discussed next.

Queries to Probablistic Datalog+/− Ontologies
There are three kinds of queries that have been proposed in this model [29,35]. We now present a brief introduction to each of them, assuming we are given a probabilistic Datalog+/− ontology $\Phi = (O, M)$.

- *Threshold queries* ask for the set of all ground atoms that have a probability of at least p, where p is specified as an input of the query. So, the answers to *threshold query* $Q = (\Phi, p)$ (with $p \in [0, 1]$) is the set of all ground atoms a with $Pr^{\Phi}(a) \geqslant p$.
- *Ranking queries* request the ranking of atomic consequences based on their probability values. So, the *answer* to a *ranking query* $Q = rank(KB)$ is a tuple $ans(Q) = \langle a_1, \ldots, a_n \rangle$ such that $\{a_1, \ldots, a_n\}$ are all of the atomic consequences of O_λ for any $\lambda \in Worlds(M)$, and $i < j \Rightarrow Pr^{KB}(a_i) \geqslant Pr^{KB}(a_j)$.
- Finally, probabilistic *conjunctive queries* are exactly as defined in Sect. 2, except that its answers are accompanied by the probability value with which it is entailed by Φ.

The following example illustrates each of these queries using the running example.

Example 7. Consider the probabilistic Datalog+/– ontology $\Phi = (O, M)$ from Example 4, and the threshold query $Q = (\Phi, 0.15)$. As seen in Example 5, and referring back to Fig. 5 for the computation of the probabilities, we have that $Pr^{\Phi}(a(x_1)) = \approx 0.191$ and $Pr^{\Phi}(d(x_3)) = 0.135$. Therefore, the former belongs to the output, while the latter does not.

The answer to query $rank(KB)$ is: $\langle a(x_1), c(x_1), d(x_3), b(x_2), d(x_2) \rangle$. Finally, the answer to probabilistic conjunctive query $Q(X) = a(X) \wedge c(X)$ is $(x_1, 0.191)$. ∎

3.3   Towards Explainable Probabilistic Ontological Reasoning

As we discussed above, probabilistic Datalog+/– is an extension of "classical" Datalog+/– with labels that refer to a probabilistic model—essentially, there are two sub-models that are in charge of representing knowledge about the domain, and these models can be either loosely or tightly coupled, depending on whether or not they share objects. So, the question we would like to pose now is *"What constitutes an explanation for a query to a probabilistic Datalog+/– knowledge base?"*. As discussed in Sect. 1, the answer to this question will depend heavily on whom the explanation is intended for, the application domain, the specific formalism being used, and the kind of query. Note that there is some initial work [18] on some variants of this problem as a generalization of MAP/MPE queries (which were developed as kinds of explanations for probabilistic models) for the special case of tuple-independent probabilistic databases.

Therefore, here we will focus on foundational aspects of designing explanations for probabilistic Datalog+/–; the discussion will generally apply to all queries presented on Page 20, unless stated otherwise. The basic building blocks of reasoning with probabilities in our formalism are the following:

– **The annotated chase structure:** The extension of the *chase*, the main algorithm used to answer queries in classical Datalog+/–, in order to take into account probabilistic annotations is the *annotated chase*, a structure that essentially keeps track of the probabilistic annotations required for each step to be possible [29]. There two basic ways in which this can be done:
  • Annotate each node with a *Boolean array* of size $|Worlds(M)|$; during the execution of the chase procedure, annotations are propagated as inferences are made. This is best for cases in which: (i) the number of worlds is not excessively large, since the space used by the chase structure will grow by a factor of $|Worlds(M)|$; or (ii) when a sampling-based approach to approximate the probabilities associated with query answers is used, since in this case the size of each array can be reduced to (a function of) the number of samples.
  Another advantage of this approach is that for models under which querying the probability of a specific world is tractable (often referred to as *tractable probabilistic models*), the Boolean array representation can be used to clearly obtain either the exact or approximate probability mass associated with each node of interest.

- Annotate each node with a *logical formula* expressing the conditions that must hold for the node to be inferrable. This approach is more compact than the array-based method, since the size of the formulas are bounded by the length of the derivation (at most the depth of the deepest spanning tree associated with the chase graph) and the length of the original annotations in the probabilistic ontology. On the other hand, extracting the specific worlds that make up the probabilistic mass associated with a given atom (or set of atoms for a query) is essentially equivalent to solving a #SAT problem; for tractable probabilistic models there is a greater chance of performing feasible computations, though the structure of the resulting logical formula depends greatly on how rules interact—this is the topic of ongoing work.

By analyzing the resulting data structure, one can extract a clear map of how the probability of an atom is derived; this is discussed next.

- **Probabilities of atomic formulas:** The annotated chase yields several tools that facilitate the provision of an explanation for the probability of an atom:
  - Different derivation paths leading to the same result (or summaries).
  - Examples of branches, perhaps highlighting well-separated ones to show variety.
  - Common aspects of worlds that make up most of the probability mass (such as atoms in the probabilistic model that appear in most derivations).

  In all cases, if we wish to provide a *balanced* explanation (as discussed above) we can also focus on the dual situation, i.e., showing the cases in which the atom in question is *not* derived. Note that all of these elements are available independently of the specific probabilistic model used in the KB—depending on the characteristics of the chosen model, other data might be available as well.

- **Probabilities of more complex queries:** The previous point covered the most basic probabilistic query (probability of atomic formulas); clearly, the same approach is useful for threshold queries, which can be answered simply by computing the probabilities of all atomic consequences and checking if their associated probabilities exceed the threshold.

  As discussed in the previous section, we have two other kinds of queries:
  - *Probabilistic conjunctive queries:* The basic building blocks described for atomic queries can be leveraged for the more complex case of conjunctive queries. Depending on the kind of annotated chase graph used (as discussed above), the probability of a set of atoms that must be true at once can be derived from that of each individual member. Opportunities for explanations of why a query is derived or not derived may also include, for instance, selecting one or more elements of the conjunction that are responsible for lowering the resulting probability of the query.
  - *Ranking queries:* The fundamental component of the result of a ranking query is the *relationship* between the probabilities of atoms—the most important question to answer regarding explanations of such results is: for a given *pair* of atoms (a, b) such that a is ranked above b, why is it

$a > b$ and not $b > a$? The basic elements discussed above can be used to shed light on this aspect.

Finally, sampling-based methods (for instance, taking into account a subset of the worlds chosen at random) yield *probability intervals* instead of point probabilities—the width of the resulting interval will be a function of the number and probability mass of the worlds taken into account vs. those left out [35]. So, explanations can involve examples or summaries of how the probability mass gets to a minimum (lower bound) and, conversely, why the maximum (upper bound) is not higher.

There is much work to be done in developing effective algorithms to leverage these and other building blocks for deriving explanations for probabilistic queries. Furthermore, developing adequate user interfaces so that the resulting explanations are useful is also a highly non-trivial task.

4   Inconsistency-Tolerant Query Answering with Datalog+/−

In this section we discuss a general approach to inconsistency-tolerant query answering in Datalog+/−; the material in this section is based mainly on [36].

We now discuss semantics for inconsistency-tolerant query answering that are based on the ideas of [2] but from the perspective of the area of *belief change*, which is an area of AI that is closely related to the management of inconsistent information, aiming to adequately model the dynamics of the knowledge that constitutes the set of beliefs of an agent when new information comes up. In [31], *kernel* consolidations are defined based on the notion of an *incision function*. Given a knowledge base KB that needs to be consolidated (i.e., KB is inconsistent), the set of kernels is defined as the set of all minimal inconsistent subsets of KB. For each kernel, a set of sentences is removed (i.e., an "incision" is made) such that the remaining formulas in the kernel are consistent; note that it is enough to remove any single formula from the kernel because they are minimal inconsistent sets. The result of consolidating KB is then the set of all formulas in KB that are not removed by the incision function. In this work, we present a framework based on a similar kind of functions to provide alternative query answering semantics in inconsistent Datalog+/− ontologies. The main difference in our proposal is that incisions are performed over inconsistent subsets of the ontology that are not necessarily minimal.

We analyze three types of incision functions that correspond to three different semantics for query answering in inconsistent Datalog+/− ontologies: (i) *consistent answers* or AR semantics [2,32], widely adopted in relational databases and DLs, (ii) *intersection semantics* or IAR, which is a sound approximation of AR [32], and (iii) a semantics first proposed in [36] that relaxes the requirements of AR semantics, allowing it to be computed in polynomial time for some fragments of Datalog+/−, without compromising the quality of the answers as much as the IAR semantics does, by allowing a certain *budget* within which the answers can be computed.

We first define the notion of a *culprit* relative to a set of constraints, which is informally a minimal (under set inclusion) inconsistent subset of the database relative to the constraints. Note that we define culprits relative to both negative constraints and EGDs, as Σ_{NC} contains all EGDs written as NCs, as we mentioned above.

Definition 10 (Culprit). Given a Datalog+/– ontology $KB = (D, \Sigma_T \cup \Sigma_E \cup \Sigma_{NC})$, a *culprit* in KB relative to $\Sigma_E \cup \Sigma_{NC}$ is a set $c \subseteq D$ such that $mods(c, \Sigma_T \cup IC) = \emptyset$ for some $IC \subseteq \Sigma_E \cup \Sigma_{NC}$, and there is no $c' \subset c$ such that $mods(c', \Sigma_T \cup IC) = \emptyset$. We denote by $culprits(KB)$ the set of culprits in KB relative to $\Sigma_E \cup \Sigma_{NC}$.

Note that we may also refer to $culprits(KB, IC)$ whenever we want to make the point that IC is an arbitrary set of constraints or to identify a specific subset of $\Sigma_E \cup \Sigma_{NC}$. The following example shows a Datalog+/– ontology that we will use as a running example through out the chapter.

The following example shows a simple Datalog+/– ontology; the language and standard semantics for query answering in Datalog+/– ontologies is recalled in the next section.

Example 8. A (guarded) Datalog+/– ontology $KB = (D, \Sigma_T \cup \Sigma_E \cup \Sigma_{NC})$ is given below. Here, the formulas in Σ_T are tuple-generating dependencies (TGDs), which say that each person working for a department is an employee (σ_1), each person that directs a department is an employee (σ_2), and that each person that directs a department and works in that department is a manager (σ_3). The formulas in Σ_{NC} are negative constraints, which say that if X supervises Y, then Y cannot be a manager (υ_1), and that if Y is supervised by someone in a department, then Y cannot direct that department (υ_2). The formula $\upsilon_3 \in \Sigma_E$ is an equality-generating dependency (EGD), saying that the same person cannot direct two different departments.

$$
\begin{aligned}
D \;=\; & \{ directs(john, d_1),\ directs(tom, d_1),\ directs(tom, d_2), \\
& supervises(tom, john),\ works\_in(john, d_1),\ works\_in(tom, d_1) \}; \\
\Sigma_T \;=\; & \{ \sigma_1 : works\_in(X, D) \rightarrow emp(X),\ \sigma_2 : directs(X, D) \rightarrow emp(X), \\
& \sigma_3 : directs(X, D) \wedge works\_in(X, D) \rightarrow manager(X) \}; \\
\Sigma_{NC} \;=\; & \{ \upsilon_1 : supervises(X, Y) \wedge manager(Y) \rightarrow \bot, \\
& \upsilon_2 : supervises(X, Y) \wedge works\_in(X, D) \wedge directs(Y, D) \rightarrow \bot \}; \\
\Sigma_E \;=\; & \{ \upsilon_3 : directs(X, D) \wedge directs(X, D') \rightarrow D = D' \}.
\end{aligned}
$$

We can easily see that this ontology is inconsistent. For instance, the atoms $directs(john, d_1)$ and $works\_in(john, d_1)$ trigger the application of σ_3, producing $manager(john)$, but that together with $supervises(tom, john)$ (which belongs to D) violates υ_1. The set of culprits relative to $\Sigma_E \cup \Sigma_{NC}$ are:

$$
\begin{aligned}
c_1 =\ & \{ supervises(tom, john),\ directs(john, d_1),\ works\_in(john, d_1) \}, \\
c_2 =\ & \{ supervises(tom, john),\ directs(john, d_1),\ works\_in(tom, d_1) \}, \\
c_3 =\ & \{ directs(tom, d_1),\ directs(tom, d_2) \}\ . \qquad\qquad\qquad \blacksquare
\end{aligned}
$$

We construct *clusters* by grouping together all culprits that share elements. Intuitively, clusters contain only information involved in some inconsistency relative to Σ, i.e., an atom is in a cluster relative to Σ iff it is in contradiction with some other set of atoms in D.

Definition 11 (Cluster [38]). Given a Datalog+/− ontology $KB = (D, \Sigma_T \cup \Sigma_{NC})$ and $IC \subseteq \Sigma_{NC}$, two culprits $c, c' \in culprits(KB, IC)$ *overlap*, denoted $c \ominus c'$, iff $c \cap c' \neq \emptyset$. Denote by Θ^* the equivalence relation given by the reflexive and transitive closure of Θ. A *cluster* is a set $cl = \bigcup_{c \in e} c$, where e is an equivalence class of Θ^*. We denote by $clusters(KB, IC)$ (resp., $clusters(KB)$) the set of all clusters in KB relative to IC (resp., $IC = \Sigma_{NC}$).

Example 9. The clusters for KB in the running example are $cl_1 = c_3$ and $cl_2 = c_1 \cup c_2$ (cf. Example 8 for culprits c_1, c_2, and c_3). ∎

We now recall the definition of *incision function* from [31], adapted for Datalog+/− ontologies. Intuitively, an incision function selects from each cluster a set of atoms to be discarded such that the remaining atoms are consistent relative to Σ.

Definition 12 (Incision Function). Given a Datalog+/− ontology $KB = (D, \Sigma)$, an *incision function* is a function χ that satisfies the following properties:

(1) $\chi(clusters(KB)) \subseteq \bigcup_{cl \in clusters(KB)} cl$, and
(2) $mods(D - \chi(clusters(KB)), \Sigma) \neq \emptyset$.

Note that incision functions in [31] do not explicitly require condition (2) from Definition 12; instead, they require the removal of at least one sentence from each α-kernel. The notion of α-kernel [31] translates in our framework to a minimal set of sentences in D such that, together with Σ, entails the sentence α, where $KB = (D, \Sigma)$. Culprits are then, no more than minimal subsets of D that, together with Σ, entail $\bot$. Here, χ produces incisions over clusters instead, therefore, condition (2) is necessary to ensure that by making the incision, the inconsistency is resolved.

4.1   Relationship with (Classical) Consistent Answers

In the area of relational databases, the notion of *repair* was used in order to identify the consistent part of a possibly inconsistent database. A repair is a model of the set of integrity constraints that is maximally close, i.e., "as close as possible" to the original database. Repairs may not be unique, and in the general case, there can be a very large number of them. The most widely accepted semantics for querying a possibly inconsistent database is that of *consistent answers* [2].

We now adapt one notion of *repair* from [32] to Datalog+/− ontologies $KB = (D, \Sigma)$. Intuitively, repairs are maximal consistent subsets of D. We also show that BCQ answering under the consistent answer semantics is co-NP-complete for guarded and linear Datalog+/− in the data complexity.

Definition 13 (Repair). A *repair* for $KB = (D, \Sigma)$ is a set D' such that (i) $D' \subseteq D$, (ii) $mods(D', \Sigma) \neq \emptyset$, and (iii) there is no $D'' \subseteq D$ such that $D' \subset D''$ and $mods(D'', \Sigma) \neq \emptyset$. We denote by $DRep(KB)$ the set of all repairs for KB.

Example 10. The Datalog$+/-$ ontology KB in Example 8 has six repairs:
$r_1 = \{ directs(john, d_1), supervises(tom, john), directs(tom, d_1), manager(tom, d_1) \}$,
$r_2 = \{ directs(john, d_1), supervises(tom, john), directs(tom, d_2), manager(tom, d_1) \}$,
$r_3 = \{ directs(john, d_1), directs(tom, d_1), works\_in(john, d_1), works\_in(tom, d_1),$
    $manager(tom, d_2) \}$,
$r_4 = \{ directs(john, d_1), directs(tom, d_2), works\_in(john, d_1), works\_in(tom, d_1),$
    $manager(tom, d_2) \}$,
$r_5 = \{ supervises(tom, john), directs(tom, d_1), works\_in(john, d_1),$
    $works\_in(tom, d_1), manager(tom, d_1) \}$,
$r_6 = \{ supervises(tom, john), directs(tom, d_2), works\_in(john, d_1),$
    $works\_in(tom, d_1), manager(tom, d_1) \}$. ∎

Repairs play a central role in the notion of *consistent answer* for a query to an ontology, which are intuitively the answers relative to each ontology built from a repair. The following definition adapts the notion of consistent answers, defined in [32] for Description Logics, for Datalog$+/-$ ontologies.

Definition 14 (Consistent Answers – AR Semantics). Let $KB = (D, \Sigma)$ be a Datalog$+/-$ ontology, and Q be a BCQ. Then, *Yes* is a *consistent answer* for Q to KB, denoted $KB \models_{AR} Q$, iff it is an answer for Q to each $KB' = (D', \Sigma)$ with $D' \in DRep(KB)$.

Example 11. Consider the ontology KB from our running example. The atom $emp(john)$ can be derived from every repair, as each contains either the atom $works\_in(john, d_1)$ or the atom $directs(john, d_1)$. Thus, BCQ $Q = emp(john)$ is true under the consistent answer semantics. ∎

In accordance with the principle of *minimal change*, incision functions that make as few changes as possible when applied the set of clusters are called *optimal* incision functions.

Definition 15 (Optimal Incision Function). Given a Datalog$+/-$ ontology $KB = (D, \Sigma)$, an incision function χ is *optimal* iff for every $B \subset \chi(clusters(KB))$, it holds that $mods(D - B, \Sigma) = \emptyset$.

The following theorem shows the relationship between an optimal incision function and repairs for a Datalog$+/-$ ontology $KB = (D, \Sigma)$. More concretely, every repair corresponds to the result of removing from D all ground atoms according to some optimal incision $\chi(clusters(KB))$ and vice versa.

Theorem 1. Let $KB = (D, \Sigma)$ be a Datalog$+/-$ ontology. Then, D' is a repair, i.e., $D' \in DRep(KB)$, iff there exists an optimal incision function χ_{opt} such that $D' = D - \chi_{opt}(clusters(KB))$.

4.2   Relationship with IAR Semantics

An alternative semantics that considers only the atoms that are in the *intersection* of all repairs was presented in [32] for *DL-Lite* ontologies. This semantics yields a unique way of repairing inconsistency; the consistent answers are intuitively the answers that can be obtained from that unique set. Here, we define IAR for Datalog+/− ontologies.

Definition 16 (Intersection Semantics − IAR). Let $KB = (D, \Sigma)$ be a Datalog+/− ontology, and Q be a BCQ. Then, *Yes* is a *consistent answer* for Q to *KB under IAR*, denoted $KB \models_{IAR} Q$, iff it is an answer for Q to $KB_I = (D_I, \Sigma)$, where $D_I = \bigcap \{D' \mid D' \in DRep(KB)\}$.

Example 12. Consider the ontology $KB = (D, \Sigma)$ from the running example. Analyzing the set of all its repairs, it is easy to verify that $D_I = \{manager(tom, d_1)\}$.   ■

The following theorem shows the relationship between the incision function χ_{all}, which is defined by $\chi_{all}(clusters(KB)) = \bigcup_{cl \in clusters(KB)} cl$, and consistent answers under the IAR semantics. Intuitively, answers relative to IAR can be obtained by removing from D all atoms participating in some cluster, and answering the query using the resulting database.

Theorem 2. Let $KB = (D, \Sigma)$ be a Datalog+/− ontology, and Q be a BCQ. Then, $KB \models_{IAR} Q$ iff $(D − \chi_{all}(clusters(KB))) \cup \Sigma \models Q$.

4.3   Lazy Answers

In the following, we present a different semantics for consistent query answering in Datalog+/− ontologies. The motivation to seek for a different semantics comes from different reasons: first is the fact that computing the AR semantics is too expensive for any reasonable-sized Datalog+/− ontology, and second, the IAR semantics is unnecessarily restrictive in the set of answers that can be obtained from a query. The *k-lazy* semantics is an alternative to classical consistent query answering in Datalog+/− ontologies; the intuition behind lazy answers is that, given a budget (the k parameter), a maximal set of consistent answers (maximal relative to the k) can be computed, which are at least as complete as those that can be obtained under IAR.

We first define the notion of *k-cut* of clusters. Let $\chi_{k\text{-}cut}$ be a function defined as follows for $cl \in cluster(KB)$:

$$\chi_{k\text{-}cut}(cl) = \begin{cases} \{C_1, \dots, C_m\} & m \geqslant 1, C_i \subset cl, |C_i| \leqslant k, \\ & s.t.\ mods(cl − C_i, \Sigma) \neq \emptyset \\ & \text{and}\ \not\exists C_i'\ s.t.\ C_i' \subset C\ and \\ & mods(cl − C_i', \Sigma) \neq \emptyset\}; \\ \{cl\} & \text{if no such } C_i \text{ exists.} \end{cases} \quad (1)$$

Intuitively, given a cluster cl, its k-cut $\chi_{k\text{-}cut}(cl)$ is the set of minimal subsets of cl of cardinality at most k, such that if they are removed from cl, what is left is consistent with respect to Σ.

We next use the k-cut of clusters to define a new type of incision functions, called k-*lazy functions*, as follows.

Definition 17. Let $KB = (D, \Sigma)$ be a Datalog$+/-$ ontology, and $k \geqslant 0$. A k-*lazy function* for KB is defined as $\chi_{lazy}(k, clusters(KB)) = \bigcup_{cl \in clusters(KB)} c_{cl}$, where $c_{cl} \in \chi_{k\text{-}cut}(cl)$.

The above k-lazy functions are indeed incision functions.

Proposition 1. Let $KB = (D, \Sigma)$ be a Datalog$+/-$ ontology, and $k \geqslant 0$. All k-lazy functions for KB are incision functions.

The function χ_{lazy} is the basis of *lazy repairs*, as defined next. Intuitively, k-lazy repairs are built by analyzing ways in which to remove at most k atoms in every cluster.

Definition 18 (k-Lazy Repair). Let $KB = (D, \Sigma)$ be a Datalog$+/-$ ontology, and $k \geqslant 0$. A k-*lazy repair* for KB is any set $D' = D - \chi_{lazy}(k, clusters(KB))$, where $\chi_{lazy}(k, clusters(KB))$ is a k-lazy function for KB. $LRep(k, KB)$ denotes the set of all such repairs.

Example 13. Consider again our running example and the clusters in KB from Example 9. Let $k = 1$, then we have that

$$\chi_{1\text{-}cut}(cl_1) = \{d_1 : \{directs(tom, d_1)\}, d_2 : \{directs(tom, d_2)\}\},$$

and that

$$\chi_{1\text{-}cut}(cl_2) = \{e_1 : \{supervises(tom, john)\}, e_2 : \{directs(john, d_1)\}\}.$$

There are four possible incisions: $ins_1 = d_1 \cup e_1$, $ins_2 = d_1 \cup e_2$, $ins_3 = d_2 \cup e_1$, and $ins_4 = d_2 \cup e_2$. Thus, there are four 1-lazy repairs, with $lrep_i = D - ins_i$; for example,

$$lrep_1 = \{directs(john, d_1), directs(tom, d_2), works\_in(john, d_1),$$

$$works\_in(tom, d_1), manager(tom, d_1)\}.$$

∎

We can now define k-*lazy answers* for a query Q as the set of atoms that are derived from every k-lazy repair.

Definition 19 (k-Lazy Answers). Let $KB = (D, \Sigma)$ be a Datalog$+/-$ ontology, Q be a BCQ, and $k \geqslant 0$. Then, *Yes* is a k-*lazy answer* for Q to KB, denoted $KB \models_{k\text{-}LCons} Q$, iff it is an answer for Q to each $KB' = (D', \Sigma)$ with $D' \in LRep(k, KB)$.

Note that k-$LCons$ is used to identify the consistency-tolerant query answering semantics corresponding to k-lazy repairs. The following proposition states some properties of k-lazy repairs and lazy answers: each lazy repair is consistent relative to Σ, and only atoms that contribute to an inconsistency are removed by a k-lazy function for KB.

Proposition 2. *Let* $KB = (D, \Sigma)$ *be a Datalog+/− ontology, and* $k \geqslant 0$. *Then, for every* $D' \in LRep(k, KB)$, *(a)* $mods(D', \Sigma) \neq \emptyset$, *and (b) if* $\beta \in D$ *and* $\beta \notin D'$, *then there exists some* $B \subseteq D$ *such that* $mods(B, \Sigma) \neq \emptyset$ *and* $mods(B \cup \{\beta\}, \Sigma) = \emptyset$.

Proposition 2 shows that lazy repairs satisfy properties that are desirable for any belief change operator to have [31]. However, the incisions performed by function $\chi_{lazy}(k, clusters(KB))$ are not always minimal relative to set inclusion; i.e., if there is no subset of a cluster of size at most k that satisfies the conditions in Definition 1, then the whole cluster is removed, and therefore not every lazy repair is a repair.

The AR semantics from Definition 14 is a *cautious semantics* to query answering, since only answers that can be entailed from *every* repair are deemed consistent. Traditionally, the alternative to this semantics is a *brave* approach, which in our framework would consider an answer as consistent if it can be entailed from *some* repair. In the case of Example 13 with $k = 1$, $works\_in(john, d_1)$ and $works\_in(tom, d_1)$ are lazy consequences of KB, which are clearly not consistent consequences of KB. However, a brave approach for query answering would allow both $supervise(tom, john)$ and $directs(john, d_1)$ as answers. In this respect, lazy answers are a compromise between brave and cautious approaches: although it is "braver" than the cautious approach, *it does not allow to derive mutually inconsistent answers*.

Proposition 3. *Let* $KB = (D, \Sigma)$ *be a Datalog+/− ontology, Q be a CQ, and* $ans_{LCons}(k, Q, D, \Sigma)$ *be the set of lazy answers for Q given k. Then, for any* $k \geqslant 0$, $mods(ans_{LCons}(k, Q, D, \Sigma), \Sigma) \neq \emptyset$.

The next proposition shows that the same property holds if we consider the union of k-lazy answers for different values of k.

Theorem 3. *Let* $KB = (D, \Sigma)$ *be a Datalog+/− ontology and Q be a CQ. Then, for any* $k \geqslant 0$, $mods(\bigcup_{0 \leqslant i \leqslant k} ans_{LCons}(i, Q, D, \Sigma), \Sigma) \neq \emptyset$.

Theorem 3 shows that lazy answers can be used to obtain answers that are not consistent answers but are nevertheless consistent *as a whole*. We refer to this as the *union-k-lazy semantics*.

The next proposition shows the relationships between AR, IAR, and the lazy semantics.

Proposition 4. *Let* $KB = (D, \Sigma)$ *be a Datalog+/− ontology, and Q be a BCQ. Then, (a) if* $KB \models_{IAR} Q$, *then* $KB \models_{k\text{-}LCons} Q$, *for any* $k \geqslant 0$, *and (b)* $KB \models_{IAR} Q$ *iff* $KB \models_{0\text{-}LCons} Q$. *Furthermore, there is* $k \geqslant 0$ *such that* $KB \models_{AR} Q$ *iff* $KB \models_{k\text{-}LCons} Q$.

Clearly, Proposition 4 entails that if we take the union of the lazy answers up to the k from the proposition, then the resulting set of lazy answers is complete with respect to AR. Example 14 shows that, in our running example, the 2-lazy answers correspond exactly to the consistent answers.

Example 14. In Example 13, if $k = 2$, then we have that $\chi_{2\text{-}cut}(cl_1) = \chi_{1\text{-}cut}(cl_1)$ and $\chi_{2\text{-}cut}(cl_2) = \chi_{1\text{-}cut}(cl_2) \cup \{\{works\_in(tom, d_1),\ works\_in(john, d_1)\}\}$. We can easily see that $LRep(2, KB) = DRep(KB)$.   ∎

The following (simpler) example shows the effects of changing the value of k as well as the results from Theorem 3.

Example 15. Consider the CQ $Q(X, Y) = p(X) \wedge q(Y)$ and the following Datalog+/− ontology $KB = (D, \Sigma)$:

$$D \ \ = \{p(a),\, p(b),\, p(c),\, p(d),\, p(e),\, p(f),\, q(g),\, q(h),\, q(i),\, q(j)\};$$
$$\Sigma_T \ = \{\};$$
$$\Sigma_{NC} = \{p(a) \wedge p(b) \to \bot,\, p(b) \wedge p(d) \to \bot,\, p(d) \wedge p(e) \to \bot,\, p(d) \wedge p(f) \to \bot$$
$$q(g) \wedge q(h) \to \bot,\, q(h) \wedge q(i) \to \bot\}$$

The set of clusters in KB is $clusters(KB, \Sigma) = \{cl_1 : \{p(a), p(b), p(d), p(e), p(f)\}$, $cl_2 : \{q(g), q(h), q(i)\}$. For $k = 0$, the only 0-lazy repair is $lrep_0 = \{p(c), q(j)\}$, which coincides with D_I; the set of 0-lazy answers (and the answers under IAR) to $Q(X, Y)$ is $\{p(c), q(j)\}$.

For $k = 1$, note that there is no way of removing one element from cl_1 making the rest consistent; therefore, the only possible cut removes the whole cluster. On the other hand, there is one 1-cut for cl_2, namely $\{q(h)\}$. Therefore, we have only one 1-lazy repair $lrep_1 = \{p(c), q(j), q(i), q(g)\}$. The set of 1-lazy answers to $Q(X, Y)$ is $\{p(c), q(j), q(i), q(g)\}$.

With $k = 2$, we have two possible 2-cuts for cl_1 and two for cl_2, this is, $\chi_{2\text{-}cut}(cl_1) = \{\{p(a), p(d)\}, \{p(b), p(d)\}\}$ and $\chi_{2\text{-}cut}(cl_2) = \{\{q(h)\}, \{q(g), q(i)\}\}$. In this case there are four 2-lazy repairs and the set of 2-lazy answers to $Q(X, Y)$ is $\{p(c), p(e), p(f), q(j)\}$.

For $k = 3$, we have $\chi_{3\text{-}cut}(cl_1) = \{\{p(a), p(d)\}, \{p(b), p(d)\}, \{p(b), p(e), p(f)\}\}$ and $\chi_{3\text{-}cut}(cl_2) = \chi_{2\text{-}cut}(cl_2) = \{\{q(h)\}, \{q(g), q(i)\}\}$. The set of 3-lazy repairs coincide with the set of repairs and therefore the set of 3-lazy answers to $Q(X, Y)$ is the set of consistent answers, namely $\{p(c), q(j)\}$.

Finally, $\bigcup_{0 \leqslant i \leqslant 3} ans_{LCons}(i, Q, D, \Sigma) = \{p(c), q(j), q(i), q(g), p(e), p(f)\}$, which is clearly consistent relative to Σ_{NC}.   ∎

After the formal presentation of lazy answers, based on the concept of incision functions, we can now provide an algorithm that computes lazy answers to conjunctive queries to Datalog+/− ontologies. In [36], an algorithm to compute lazy answers to conjunctive queries is provided; the algorithm uses the concept of *finite chase graph* [17] for a given ontology $KB = (D, \Sigma)$, which is a graph consisting of the necessary finite part of $chase(D, \Sigma)$ relative to query Q, i.e., the finite part of the chase graph for D and Σ such that $chase(D, \Sigma) \models Q$. The idea of the algorithm is pretty straight forward, it first computes the set of clusters in KB, and next, for each cluster, function $\chi_{k\text{-}cut}$ is constructed by removing each possible subset (of size at most k) of the cluster in turn and checking if the remaining tuples are consistent (and that the subset in question is not a superset of an incision already found). A lazy repair then arises from each such possible combination by removing the incisions from D. The answer is finally computed using these repairs. Thought [36] proposes also an algorithm to compute clusters and kernels, there exists several algorithms in the literature to efficiently compute kernels in propositional logic that can be leveraged for Datalog+/− ontologies [43,51,52].

Lazy answers are based on a budget that restricts the size of removals that need to be made in a set of facts in order to make it consistent—if the budget is large enough, then we go to the trouble of considering all possible ways of solving the conflicts within the budget, but if it is not enough then we get rid of all the sentences that are involved in that particular conflict. If we think of the problem of querying inconsistent KBs as a reasoning task for an intelligent agent, then the value of the budget would be a bound on its reasoning capabilities (more complex reasoning can thus be afforded with higher budgets). On the other hand, considering clusters instead of culprits (or kernels) allows to identify a class of incision functions that solve conflicts from a global perspective; for more details on the relation of cluster incision functions and kernel incision functions cf. [20].

The key points that differentiates the k-lazy semantics from the Ar and its approximations is reflected in the fact that the set of repairs and the set of k-lazy repairs do not coincide in general, unless k is such that it forces to consider all the possible ways to solve the conflicts. This allows to consider answers that are not consistent answers in the sense of AR semantics but that are consistent with respect to the way conflicts are solved given the provided budget.

4.4   Towards Explainable Inconsistency-Tolerant Query Answering

Inconsistency-tolerant semantics for query answering provide a way to reason in logical knowledge bases in the presence of inconsistency. This is an important advantage over classical query answering processes, where answers may become meaningless. In this sense, the presence of inconsistency in the knowledge base remains transparent to the user that is issuing the query, which is, arguably, a good property as it does not disrupt the process. However, as these tools are often used to aid in the process of making decisions for different application domains (it may be an automated system itself the one that makes decisions based on the answers obtained from the knowledge base), it seems reasonable to try to provide information that complements the set of answers and helps the user understand *why* they obtained that set of answers and, particularly, if there was some piece of information, related to their query, that is subject to logical conflicts and how that affected the computed answers, especially if the answer was negative or did not include an individual that was expected.

For instance, suppose the user asks if the query $q() = \exists X p(X)$ is true and they get the answer **No**. It is only natural to pose the following question:

"*Was it the case that there is no possible way to derive $p(X)$ from the knowledge base, so the answer is* **No** *in every possible repair, or was it the case that q is true in some repairs but false in others, such that the semantics cannot assure its truth value*".

This distinction may be significant depending on the implications of the answer and how it is used. With this example in mind, given a Datalog+/− ontology (D, Σ) and a query Q, a natural question that one may be interested in asking for explanatory purposes is:

"*What makes Q true under some semantics S?*", or alternatively
"*What makes Q false under some semantics S?*".

The work of [11] proposes the notion of explanation for positive and negative query answers under the *brave*, AR, and IAR semantics, for Description Logics. An explanation for a query Q in this case is based on *causes* for Q, which are sets of facts from the original knowledge base that yield Q; this means that, in terms of Datalog+/−, causes are subsets of the D that together with Σ yield Q. Positive explanations for the *brave* semantics (the answer is true in some repair) is any cause for Q, that is, any consistent subset of D that entails the query by means of Σ. For the IAR semantics, an explanation is any cause of Q that does not participate in any contradiction. In the case of AR it is not enough to provide just one set of facts that are a cause of the query, as different repairs may use different causes. Therefore, they provide explanations in the form of (minimal) disjunctions of causes that cover all repairs (every cause belongs to at least one repair and for each repair there is one cause in the set).

Explanations for negative answers for Q under AR are minimal subsets of D such that together with any cause for Q yield an inconsistency. On the other hand, explanations for negative answers under IAR it is only necessary to ensure that every cause is contradicted by some consistent subset of D, which is enough to show that no cause belongs to all repairs. The proposal is accompanied by a computational complexity study of the difficulty of the decision problems related to checking and computing the different types of explanations. Most of these problems are polynomial for the case of explanations for positive and negative answers under *brave* and IAR. Not surprisingly, explanations in both cases under the AR semantics are intractable.

The notion of explanation, both for positive and negative answers, are directly translatable for k-Lazy answers. In particular, incisions correspond to explanations for negative answers. If we look at Example 14 and take query $Q() = p(e)$ for $k = 1$, we have that $KB \not\models_{1\text{-}LCons} Q$ and the reason for that is that the only possible 1-cut for cl_1 includes $p(e)$ (in the general case we can actually see that the incision contradicts every possible cause of $p(e)$).

In addition, this proposal can provide other types of answers in relation to explaining the behavior of the semantics. Other interesting questions may include:

1. What is the smallest k needed to make Q true under both k-lazy and union-k-lazy semantics?
2. What are the causes that make Q change its truth value from k to $k+1$ under the k-lazy semantics (either from true to false or the other way around)?
3. If Q is true under (union-) k-lazy semantics for some $k \geqslant 0$ but it is not a consistent answer, what are the causes for this behavior? This question

actually elaborates on the previous one, as we can try to find for which $k' \geqslant k$ the truth value of Q changes, and find the reason by comparing k'-cuts against $k' + 1$-cuts.

If we try to answer question (1) for Example 14 and $Q() = p(e)$, we find that the smallest k is 2, which means that $p(e)$ is involved in some inconsistency (it belongs or can be derived from a cluster) and that the conflict is such that it is necessary to remove two atoms at the same time from the cluster, so that $p(e)$ appears or survives. Note that this number is actually related to the fact that there is a *chain of conflicts*: $p(a) \wedge p(b) \rightarrow \bot$, $p(b) \wedge p(d) \rightarrow \bot$, and $p(d) \wedge p(e) \rightarrow \bot$. In this way we can use causes, together with NCs and incisions, to complement the required explanations and can show explanations that are local to specific k's.

Finally, we will show examples of how argumentation theory can help in the construction of meaningful explanations for inconsistency-tolerant query answering. As mentioned in the introduction, argumentation provides a natural dialogic structure and mechanism as part of the reasoning process and that can, in principle, be examined by a user in order to understand both *why* and *how* conclusions (answers) are reached.

The work in [4], proposes explanations as a set of logical arguments supporting the query. Without going into the fine details, an argument can be seen as a set of premises (facts) that derives a conclusion by means of a logical theory, or in the case of Datalog+/− a set of TGDs. In this context, we can think of causes of a query, defined by [11], as arguments that entail or support the entailment of the query. Conversely, we can build arguments that contradict some sentence, and these can be used as reasons against a query or, more generally, explanations for negative answers. For more details on argumentation we refer the reader to [21, 25]. All the examples of explanation proposals mentioned so far can be considered as argument-based explanations, depending on the richness of the language and the framework, different notions of argument and counterarguments can be constructed as a means for explanations.

The proposals mentioned above provide arguments for and against conclusions but in a static way, after the user inquiry for explanations the system retrieves and shows the set of explanations. Alternative, it is possible to exploit the dynamical characteristics of argumentation frameworks in order to create an interactive explanation mechanism. In [3], the notion of *dialectical explanations* is developed where it is assumed that the explanation is an interactive process with the system where a dialogue is established. The idea is that the explainer (e.g., the system) aims to make an explainee (e.g., the user) understand why a query Q is or is not entailed by the query answering semantics. It is shown that the query answering process can be represented as such a dialogue, in which arguments for and against the entailment of the query are identified, analyzed, and weighed among each other. A query is entailed under a specific semantics if and only if the dialectical process ends with a winning argument in favor of the query. That work develops this dialectical process for the *ICR* semantics [10], which is a sound approximation of *AR* and generalizes *IAR*. In [3], the proposal is extended for the *brave* and *IAR* semantics.

In this same spirit, [37] introduces an inconsistency-tolerant semantics for Datalog+/− ontologies query asnwering based on defeasible argumentative reasoning, which allows consequences to represent statements whose truth can be challenged. The proposal incorporates argumentation theory within the Datalog+/− query answering process itself. This process has the ability of considering reasons for and against potential conclusions and deciding which are the ones that can be obtained (warranted) from the knowledge base. This provides the possibility of implementing different inconsistency-tolerant semantics depending on the comparison criterion selected, all within the same framework, and as part of the query answering process. Indeed, the paper shows that most inconsistency-tolerant semantics that are based on the notion of repair (AR and the family of semantics that approximates it), as well as other such as the k-support and k-defeater semantics [12], can be obtained within this framework. This proposal has two advantages; first, it is not necessary to use and compute elements that are outside of the logic, such as repairs, kernels, clusters, incisions, etc., as the query answering engine is inconsistency-tolerant in itself. Second, the argumentative process underlying the query answering task allows to compute the answers and the required explanations at the same time. This means that there is, in principle, not extra cost for computing explanations, as happens also in [3]. Of course, there is the potential of creating a more complex explanatory mechanism exploiting other elements that are explicitly built within the argumentative process.

5   Discussion and Future Research Directions

Querying and managing incomplete and inconsistent information in an automatic and systematic way is becoming more and more necessary in order to cope with the amount of information that feeds the systems that are used to make decisions in a wide variety of domains, from product or service recommendation systems to medical or political applications. In order to build automated systems that aid humans in the process of making decisions in such a way that they improve their performance and understanding, proper explanations and interpretability of results are of the utmost importance. As we mentioned before, what is considered a good or reasonable explanation—or explanation process—strongly depends on the application domain and the particular problem the user is trying to solve based on the system's results.

Being able to produce adequate and meaningful explanations from automated systems is about being able to trust their results. This kind of trust is important from the system's functional point of view, but also important from a regulatory perspective. As it has already being discussed in different forums, such as the European General Data Protection Regulation,[3] users (or subjects) of automated data processing have the right "to obtain human intervention, to express his or her point of view, to obtain an explanation of the decision reached after

---

[3] https://eugdpr.org/.

such assessment and to challenge the decision". Such legal (and social) requirements clearly set the stage for discussions and further research efforts regarding adequate explanation models, mechanisms, and tools.

In this work we drafted some ideas on how to exploit the potential of knowledge based AI systems in order to produce meaningful explanations for query answering in the presence of uncertain information, where the uncertainty may arise from a probabilistic model or from the presence of inconsistency. After these first steps, it becomes clear that the road to designing and implementing explainable tools based on these and other formalisms is long; the roadmap includes the following activities, among others:

- Research different *kinds of explanations* for each type of query that can be posed to the system, making full use of the knowledge encoded in the models.
- Related to the previous point, take into account the actual users of the system, for whom the explanations are generated. This includes many different *human-centered aspects*, such as effective interfaces that don't overwhelm the user, and conveying full transparency in order to gain the user's trust.
- Design explanation techniques that allow a *level of detail* to be set, so as to support the wide range between novice and expert users, as well as different levels of privacy and clearance (in terms of security).
- Study the relationship between explainability and *human-in-the-loop* systems—for instance, it is possible for users to only require explanations for certain parts of a result, or only an explanation of why the result wasn't one that they were expecting.
- Explore how explainability relates to the vast body of work in *software auditing*—clearly, explanations might not only be required at query time, but at a later stage when other interested parties are reviewing the system's outputs.
- Ensure *computational tractability*: producing explanations should not be an excessive computational burden.

Each of these tasks can be considered a research and development program in its own right; we envision progress to continue being made slowly but steadily from ad hoc approaches to well-founded developments in the future.

Acknowledgments. This work was partially supported by funds provided by CONICET, Agencia Nacional de Promoción Científica y Tecnológica, Universidad Nacional del Sur (UNS), Argentina, and by the EU H2020 research and innovation programme under the Marie Sklodowska-Curie grant agreement 690974 for the project "MIREL: MIning and REasoning with Legal texts".

References

1. Alonso, J.M., Castiello, C., Mencar, C.: A bibliometric analysis of the explainable artificial intelligence research field. In: Medina, J., et al. (eds.) IPMU 2018. CCIS, vol. 853, pp. 3–15. Springer, Cham (2018). https://doi.org/10.1007/978-3-319-91473-2_1

2. Arenas, M., Bertossi, L.E., Chomicki, J.: Consistent query answers in inconsistent databases. In: Proceedings of PODS, pp. 68–79 (1999)
3. Arioua, A., Croitoru, M.: Dialectical characterization of consistent query explanation with existential rules. In: FLAIRS: Florida Artificial Intelligence Research Society (2016)
4. Arioua, A., Tamani, N., Croitoru, M.: Query answering explanation in inconsistent datalog+/− knowledge bases. In: Chen, Q., Hameurlain, A., Toumani, F., Wagner, R., Decker, H. (eds.) DEXA 2015. LNCS, vol. 9261, pp. 203–219. Springer, Cham (2015). https://doi.org/10.1007/978-3-319-22849-5_15
5. Atserias, A., Dawar, A., Kolaitis, P.G.: On preservation under homomorphisms and unions of conjunctive queries. J. ACM (JACM) **53**(2), 208–237 (2006)
6. Baget, J., Mugnier, M., Rudolph, S., Thomazo, M.: Walking the complexity lines for generalized guarded existential rules. In: Proceedings of the International Joint Conference on Artificial Intelligence (IJCAI), pp. 712–717 (2011)
7. Baral, C., Gelfond, M., Rushton, N.: Probabilistic reasoning with answer sets. In: Lifschitz, V., Niemelä, I. (eds.) LPNMR 2004. LNCS (LNAI), vol. 2923, pp. 21–33. Springer, Heidelberg (2003). https://doi.org/10.1007/978-3-540-24609-1_5
8. Barber, D.: Bayesian Reasoning and Machine Learning. Cambridge University Press, Cambridge (2012)
9. Beeri, C., Vardi, M.Y.: The implication problem for data dependencies. In: Even, S., Kariv, O. (eds.) ICALP 1981. LNCS, vol. 115, pp. 73–85. Springer, Heidelberg (1981). https://doi.org/10.1007/3-540-10843-2_7
10. Bienvenu, M.: Inconsistency-tolerant conjunctive query answering for simple ontologies. In: Kazakov, Y., Lembo, D., Wolter, F. (eds.) Proceedings of DL, vol. 846. CEUR-WS.org (2012)
11. Bienvenu, M., Bourgaux, C., Goasdoue, F.: Explaining inconsistency-tolerant query answering over description logic knowledge bases. In: Proceedings of AAAI 2016, pp. 900–906. AAAI Press (2016)
12. Bienvenu, M., Rosati, R.: Tractable approximations of consistent query answering for robust ontology-based data access. In: Proceedings of IJCAI, pp. 775–781 (2013)
13. Calì, A., Gottlob, G., Kifer, M.: Taming the infinite chase: query answering under expressive relational constraints. In: Proceedings of the International Conference on Principles of Knowledge Representation and Reasoning (KR), pp. 70–80 (2008)
14. Calì, A., Gottlob, G., Kifer, M.: Taming the infinite chase: query answering under expressive relational constraints. J. Artif. Intell. Res. (JAIR) **48**, 115–174 (2013)
15. Calì, A., Gottlob, G., Lukasiewicz, T.: A general Datalog-based framework for tractable query answering over ontologies. In: Proceedings of the ACM SIGMOD-SIGACT-SIGAI Symposium on Principles of Database Systems (PODS), pp. 77–86 (2009)
16. Calì, A., Gottlob, G., Pieris, A.: Towards more expressive ontology languages: the query answering problem. Artif. Intell. J. (AIJ) **193**, 87–128 (2012)
17. Calì, A., Gottlob, G., Lukasiewicz, T.: A general Datalog-based framework for tractable query answering over ontologies. J. Web Sem. **14**, 57–83 (2012)
18. Ceylan, İ.İ., Borgwardt, S., Lukasiewicz, T.: Most probable explanations for probabilistic database queries. In: Proceedings of IJCAI, pp. 950–956 (2017)
19. Chandra, A.K., Merlin, P.M.: Optimal implementation of conjunctive queries in relational data bases. In: Proceedings of the ACM Symposium on Theory of Computing (STOC), pp. 77–90 (1977)
20. Deagustini, C.A.D., Martínez, M.V., Falappa, M.A., Simari, G.R.: Improving inconsistency resolution by considering global conflicts. In: Straccia, U., Calì, A.

(eds.) SUM 2014. LNCS (LNAI), vol. 8720, pp. 120–133. Springer, Cham (2014). https://doi.org/10.1007/978-3-319-11508-5_11

21. Dung, P.M.: On the acceptability of arguments and its fundamental role in non-monotonic reasoning, logic programming and n-person games. Artif. Intell. **77**, 321–357 (1995)

22. Fagin, R., Kolaitis, P.G., Miller, R.J., Popa, L.: Data exchange: semantics and query answering. In: Calvanese, D., Lenzerini, M., Motwani, R. (eds.) ICDT 2003. LNCS, vol. 2572, pp. 207–224. Springer, Heidelberg (2003). https://doi.org/10.1007/3-540-36285-1_14

23. Fagin, R., Kolaitis, P.G., Miller, R.J., Popa, L.: Data exchange: semantics and query answering. Theoret. Comput. Sci. **336**(1), 89–124 (2005)

24. Falappa, M.A., Kern-Isberner, G., Simari, G.R.: Explanations, belief revision and defeasible reasoning. Artif. Intell. **141**(1–2), 1–28 (2002)

25. García, A.J., Simari, G.R.: Defeasible logic programming: delp-servers, contextual queries, and explanations for answers. Argument Comput. **5**(1), 63–88 (2014)

26. García, A.J., Simari, G.R.: Defeasible logic programming: an argumentative approach. TPLP **4**(1–2), 95–138 (2004)

27. Gottlob, G., Manna, M., Pieris, A.: Combining decidability paradigms for existential rules. Theory Practice Logic Program. (TPLP) **13**(4–5), 877–892 (2013)

28. Gottlob, G., Orsi, G., Pieris, A., Šimkus, M.: Datalog and its extensions for semantic web databases. In: Eiter, T., Krennwallner, T. (eds.) Reasoning Web 2012. LNCS, vol. 7487, pp. 54–77. Springer, Heidelberg (2012). https://doi.org/10.1007/978-3-642-33158-9_2

29. Gottlob, G., Lukasiewicz, T., Martinez, M.V., Simari, G.I.: Query answering under probabilistic uncertainty in datalog+/- ontologies. Ann. Math. Artif. Intell. **69**(1), 37–72 (2013)

30. Grover, S., Pulice, C., Simari, G.I., Subrahmanian, V.S.: BEEF: balanced English explanations of forecasts. IEEE Trans. Comput. Soc. Syst. **6**(2), 350–364 (2019)

31. Hansson, S.O.: Semi-revision. J. Appl. Non-Classical Logic **7**, 151–175 (1997)

32. Lembo, D., Lenzerini, M., Rosati, R., Ruzzi, M., Savo, D.F.: Inconsistency-tolerant semantics for description logics. In: Hitzler, P., Lukasiewicz, T. (eds.) RR 2010. LNCS, vol. 6333, pp. 103–117. Springer, Heidelberg (2010). https://doi.org/10.1007/978-3-642-15918-3_9

33. Leone, N., Manna, M., Terracina, G., Veltri, P.: Efficiently computable Datalog programs. In: Proceedings of the International Conference on Principles of Knowledge Representation and Reasoning (KR), pp. 13–23 (2012)

34. Lukasiewicz, T., Martinez, M.V., Orsi, G., Simari, G.I.: Heuristic ranking in tightly coupled probabilistic description logics. In: Proceedings of UAI 2012, pp. 554–563 (2012)

35. Lukasiewicz, T., Martinez, M.V., Orsi, G., Simari, G.I.: Exact and approximate query answering in tightly coupled probabilistic datalog+/-. Forthcoming (2019)

36. Lukasiewicz, T., Martinez, M.V., Simari, G.I.: Inconsistency handling in Datalog+/- ontologies. In: Proceedings of ECAI, pp. 558–563 (2012)

37. Martinez, M.V., Deagustini, C.A.D., Falappa, M.A., Simari, G.R.: Inconsistency-tolerant reasoning in datalog$^\pm$ ontologies via an argumentative semantics. In: Bazzan, A.L.C., Pichara, K. (eds.) IBERAMIA 2014. LNCS (LNAI), vol. 8864, pp. 15–27. Springer, Cham (2014). https://doi.org/10.1007/978-3-319-12027-0_2

38. Martinez, M.V., Pugliese, A., Simari, G.I., Subrahmanian, V.S., Prade, H.: How dirty is your relational database? An axiomatic approach. In: Proceedings of ECSQARU, pp. 103–114 (2007)

39. Milani, M., Bertossi, L.: Tractable query answering and optimization for extensions of weakly-sticky Datalog+/-. arXiv:1504.03386 (2015)
40. Miller, T.: Explanation in artificial intelligence: insights from the social sciences. Artif. Intell. **267**, 1–38 (2019)
41. Pearl, J.: Probabilistic Reasoning in Intelligent Systems: Networks of Plausible Inference (1988)
42. Poole, D.: The independent choice logic for modelling multiple agents under uncertainty. Artif. Intell. **94**(1–2), 7–56 (1997)
43. Ribeiro, M.M., Wassermann, R.: Minimal change in AGM revision for non-classical logics. In: Principles of Knowledge Representation and Reasoning: Proceedings of the Fourteenth International Conference, KR 2014, 20–24 July 2014, Vienna, Austria (2014)
44. Richardson, A., Rosenfeld, A.: A survey of interpretability and explainability in human-agent systems. In: XAI Workshop on Explainable Artificial Intelligence, pp. 137–143 (2018)
45. Richardson, M., Domingos, P.: Markov logic networks. Mach. Learn. **62**, 107–136 (2006)
46. Shortliffe, E.H., Davis, R., Axline, S.G., Buchanan, B.G., Green, C.C., Cohen, S.N.: Computer-based consultations in clinical therapeutics: explanation and rule acquisition capabilities of the mycin system. Comput. Biomed. Res. **8**(4), 303–320 (1975)
47. Simari, G.I., Molinaro, C., Martinez, M.V., Lukasiewicz, T., Predoiu, L.: Ontology-Based Data Access Leveraging Subjective Reports. Springer, Heidelberg (2017). https://doi.org/10.1007/978-3-319-65229-0
48. Tifrea-Marciuska, O.: Personalised search for the social semantic web. Ph.D. thesis, Department of Computer Science, University of Oxford (2016)
49. Vardi, M.Y.: The complexity of relational query languages (extended abstract). In: Proceedings of the ACM Symposium on Theory of Computing (STOC), pp. 137–146 (1982)
50. Wainwright, M.J., Jordan, M.I., et al.: Graphical models, exponential families, and variational inference. Found. Trends® Mach. Learn. **1**(1–2), 1–305 (2008)
51. Wang, S., Pan, J.Z., Zhao, Y., Li, W., Han, S., Han, D.: Belief base revision for datalog+/- ontologies. In: Kim, W., Ding, Y., Kim, H.-G. (eds.) JIST 2013. LNCS, vol. 8388, pp. 175–186. Springer, Cham (2014). https://doi.org/10.1007/978-3-319-06826-8_14
52. Wassermann, R.: An algorithm for belief revision. In: Proceedings of the Seventh International Conference Principles of Knowledge Representation and Reasoning, KR 2000, 11–15 April 2000, Breckenridge, Colorado, USA, pp. 345–352 (2000)
53. White, C.C.: A survey on the integration of decision analysis and expert systems for decision support. IEEE Trans. Syst. Man Cybern. **20**(2), 358–364 (1990)

Provenance in Databases: Principles and Applications

Pierre Senellart[1,2,3]([⊠]) (ID)

[1] DI ENS, ENS, CNRS, PSL University, Paris, France
pierre@senellart.com
[2] Inria, Paris, France
[3] LTCI, Télécom Paris, IP Paris, Paris, France

Abstract. Data provenance is extra information computed during query evaluation over databases, which provides additional context about query results. Several formal frameworks for data provenance have been proposed, in particular based on provenance semirings. The provenance of a query can be computed in these frameworks for a variety of query languages. Provenance has applications in various settings, such as probabilistic databases, view maintenance, or explanation of query results. Though the theory of provenance semirings has mostly been developed in the setting of relational databases, it can also apply to other data representations, such as XML, graph, and triple-store databases.

Keywords: Provenance · Databases

1 Introduction

This short paper provides a very high-level overview of the principles and applications of provenance in databases. A more in-depth but still accessible presentation of the same concepts can be found in [21]; we also refer the reader to the other references listed in this paper.

We first briefly define data provenance in Sect. 2, then highlight a few example applications in Sect. 3 before discussing provenance over databases that are not in the classical relational setting in Sect. 4.

2 Provenance

The main task in data management is *query evaluation*: given a database D (in some structured form) and a query q (from some class), compute the result of the query over the database, $q(D)$. In the most commonly used setting of relational databases [1], for example, a database is a collection of named tables, a query can be expressed in the SQL query language, and the result of a query is a table.

However, in a number of applications (see examples in Sect. 3), knowing the query result is not enough: it is also useful to obtain extra information about

© Springer Nature Switzerland AG 2019
M. Krötzsch and D. Stepanova (Eds.): Reasoning Web 2019, LNCS 11810, pp. 104–109, 2019.
https://doi.org/10.1007/978-3-030-31423-1_3

this result, where it comes from, or how it has been computed. We call this extra information *data provenance* [5,8]. *Provenance management* deals with the computation of data provenance.

Data provenance can take multiple forms, depending on what kinds of information is required. A good and simple example of this is *Boolean provenance*, a notion introduced in [19] under a different terminology. Let X be a set of Boolean variables (variables that can be set to the values 0 or 1). We assume that every valuation ν of the variables of X, when applied to the database D, defines a new database $\nu(D)$. For example, if D is a relational database, every tuple of D can be associated with a different variable of X, and then $\nu(D)$ is simply the subdatabase of D formed only of tuples whose associated variable is set to 1 by ν. Then, by definition, the *provenance* of an element t in the query result $q(D)$ (e.g., in the relational setting, a tuple $t \in q(D)$) is the function from valuations of X to $\{0,1\}$:

$$\nu \mapsto \begin{cases} 1 \text{ if } t \in q(\nu(D)) \\ 0 \text{ otherwise.} \end{cases}$$

Boolean provenance is useful because the Boolean provenance of t in $q(D)$ is sufficient to determine the presence of t in any database of the form $\nu(D)$. In other words, if the Boolean provenance can be efficiently computed, it is possible to answer many kinds of hypothetical questions about what the output of the query q is over other databases than the database D.

Boolean provenance is special in that it can be defined quite abstractly, independently of a query language or even a precise data model. This definition, however, does not yield an efficient computation. A seminal paper on data provenance [17] has shown that, if we restrict the data model to relational databases and the query language to the *positive relational algebra* (the SELECT-FROM-WHERE core of SQL), Boolean provenance is simply a particular case of *semiring provenance*, and all forms of semiring provenances can be computed efficiently under the same restrictions. A semiring is an algebraic structure with two operators, $\oplus$ and $\otimes$, verifying some axioms; when semirings are used for provenance, the $\oplus$ operator corresponds to different possible ways of producing a given result (e.g., with union and duplicate elimination in the relational algebra), while $\otimes$ is used to indicate different information that need to be combined to produce a result (e.g., with joins and cross products). Semiring provenance, which is parameterized by an arbitrary commutative semiring – Boolean provenance corresponds to a parameterization by the semiring of Boolean functions –, captures most existing provenance formalisms, and yields multitude applications. See [17,21] for precise definitions.

3 Example Applications

We now list a few important applications of different forms of data provenance. The list is by no means restrictive, see, e.g., [21] for other examples.

Probabilistic databases. *Probabilistic databases* [23] are probability distributions over regular databases, these distributions being represented in some compact format. The central question in probabilistic databases is *probabilistic query evaluation*, namely computing the probability that a query is satisfied over a database. It turns out [18,23] that this problem can be solved using Boolean provenance: first, assign Boolean variables to the input database, and assign probabilities to these variables in a way consistent with the probability distribution; second, compute the Boolean provenance of the query; third, compute the probability that the Boolean provenance, seen as a Boolean function, evaluates to 1. This last part is intractable in general (#P-hard) but is amenable to techniques from the field of *knowledge compilation* [10,22].

View maintenance and view update. In databases, materialized views are stored representations of the result of a given query. If the original database is updated (e.g., through the deletion of some tuples), the materialized view needs to be maintained so as to reflect the new output of the query, hopefully without fully recomputing it; this is the *view maintenance* problem. Conversely, it should be possible (at least in simple situations) to issue an update (e.g., a deletion) over the content of the materialized view, and that this update be propagated to the original database; this is the *view update* problem. Both these problems can be solved using data provenance: View maintenance under deletions can be solved by maintaining the Boolean provenance of the view, and deleting tuples whose provenance evaluates to 0 once the variables associated to original deleted tuples are set to 0. View update under deletions can be solved using why-provenance [6], a form of semiring provenance.

Explanation of query results. Different forms of provenance can also be used to present a user with explanation of query results: *where-provenance* [5] can explain where a particular data value in the output comes from; *why-provenance* [5] which data inputs have been combined to produce a query result; *how-provenance* [17] how the entire result was constructed; *why-not provenance* [7] why a particular result was not produced. Though why- and how- provenance can be computed in the framework of semiring provenance, where- and why-not provenance require different techniques.

Provenance Management Systems. In order to support such applications, a number of provenance management systems have been designed. We restrict the discussion here to general-purpose provenance management in database systems, and not in other settings, such as scientific workflows [11]. Perm [16] modifies the internals of a now-obsolete PostgreSQL relational database management system to add support for computation of provenance. This design, unfortunately, had made it hard to maintain the system and to deploy it in modern environments. GProM [4] and ProvSQL [22] are two more recent provenance management systems which address this issue in two different ways: GProM is implemented as a middleware between the user and a database system, queries being rewritten on the fly to compute provenance annotations; ProvSQL is implemented as a lightweight add-on to PostgreSQL, which can be deployed on an existing PostgreSQL installation. GProM and ProvSQL both support provenance computation in various provenance semirings; ProvSQL also is a probabilistic database system, computing probabilities from the Boolean provenance. See the discussion in [22] for a comparison of the main features of GProM and ProvSQL.

4   Beyond Relational Provenance

Most research on provenance (and in particular on semiring provenance) has been carried out in the common setting of relational databases for the simple query language of the positive relational algebra. Extensions to richer query languages, and to different data models, are possible, though sometimes with different approaches.

Non-monotone queries. Semiring provenance can only capture the provenance of monotone queries, such as those of the positive relational algebra. Moving to non-monotone queries and the full relational algebra requires considering *semirings with monus* [2,14], where the monus $\ominus$ operator is used to represent negative information.

Aggregation. In order to capture the provenance of aggregation operators (such as sum or count), it is necessary to move from semirings to semimodules over the scalars that are the aggregation values [3].

Recursive queries. To add support for query languages involving recursion (such as Datalog), it is necessary to add constraints on to which semirings are considered: depending on these constraints (e.g., ω-continuity [17], absorptivity [12], existence of a * operator [20]), different algorithms can be used to compute the provenance.

XML databases. XML databases organize information in a hierarchical, tree-like manner. Queries over XML databases typically resemble tree patterns to be matched over the tree database. Semiring provenance concepts can be extended to this setting in a quite straightforward manner [13].

Graph databases. In graph databases, data is represented as a labeled, annotated graph, and queries make it possible to ask for the existence of a path between two nodes with constraints on its labels. Graph queries are inherently recursive, and require similar techniques as to support Datalog queries over relational databases [20].

Triple stores. Triple stores model information using the Semantic Web standard of subject–predicate–object triples. Queries, for example expressed in the standard SPARQL query language, represent complex patterns of triples. Negation is an important feature of Semantic Web languages, so semirings with monus are also deployed in this setting [9]; these semirings with monus must also verify additional axioms imposed by the semantics of SPARQL [15].

5   Outlook

The principles of provenance management in databases are now well-understood. The framework of provenance semirings, in particular, has revealed to be very fruitful. It also lends itself to a number of extensions beyond the positive relational algebra, as discussed in Sect. 4; some of these extensions are not fully fleshed out, however, and still require more work. Some other areas are in need of more research: for instance, on how updates in databases should interact with provenance annotations; or on how to combine provenance computation with efficient query processing. However, there are now enough foundations to build and optimize concrete provenance management systems (starting with the existing software, in particular, GProM and ProvSQL), and to apply them to real-world use cases.

References

1. Abiteboul, S., Hull, R., Vianu, V.: Foundations of Databases. Addison-Wesley, Boston (1995)
2. Amer, K.: Equationally complete classes of commutative monoids with monus. Algebra Universalis **18**(1), 129–131 (1984)
3. Amsterdamer, Y., Deutch, D., Tannen, V.: Provenance for aggregate queries. In: PODS (2011)
4. Arab, B.S., Feng, S., Glavic, B., Lee, S., Niu, X., Zeng, Q.: GProM - a swiss army knife for your provenance needs. IEEE Data Eng. Bull. **41**(1), 51–62 (2018)
5. Buneman, P., Khanna, S., Tan, W.C.: Why and where: a characterization of data provenance. In: ICDT (2001)
6. Buneman, P., Khanna, S., Tan, W.C.: On propagation of deletions and annotations through views. In: PODS (2002)
7. Chapman, A., Jagadish, H.V.: Why not? In: SIGMOD (2009)
8. Cheney, J., Chiticariu, L., Tan, W.C.: Provenance in databases: why, how, and where. Found. Trends Databases **1**(4), 379–474 (2009)
9. Damásio, C.V., Analyti, A., Antoniou, G.: Provenance for SPARQL queries. In: Cudré-Mauroux, P., et al. (eds.) ISWC 2012. LNCS, vol. 7649, pp. 625–640. Springer, Heidelberg (2012). https://doi.org/10.1007/978-3-642-35176-1_39
10. Darwiche, A., Marquis, P.: A knowledge compilation map. J. Artif. Intell. Res. **17**(1), 229–264 (2002)
11. Davidson, S.B., et al.: Provenance in scientific workflow systems. IEEE Data Eng. Bull. **30**(4), 44–50 (2007)
12. Deutch, D., Milo, T., Roy, S., Tannen, V.: Circuits for Datalog provenance. In: ICDT (2014)
13. Foster, J.N., Green, T.J., Tannen, V.: Annotated XML: queries and provenance. In: PODS (2008)
14. Geerts, F., Poggi, A.: On database query languages for K-relations. J. Appl. Logic **8**(2), 173–185 (2010)
15. Geerts, F., Unger, T., Karvounarakis, G., Fundulaki, I., Christophides, V.: Algebraic structures for capturing the provenance of SPARQL queries. J. ACM **63**(1), 7 (2016)
16. Glavic, B., Alonso, G.: Perm: processing provenance and data on the same data model through query rewriting. In: ICDE, pp. 174–185 (2009)
17. Green, T.J., Karvounarakis, G., Tannen, V.: Provenance semirings. In: PODS (2007)
18. Green, T.J., Tannen, V.: Models for incomplete and probabilistic information. IEEE Data Eng. Bull. **29**(1), 17–24 (2006)

19. Imielinski, T., Lipski Jr., W.: Incomplete information in relational databases. J. ACM **31**(4), 761–791 (1984)
20. Ramusat, Y., Maniu, S., Senellart, P.: Semiring provenance over graph databases. In: TaPP (2018)
21. Senellart, P.: Provenance and probabilities in relational databases: from theory to practice. SIGMOD Rec. **46**(4), 5–15 (2017)
22. Senellart, P., Jachiet, L., Maniu, S., Ramusat, Y.: ProvSQL: provenance and probability management in PostgreSQL. PVLDB **11**(12), 2034–2037 (2018)
23. Suciu, D., Olteanu, D., Ré, C., Koch, C.: Probabilistic Databases. Morgan & Claypool (2011)

Knowledge Representation and Rule Mining in Entity-Centric Knowledge Bases

Fabian M. Suchanek[1(✉)], Jonathan Lajus[1], Armand Boschin[1], and Gerhard Weikum[2]

[1] Telecom Paris, Institut Polytechnique de Paris, Paris, France
{suchanek,jlajus,aboschin}@enst.fr
[2] Max Planck Institute for Informatics, Saarbrücken, Germany
weikum@mpi-inf.mpg.de

Abstract. Entity-centric knowledge bases are large collections of facts about entities of public interest, such as countries, politicians, or movies. They find applications in search engines, chatbots, and semantic data mining systems. In this paper, we first discuss the knowledge representation that has emerged as a pragmatic consensus in the research community of entity-centric knowledge bases. Then, we describe how these knowledge bases can be mined for logical rules. Finally, we discuss how entities can be represented alternatively as vectors in a vector space, by help of neural networks.

1 Introduction

1.1 Knowledge Bases

When we send a query to Google or Bing, we obtain a set of Web pages. However, in some cases, we also get more information. For example, when we ask "When was Steve Jobs born?", the search engine replies directly with "February 24, 1955". When we ask just for "Steve Jobs", we obtain a short biography, his birth date, quotes, and spouse. All of this is possible because the search engine has a huge repository of knowledge about people of common interest. This knowledge takes the form of a knowledge base (KB).

The KBs used in such search engines are entity-centric: they know individual entities (such as *Steve Jobs*, the *United States*, the *Kilimanjaro*, or the *Max Planck Society*), their semantic classes (such as *SteveJobs is-a computer-Pioneer*, *SteveJobs is-a entrepreneur*), relationships between entities (e.g., *Steve-Jobs founded AppleInc*, *SteveJobs hasInvented iPhone*, *SteveJobs hasWonPrize NationalMedalOfTechnology*, etc.) as well as their validity times (e.g., *SteveJobs wasCEOof Pixar [1986,2006]*).

The idea of such KBs is not new. It goes back to seminal work in Artificial Intelligence on universal knowledge bases in the 1980s and 1990s, most notably, the Cyc project [41] at MCC in Austin and the WordNet project [19] at Princeton

M. Krötzsch and D. Stepanova (Eds.): Reasoning Web 2019, LNCS 11810, pp. 110–152, 2019.
https://doi.org/10.1007/978-3-030-31423-1_4

University. These knowledge collections were hand-crafted and manually curated. In the last ten years, in contrast, KBs are often built automatically by extracting information from the Web or from text documents. Salient projects with publicly available resources include KnowItAll (UW Seattle, [17]), ConceptNet (MIT, [44]), DBpedia (FU Berlin, U Mannheim, & U Leipzig, [40]), NELL (CMU, [9]), BabelNet (La Sapienza, [58]), Wikidata (Wikimedia Foundation, [77]), and YAGO (Telecom Paris & Max Planck Institute, [70]). Commercial interest in KBs has been strongly growing, with projects such as the Google Knowledge Graph [15] (including Freebase [6]), Microsoft's Satori, Amazon's Evi, LinkedIn's Knowledge Graph, and the IBM Watson KB [20]. These KBs contain many millions of entities, organized in hundreds to hundred thousands of semantic classes, and hundred millions of relational facts between entities. Many public KBs are interlinked, forming the Web of Linked Open Data [5].

1.2 Applications

KBs are used in several applications, including the following:

Semantic Search and Question Answering. Both the Google search engine [15] and Microsoft Bing[1] use KBs to give intelligent answers to queries, as we have seen above. They can answer simple factual questions, provide movie showtimes, or show a list of "best things to do" at a travel destination. Wolfram Alpha[2] is another prominent example of a question answering system. The IBM Watson system [20] used knowledge from a KB to win against human champions in the TV quiz show Jeopardy.

Intelligent Assistants. Chatbots such as Apple's Siri, Amazon's Alexa, Google's Allo, or Microsoft's Cortana aim to help a user achieve daily tasks. The bots can, e.g., suggest restaurants nearby, answer simple factual questions, or manage calendar events. The background knowledge that the bots need for this work usually comes from a KB. With embodiments such as Amazon's Echo system or Google Home, such assistants will share more and more people's homes in the future. Other companies, too, are experimenting with chat bots that treat customer requests or provide help to users.

Semantic Data Mining. Daily news, social media, scholarly publications, and other Web contents are the raw inputs for analytics to obtain insights on business, politics, health, and more. KBs can help to discover and track entities and relationships in order to generate opinion maps, informative recommendations, and other kinds of intelligence towards decision making. For example, we can mine the gender bias from newspapers, because the KB knows the gender of people (see [71] for a survey). There is an entire domain of research dedicated to "predictive analytics", i.e., the prediction of events based on past events.

---

[1] http://blogs.bing.com/search/2013/03/21/understand-your-world-with-bing/.
[2] https://wolframalpha.com.

1.3   Knowledge Representation and Rule Mining

In this article, we first discuss how the knowledge is usually represented in entity-centric KBs. The field of knowledge representation has a long history, and goes back to the early days of Artificial Intelligence. It has developed numerous knowledge representation models, from frames and KL-ONE to recent variants of description logics. The reader is referred to survey works for comprehensive overviews of historical and classical models [62,67]. In this article, we discuss the knowledge representation that has emerged as a pragmatic consensus in the research community of entity-centric knowledge bases.

In the second part of this article, we discuss logical rules on knowledge bases. A logical rule can tell us, e.g., that if two people are married, then they (usually) live in the same city. Such rules can be mined automatically from the knowledge base, and they can serve to correct the data or fill in missing information. We discuss first classical Inductive Logic Programming approaches, and then show how these can be applied to the case of knowledge bases.

In the third part of this article, we discuss an alternative way to represent entities: as vectors in a vector space. Such so-called embeddings can be learned by neural networks from a knowledge base. The embeddings can then help deduce new facts – much like logical rules.

2   Knowledge Representation

2.1   Entities

2.1.1   Entities of Interest

The most basic element of a KB is an *entity*. An entity is any abstract or concrete object of fiction or reality, or, as Bertrand Russell puts it in his *Principles of Mathematics* [81]:

Definition 1 (Entity): *An entity is whatever may be an object of thought.*

This definition is completely all-embracing. Steve Jobs, the Declaration of Independence of the United States, the Theory of Relativity, and a molecule of water are all entities. Events (such as the French Revolution), are entities, too. An entity does not even have to exist: Harry Potter, e.g., is a fictional entity. Phlogiston was presumed to be the substance that makes up heat. It turned out to not exist – but it is still an entity.

KBs model a part of reality. This means that they choose certain entities of interest, give them names, and put them into a structure. Thus, a KB is a structured view on a selected part of the world. KBs typically model only distinct entities. This cuts out a large portion of the world that consists of variations, flows and transitions between entities. Drops of rain, for instance, fall down, join in a puddle and may be splattered by a passing car to form new drops [66]. KBs will typically not model these phenomena. This choice to model only discrete entities is a projection of reality; it is a grid through which we see only distinct things. Many entities consist of several different entities. A car, for example, consists of wheels, a bodywork, an engine, and many other pieces. The engine consists of the pistons, the valves, and the spark plug. The valves consist again of several parts, and so on, until we ultimately arrive at the level of atoms or below. Each of these components is an entity. However, KBs will typically not be concerned with the lower levels of granularity. A KB might model a car, possibly its engine and its wheels, but most likely not its atoms. In all of the following, we will only be concerned with the entities that a KB models.

Entities in the real world can change gradually. For example, the Greek philosopher Eubilides asks: If one takes away one molecule of an object, will there still be the same object? If it is still the same object, this invites one to take away more molecules until the object disappears. If it is another object, this forces one to accept that two distinct objects occupy the same spatio-temporal location: The whole and the whole without the molecule. A related problem is the question of identity. The ancient philosopher Theseus uses the example of a ship: Its old planks are constantly being substituted. One day, the whole ship has been replaced and Theseus asks, "Is it still the same ship?". To cope with these problems, KBs typically model only atomic entities. In a KB, entities can only be created and destroyed as wholes.

2.1.2   Identifiers and Labels

In computer systems (as well as in writing of any form), we refer to entities by *identifiers*.

Definition 2 (Identifier): *An identifier for an entity is a string of characters that represents the entity in a computer system.*

Typically, these identifiers take a human-readable form, such as *ElvisPresley* for the singer Elvis Presley. However, some KBs use abstract identifiers. Wikidata, e.g., refers to Elvis Presley by the identifier *Q303*, and Freebase by */m/02jq1*. This choice was made so as to be language-independent, and so as to provide an identifier that is stable in time. If, e.g., Elvis Presley reincarnates in the future, then *Q303* will always refer to the original Elvis Presley. It is typically assumed that there exists exactly one identifier per entity in a KB. For what follows, we will not distinguish identifiers from entities, and just talk of entities instead.

Entities have names. For example, the city of New York can be called "city of New York", "Big Apple", or "Nueva York". As we see, one entity can have several names. Vice versa, the same name can refer to several entities. "Paris", e.g., can refer to the city in France, to a city of that name in Texas, or to a hero of Greek mythology. Hence, we need to carefully distinguish *names* – single words or entire phrases – from their *senses* – the entities that they denote. This is done by using labels.

Definition 3 (Label): *A label for an entity is a human-readable string that names the entity.*

If an entity has several labels, the labels are called synonymous. If the same label refers to several entities, the label is polysemous. Not all entities have labels. For example, your kitchen chair is clearly an entity, but it probably does not have any particular label. An entity that has a label is called a *named entity*. KBs typically model mainly named entities. There is one other type of entities that appears in KBs: literals.

Definition 4 (Literal): *A literal is a fixed value that takes the form of a string of characters.*

Literals can be pieces of text, but also numbers, quantities, or timestamps. For example, the label "Big Apple" for the city of New York is a literal, as is the number of its inhabitants (8,175,133).

2.2   Classes

2.2.1   Classes and Instances

KBs model entities of the world. They usually group entities together to form a *class*:

Definition 5 (Class): *A class (also: concept, type) is a named set of entities that share a common trait. An element of that set is called an instance of the class.*

Under this definition, the following are classes: The class of singers (i.e., the set of all people who sing professionally), the class of historical events in Latin America, and the class of cities in Germany. Some instances of these classes are, respectively, Elvis Presley, the independence of Argentina, and Berlin. Since everything is an entity, a class is also an entity. It has (by definition) an identifier and a label.

Theoretically, KBs can form classes based on arbitrary traits. We can, e.g., construct the class of singers whose concerts were the first to be broadcast by satellite. This class has only one instance (Elvis Presley). We can also construct the class of left-handed guitar players of Scottish origin, or of pieces of music that the Queen of England likes. There are several theories as to whether humans actually build and use classes, too [46]. Points of discussion are whether humans form crisp concepts, and whether all elements of a concept have the same degree of membership. For the purpose of KBs, however, classes are just sets of entities.

It is not always easy to decide whether something should be modeled as an instance or as a class. We could construct, e.g., for every instance a singleton class that contains just this instance (e.g., the class of all Elvis Presleys). Some things of the world can be modeled both as instances and as classes. A typical example is *iPhone*. If we want to designate the type of smartphone, we can model it as an instance of the class of smartphone brands. However, if we are interested in the iPhones owned by different people and want to capture them individually, then *iPhone* should be modeled as a class. A similar observation holds for abstract entities such as *love*. Love can be modeled as an instance of the class *emotion*, where it resides together with the emotions of *anger*, *fear*, and *joy*. However, when we want to model individual feelings of love, then *love* would be a class. Its instances are the different feelings of love that different people have. It is our choice how we wish to model reality.

A pragmatic test of whether something should be modeled as a class is as follows: If we are interested in the plural form of a word or phrase, then we should model it as a class. If we talk, e.g., about "iPhones", then we model several instances of iPhones, and hence *iPhone* should be a class. If we only talk about "iPhone" along with other brand names (such as "HTC One"), then *iPhone* may well be considered an instance. Analogously, if we talk of "love" only in singular, then we may model it as an instance, along with other emotions. If we talk of "loves" (as in "Elvis had many loves during his time as a star"), then *love* is the set of all love affairs – and thus a class. The reason for this test is that only countable nouns can be classes, and only countable nouns can be put into plural. Another method to distinguish classes from instances is to say "An X", or "Every X". If that is possible, then X is best modeled as a class, because it can have instances. For example, it is possible to say "a CEO", but not "a Steve Jobs". Hence, *ceo* should be a class, and *SteveJobs* should not. If we can say "This is X", then X is an instance – as in "This is Steve Jobs". If we can say "X is a Y", then X is an instance of Y – as in "Steve Jobs is a CEO".

A particular case are mass nouns like "milk". The word "milk" (in the sense of the liquid) does not have a plural form. Therefore, we could model it as an instance (e.g., as an instance of the class of liquids). However, if we are interested in individual servings of milk, such as bottles of milk, then we can model it as a class, *servingOfMilk*.

Some KBs do not make the distinction between classes and instances (e.g., the SKOS vocabulary, [84]). In these KBs, everything is an entity. There is, however, usually a "is more general than" link between a more special entity and a more general entity. Such a KB may contain, e.g., the knowledge that *iPhone* is more special than *smartphone*, without worrying whether one of them is a class. The distinction between classes and instances adds a layer of granularity. This granularity is used, e.g., to define the domains and ranges of relations, as we shall see in Sect. 2.3.

2.2.2 Taxonomies

Definition 6 (Subsumption): *Class A is a subclass of class B if A is a subset of B.*

For example, the class of singers is a subclass of the class of persons, because every singer is a person. We also say that the class of singers is a *specialization* of the class of persons, or that singer *is subsumed by* or *included in* person. Vice versa, we say that person is a *superclass* or a *generalization* of the class of singers. Technically speaking, two equivalent classes are subclasses of each other. This is the way the RDFS standard models subclasses [83]. We say that a class is a *proper subclass* of another class, if the second contains more entities than the first. We use the notion of subclass here to refer to *proper* subclasses only.

It is important not to confuse class inclusion with the relationship between parts and wholes. For example, an arm is a part of the human body. That does not mean, however, that every arm is a human body. Hence, *arm* is not a subclass of *body*. In a similar manner, New York is a part of the US. That does not mean that New York would be a subclass of the US. Neither New York nor the US are classes, so they cannot be subclasses of each other.

Class inclusion is transitive: If A is a subclass of B, and B is a subclass of C, then A is a subclass of C. For example, *viper* is a subclass of *snake*, and *snake* is a subclass of *reptile*. Hence, by transitivity, *viper* is also a subclass of *reptile*. We say that a class is a *direct subclass* of another class, if there is no class in the KB that is a superclass of the former and a subclass of the latter. When we talk about subclasses, we usually mean only direct subclasses. The other subclasses are *transitive subclasses*. Since classes can be included in other classes, they can form an inclusion hierarchy – a taxonomy.

Definition 7 (Taxonomy): *A taxonomy is a directed graph, where the nodes are classes and there is an edge from class X to class Y if X is a proper direct subclass of Y.*

The notion of taxonomy is known from biology. Zoological or botanic species form a taxonomy: *tiger* is a subclass of *cat*. *cat* is a subclass of *mammal*, and so on. This principle carries over to all other types of classes. We say, e.g., that *internetCompany* is a subclass of *company*, and that *company* is a subclass of *organization*, etc. Since a taxonomy models proper inclusion, it follows that the taxonomic graph is acyclic: If a class is the subclass of another class, then the latter cannot be a subclass of the former. Thus, a taxonomy is a directed acyclic graph. A taxonomy does not show the transitive subclass edges. If the graph contains transitive edges, we can always remove them. Given a finite directed acyclic graph with transitive edges, the set of direct edges is unique [2].

Transitivity is often essential in applications. For example, consider a question-answering system where a user asks for artists that are married to actors. If the KB only knew about Elvis Presley and Priscilla Presley being in the classes *rockSinger* and *americanActress*, the question could not be answered. However, by reasoning that *rockSingers* are also *singers*, who in turn are *artists* and *americanActresses* being *actresses*, it becomes possible to give this correct answer.

Usually (but not necessarily), taxonomies are connected graphs: Every node in the graph is, directly or indirectly, linked to every other node. Usually, the taxonomies have a single root, i.e., a single node that has no outgoing edges. This node identifies the most general class, of which every other class is a subclass. In zoological KBs, this may be class *animal*. In a person database, it may be the class *person*. In a general-purpose KB, this class has to be the most general possible class. In YAGO and Wordnet, the class is *entity*. In the RDF standard, it is called *resource* [82]. In the OWL standard [85], the highest class that does not include literals is called *thing*.

Some taxonomies have at most one outgoing edge per node. Then, the taxonomy forms a tree. The biological taxonomy, e.g., forms a tree, as does the Java class hierarchy. However, there can be taxonomies where a class has two distinct direct superclasses. For example, if we have the class *singer* and the classes of *woman* and *man*, then the class *femaleSinger* has two superclasses: *singer* and *woman*. Note that it would be wrong to make *singer* a subclass of *man* and *woman* (as if to say that singers can be men or women). This would actually mean that all singers are at the same time men and women.

When a taxonomy includes a "combination class" such as *FrenchFemaleSingers*, then this class can have several superclasses. *FrenchFemaleSingers*, e.g., can have as direct superclasses *FrenchPeople*, *Women*, and *Singers*. In a similar manner, one entity can be an instance of several classes. Albert Einstein, e.g., is an instance of the classes *physicist*, *vegetarian*, and *violinPlayer*.

When we populate a KB with new instances, we usually try to assign them to the most specific suitable class. For example, when we want to place *Bob Dylan* in our taxonomy, we would put him in the class *americanBluesSinger*, if we have such a class, instead of in the class *person*. However, if we lack more specific information about the instance, then we might be forced to put it into a general class. Some named entity recognizers, e.g., distinguish only between organizations, locations, and people, which means that it is hard to populate more specific classes. It may also happen that our taxonomy is not specific enough at the leaf level. For example, we may encounter a musician who plays the Arabic oud, but our taxonomy does not have any class like *oudPlayer*. Therefore, a class may contain more instances than the union of its subclasses. That is, for a class C with subclasses $C_1, \ldots, C_k$, the invariant is $\cup_{i=1..k} C_k \subseteq C$, but $\cup_{i=1..k} C_k = C$ is often false.

2.2.3 Special Cases

Some KBs assign literals to classes, too. For example, the literal "Hello" can be modeled as an instance of the class *string*. Such literal classes can also form taxonomies. For example, the class *nonNegativeIntegers* is a subclass of the class of *integers*, which is again a subclass of the more general class *numbers*.

We already observed that classes are entities. Thus, we can construct classes that contain other classes as instances. For example, we can construct the class of all classes *class* ={*car, person, scientist, ...*}. This leads to awkward questions about self-containment, reminiscent of Bertrand Russel's famous set of sets that do not include themselves. The way this is usually solved [82] is to distinguish the class (as an abstract concept) from the extension of the class (the set of its instances). For example, the class of singers is the abstract concept of people who sing. Its extension is the set {*Elvis, Madonna, ...*}. In this way, a class is not a set, but just an abstract entity. Therefore, the extension of a class can contain another class. This is, however, a rather theoretical problem, and in what follows, we will not distinguish classes from their extensions.

To distinguish classes from other entities, we call an entity that is neither a class nor a literal an *instance* or a *common entity*.

2.3   Relations

2.3.1   Relations and Statements

KBs model also *relationships* between entities:

Definition 8 (Relation): *A relationship (also: relation) over the classes $C_1, ..., C_n$ is a named subset of the Cartesian product $C_1 \times ... \times C_n$.*

For example, if we have the classes *person*, *city*, and *year*, we may construct the *birth* relationship as a subset of the cartesian product *person* × *city* × *year*. It will contain tuples of a person, their city of birth, and their year of birth. For example, $\langle ElvisPresley, Tupelo, 1935 \rangle \in birth$. In a similar manner, we can construct *tradeAgreement* as a subset of *country* × *country* × *commodity*. This relation can contain tuples of countries that made a trade agreement concerning a commodity. Such relationships correspond to classical relations in algebra or databases.

As always in matters of knowledge representation (or, indeed, informatics in general), the identifier of a relationship is completely arbitrary. We could, e.g., call the *birth* relationship *k42*, or, for that matter, *death*. Nothing hinders us to populate the *birth* relationship with tuples of a person, and the time and place where that person ate an ice cream. However, most KBs aim to model reality, and thus use identifiers and tuples that correspond to real-world relationships.

If $\langle x_1, ..., x_n \rangle \in R$ for a relationship R, we also write $R(x_1, ..., x_n)$. In the example, we write $birth(ElvisPresley, Tupelo, 1935)$. The classes of R are called the *domains* of R. The number of classes n is called the *arity* of R. $\langle x_1, ..., x_n \rangle$ is a *tuple of R*. $R(x_1, ..., x_n)$ is called a *statement*, *fact*, or *record*. The elements $x_1, ..., x_n$ are called the *arguments* of the facts. Finally, a *knowledge base*, in its simplest form, is a set of statements. For example, a KB can contain the relations *birth*, *death* and *marriage*, and thus model some of the aspects of people's lives.

2.3.2   Binary Relations

Definition 9 (Binary Relation): *A binary relation is a relation of arity 2.*

Examples of binary relations are *birthPlace*, *friendOf*, or *marriedTo*. The first argument of a binary fact is called the *subject*, and the second argument is called the *object* of the fact. The relationships are sometimes called *properties*. Relationships that have literals as objects, and that have at most one object per subject are sometimes called *attributes*. Examples are *hasBirthDate* or *hasISBN*. The *domain* of a binary relation $R \subset A \times B$ is A, i.e., the class from which the subjects are taken. B is called the *range* of R. For example, the domain of *birthPlace* is *person*, and its range is *city*. The *inverse* of a binary relation R is a relation R^{-1}, such that $R^{-1}(x, y)$ iff $R(x, y)$. For example, the inverse relation of *hasNationality* (between a person and a country) is *hasNationality⁻* (between a country and a person) – which we could also call *hasCitizen*.

Any n-ary relation R with $n > 2$ can be split into n binary relations. This works as follows. Assume that there is one argument position i that is a *key*, i.e., every fact $R(x_1, ..., x_n)$ has a different value for x_i. In the previously introduced 3-ary *birth* relationship, which contains the person, the birth place, and the birth date, the person is the key: every person is born only once at one place. Without loss of generality, let the key be at position $i = 1$. We introduce binary relationships $R_2, ..., R_n$. In the example, we introduce *birthPlace* for the relation between the person and the birth place, and *birthDate* for the relation between the person and the birth year. Every fact $R(x_1, ..., x_n)$ gets rewritten as $R_2(x_1, x_2), R_3(x_1, x_3), R_4(x_1, x_4), ..., R_n(x_1, x_n)$. In the example, the fact *birth(Elvis, Tupelo, 1935)* gets rewritten as *birthPlace(Elvis, Tupelo)* and *birthDate(Elvis, 1935)*. Now assume that a relation R has no key. As an example, consider again the *tradeAgreement* relationship. Obviously, there is no key in this relationship, because any country can make any number of trade-agreements on any commodity. We introduce binary relationships $R_1, ... R_n$ for every argument position of R. For *tradeAgreement*, these could be *country1*, *country2* and *tradeCommodity*. For each fact of R, we introduce a new entity, an *event entity*. For example, if the US and Brazil make a trade-agreement on coffee, *tradeAgreement(Brazil, US, Coffee)*, then we create *coffeeAgrBrUs*. This entity represents the fact that these two countries made this agreement. In general, every fact $R(x_1, ..., x_n)$ gives rise to an event entity $e_{x1,...,xn}$. Then, every fact $R(x_1, ..., x_n)$ is rewritten as $R_1(e_{x1,...,xn}, x_1), R_2(e_{x1,...,xn}, x_2), ..., R_n(e_{x1,...,xn}, x_n)$. In the example, *country1(coffeeAgrBrUs, Brazil), country2(coffeeAgrBrUs, US), tradeCommodity(coffeeAgrBrUs, Coffee)*. This way, any n-ary relationship with $n > 2$ can be represented as binary relationships. For $n = 1$, we can always invent a binary relation *hasProperty*, and use the relation as an additional argument. For example, instead of *male(Elvis)*, we can say *hasProperty(Elvis, male)*.

The advantage of binary relationships is that they can express facts even if one of the arguments is missing. If, e.g., we know only the birth year of Steve Jobs, but not his birth place, then we cannot make a fact with the 3-ary relation $birth \subset person \times city \times year$. We have to fill the missing arguments, e.g., with *null* values. If the relationship has a large arity, many of its arguments may have to be null values. In the case of binary relationships, in contrast, we can easily state *birthDate(SteveJobs, 1955)*, and omit the *birthPlace* fact. Another disadvantage of n-ary relationships is that they do not allow adding new pieces of information a posteriori. If, e.g., we forgot to declare the astrological ascendant as an argument to the 3-ary relation *birth*, then we cannot add the ascendant for Steve Job's birth without modifying the relationship. In the binary world, in contrast, we can always add a new relationship *birthAscendant*. Thus, binary relationships offer more flexibility. This flexibility can be a disadvantage, because it allows adding incomplete information (e.g., a birth place without a birth date). However, since knowledge bases are often inherently incomplete, binary relationships are usually the method of choice.

2.3.3 Functions

Definition 10 (Function): *A function is a binary relation that has for each subject at most one object.*

Typical examples for functions are *birthPlace* and *hasLength*: Every person has at most one birth place and every river has at most one length. The relation *ownsCar*, in contrast, is not a function, because a (rich) person can own multiple cars. In our terminology, we call a relation a function also if it has no objects for certain subjects, i.e., we include partial functions (such as *deathDate*).

Some relations are *functions in time*. This means that the relation can have several objects, but at each point of time, only one object is valid. A typical example is *isMarriedTo*. A person can go through several marriages, but can only have one spouse at a time (in most systems). Another example is *hasNumberOfInhabitants* for cities. A city can grow over time, but at any point of time, it has only a single number of inhabitants. Every function is a function in time.

A binary relation is an *inverse function*, if its inverse is a function. Typical examples are *hasCitizen* (if we do not allow double nationality) or *hasEmailAddress* (if we talk only about personal email addresses that belong to a single person). Some relations are both functions and inverse functions. These are identifiers for objects, such as the social security number. A person has exactly one social security number, and a social security number belongs to exactly one person. Functions and inverse functions play a crucial role in entity matching: If two KBs talk about the same entity with different names, then one indication for this is that both entities share the same object of an inverse function. For example, if two people share an email address in a KB about customers, then the two entities must be identical.

Some relations are "nearly functions", in the sense that very few subjects have more than one object. For example, most people have only one *nationality*, but some may have several. This idea is formalized by the notion of *functionality* [69]. The functionality of a relation r in a KB is the number of subjects, divided by the number of facts with that relation:

$$fun(r) := \frac{|\{x : \exists y : r(x,y)\}|}{|\{x,y : r(x,y)\}|}$$

The functionality is always a value between 0 and 1, and it is 1 if r is a function. It is undefined for an empty relation.

We usually have the choice between using a relation and its inverse relation. For example, we can either have a relationship *isCitizenOf* (between a person and their country) or a relationship *hasCitizen* (between a country and its citizens). Both are valid choices. In general, KBs tend to choose the relation with the higher functionality, i.e., where the subject has fewer objects. In the example, the choice would probably be *isCitizenOf*, because people have fewer citizenships than countries have citizens. The intuition is that the facts should be "facts about the subject". For example, the fact that two authors of this paper are citizens of Germany is clearly an important property of the authors (it appears on the Wikipedia page of the last author). Vice versa, the fact that Germany is fortunate enough to count these authors among its citizens is a much less important property of Germany (it does not appear on the Wikipedia page of Germany).

2.3.4   Relations with Classes

In Sect. 2.2.3, we have introduced the class *class*, which contains all classes. This allows us to introduce the relationship between an instance and its class: *type ⊂ entity × class*. We can now say *type(Elvis, singer)*.[3] We also introduce *subclassOf ⊂ class × class*, which is the relationship between a class and its super-classes. For example, *subclassOf(singer, person)*. In the same way as we have introduced the class of all classes, we can introduce the class of all relations. We call this class *property*. With this, we can define the relationship between a binary relation and its domain: *domain ⊂ property × class*. We can now say *domain(birthPlace, person)*. Analogously, we introduce *range ⊂ property × class*, so that we can say *range(birthPlace, city)*. This way, an entire KB, with its relations and schema information, can be written as binary relationships. There is no distinction between data and meta-data – the KB describes itself.

In some cases, we have the choice whether to model something as a relationship or as a class. For example, to say that Berlin is located in Germany, we can either say *locatedIn(Berlin, Germany)* or *type(Berlin, germanCity)*, or both. There is no definite agreement as to which method is the right way to go, but there are advantages and disadvantages for each of them. If the entities in question can have certain properties that other entities cannot have, then it is useful to group them into a class. Practically speaking, this means that as soon as there is a relationship that has these entities as domain or range, the entities should become a class. For example, if we model Landkreise (the German equivalent of regions), then we can have *inLandkreis ⊂ germanCity × Landkreis*. No city other than German cities can be in a Landkreis. Thus, it is useful to have the class *germanCity*. If, however, German cities behave just like all other cities in our KB, then a class for them is less useful. In this spirit, it makes sense to have a class for scientists (who have a graduation university), or digital cameras (which have a resolution), but less so for male scientists or Sony cameras.

However, if we want to express that an entity stands in a relationship with another entity, and if that other entity has itself many relationships, then it is useful to use a relational fact. This allows more precise querying. For example, German cities stand in a relationship with Germany. Germany is located in Europe, and it is one of the German speaking countries. Thus, by saying *located-In(Berlin, Germany)*, we can query for cities located in European countries and for German-speaking cities, without introducing a class for each of them. In this spirit, it makes sense to use the relational modeling for German cities or American actors, but much less so for, say, zoological categories such as mammals or reptiles. Sometimes neither choice may have strong arguments in favor, and sometimes both forms of modeling together may be useful.

---

[3] We can even say *type(class, class)*, i.e., *class* is an instance of *class*.

2.4   Knowledge Bases

2.4.1   Completeness and Correctness

Knowledge bases model only a part of the world. In order to make this explicit, one imagines a complete knowledge base $\mathcal{K}^*$ that contains all entities and facts of the real world in the domain of interest. A given KB $\mathcal{K}$ is *correct*, if $\mathcal{K} \subseteq \mathcal{K}^*$. Usually, KBs aim to be correct. In real life, however, large KBs tend to contain also erroneous statements. YAGO, e.g., has an accuracy of 95%, meaning that 95% of its statements are in $\mathcal{K}^*$ (or, rather, in Wikipedia, which is used as an approximation of $\mathcal{K}^*$). This means that YAGO still contains hundreds of thousands of wrong statements. For most other KBs, the degree of correctness is not even known.

A knowledge base is *complete*, if $\mathcal{K}^* \subseteq \mathcal{K}$ (always staying within the domain of interest). The *closed world assumption* (CWA) is the assumption that the KB at hand is complete. Thus, the CWA says that any statement that is not in the KB is not in $\mathcal{K}^*$ either. In reality, however, KBs are hardly ever complete. Therefore, KBs typically operate under the *open world assumption* (OWA), which says that if a statement is not in the KB, then this statement can be either true or false in the real world.

KBs usually do not model negative information. They may say that Caltrain serves the city of San Francisco, but they will not say that this train does not serve the city of Moscow. While incompleteness tells us that some facts may be missing, the lack of negative information prevents us from specifying which facts are missing because they are false. This poses considerable problems, because the absence of a statement does not allow any conclusion about the real world [60].

2.5   The Semantic Web

The common exchange format for knowledge bases is RDF/RDFS [82]. It specifies a syntax for writing down statements with binary relations. Most notably, it prescribes URIs as identifiers, which means that entities can be identified in a globally unique way. To query such RDF knowledge bases, one can use the query language SPARQL [86]. SPARQL borrows its syntax from SQL, and allows the user to specify graph patterns, i.e., triples where some components are replaced by variables. For example, we can ask for the birth date of Elvis by saying "SELECT ?birthdate WHERE { ⟨Elvis⟩ ⟨bornOnDate⟩ ?birthdate }".

To define semantic constraints on the data, RDF is extended by OWL [85]. This language allows specifying constraints such as functions or disjointness of classes, as well as more complex axioms. The formal semantics of these axioms is given by Description Logics [3]. These logics distinguish facts about instances from facts about classes and axioms. The facts about instances are called the *A-Box* ("Assertions"), and the class facts and axioms are called the *T-Box* ("Theory"). Sometimes, the term *ontology* is used to mean roughly the same as *T-Box*. Description Logics allow for automated reasoning on the data.

Many KBs are publicly available online. They form what is known as the *Semantic Web*. Some of these KBs talk about the same entities – with different identifiers. The Linked Open Data project [5] aims to establish links between equivalent identifiers, thus weaving all public KBs together into one giant knowledge graph.

2.6 Challenges in Knowledge Representation

Knowledge representation is a large field of research, which has received ample attention in the past, and which still harbors many open questions. Some of these open issues in the context of knowledge bases are the following.

Negative Information. For some applications (such as question answering or knowledge curation), it is important to know whether a statement is *not* true. As we have seen, KBs usually do not store negative information, and thus the mining of negative information is an active field of research. In some cases, axioms can help deducing negative information. For example, if some relation is a function, and if one object is present, then it follows that all other objects cannot be in the relation. In other cases, a variant of the closed world assumption can help [55].

Completeness. Today's KBs do not store the fact that they are complete in some domains. For example, if the KB knows all children of Barack Obama, then it would be helpful to store that the KB is complete on the children of Obama. Different techniques for storing completeness information have been devised (see [60] for a survey), and completeness can also be determined automatically to some degree [23,38,65], but these techniques are still in their infancy.

Correctness. Some KBs (e.g., NELL or YAGO) store a probability value with each statement, indicating the likelihood that the statement is correct. There is an ample corpus of scientific work on dealing with such probabilistic knowledge bases, but attaching probabilities to statements is currently not a universally adopted practice.

Provenance. Some KBs (e.g., Wikidata, NELL and YAGO) attach provenance information to their statements, i.e., the source where the statement was found, and the technique that was used to extract it. This information can be used to debug the KB, to justify the statements, or to optimize the construction process. Again, there is ample literature on dealing with provenance (see [4] for a survey of works in artificial intelligence, databases, and the Semantic Web) – although few KBs actually attach provenance information.

Time and Space. Some KBs (e.g., Wikidata and YAGO) store time and space information with their facts. Thus, they know where and when a fact happened. This is often achieved by giving each fact a fact identifier, and by making statements about that fact identifier. Other approaches abound [21,28,33,62,80]. They include, e.g., the use of 5-ary facts, the introduction of a sub-property for each temporal statement, or the attachment of time labels.

Facts about Facts. We sometimes wish to store not just the time of a statement, but more facts about that statement. For example, we may want to store the correctness or provenance of a fact, but also the authority who vouches for the fact, access rights to the fact, or beliefs or hypotheses (as in "Fabian believes that Elvis is alive"). RDF provides a mechanism called *reification* for this purpose, but it is clumsy to use. Named Graphs [10] and annotations [76] have been proposed as alternatives. Different other alternatives are surveyed in [4]. Newer approaches attach attributes to statements [37,47].

Textual Extension. The textual source of the facts often contains additional subtleties that cannot be captured in triples. It can therefore be useful to add the textual information into the KB, as it is done, e.g., in [87].

NoRDF. For some information (such as complex events, narratives, or larger contexts), the representation as triples is no longer sufficient. We call this the realm of NoRDF knowledge (in analogy to NoSQL databases). For example, it is clumsy, if not impossible, to represent with binary relations the fact that Leonardo diCaprio was baptized "Leonardo" by his mother, because she visited a museum in Italy while she was still pregnant, and felt that the baby kicked while she saw a work of Leonardo DaVinci.

Commonsense Knowledge. Properties of everyday objects (e.g. that spiders have eight legs) and general concepts are of importance for text understanding, sentiment analysis, and object recognition in images and videos. This line of knowledge representation is well covered in classical works [41,62], and is lately also enjoying attention in the KB community [72,73].

Intensional Knowledge. Commonsense knowledge can also take the form of rules. For example, if a doctoral student is advised by a professor, then the university of graduation will be the employer of the professor. Again, this type of knowledge representation is well covered in classical works [41,62], and recent approaches have turned to using it for KBs [11,25,26]. This type of intensional knowledge is what we will now discuss in the next section.

3   Rule Mining

3.1   Rules

Once we have a knowledge base, it is interesting to look out for *patterns* in the data. For example, we could notice that if some person A is married to some person B, then usually B is also married to A (symmetry of marriage). Or we could notice that, if, in addition, A is the parent of some child, then B is usually also a parent of that child (although not always).

We usually write such rules using the syntax of first-order logic. For example, we would write the previous rules as:

$$marriedTo(x, y) \Rightarrow marriedTo(y, x)$$

$$marriedTo(x, y) \land hasChild(x, z) \Rightarrow hasChild(y, z)$$

Such rules have several applications: First, they can help us complete the KB. If, e.g., we know that Elvis Presley is married to Priscilla Presley, then we can deduce that Priscilla is also married to Elvis – if the fact was missing. Second, the rules can help us disambiguate entities and correct errors. For example, if Elvis has a child Lisa, and Priscilla has a different child Lisa, then our rule could help find out that the two Lisa's are actually a single entity. Finally, those frequent rules give us insight about our data, biases in the data, or biases in the real world. For example, we may find that European presidents are usually male or that Ancient Romans are usually dead. These two rules are examples of rules that have not just variables, but also entities:

$$type(x, AncientRoman) \Rightarrow dead(x)$$

We are now interested in discovering such rules automatically in the data. This process is called Rule Mining. Let us start with some definitions. The components of a rule are called atoms:

Definition 11 (Atom): *An atom is of the form $r(t_1, \ldots, t_n)$, where r is a relation of arity n (for KBs, usually $n = 2$) and $t_1, \ldots t_n$ are either variables or entities.*

In our example, $marriedTo(x, y)$ is an atom, as is $marriedTo(Elvis, y)$. We say that an atom is *instantiated*, if it contains at least one entity. We say that it is *grounded*, if it contains only entities and no variables. A *conjunction* is a set of atoms, which we write as $\boldsymbol{A} = A_1 \wedge \ldots \wedge A_n$. We are now ready to combine atoms to rules:

Definition 12 (Rule): *A Horn rule (rule, for short) is a formula of the form $\boldsymbol{B} \Rightarrow h$, where $\boldsymbol{B}$ is a conjunction of atoms, and h is an atom. $\boldsymbol{B}$ is called the body of the rule, and h its head.*

For example, $marriedTo(x, y) \Rightarrow marriedTo(y, x)$ is a rule. Such a rule is usually read as "If x is married to y, then y is married to x". In order to apply such a rule to specific entities, we need the notion of a *substitution*:

Definition 13 (Substitution): *A substitution is a function that maps variables to entities or to other variables.*

For example, a substitution σ can map $\sigma(x) = Elvis$ and $\sigma(y) = z$ – but not $\sigma(Elvis) = z$. A substitution can be generalized straightforwardly to atoms, sets of atoms, and rules: if $\sigma(x) = Elvis$, then $\sigma(marriedTo(Priscilla, x)) = marriedTo(Priscilla, Elvis)$. With this, an *instantiation of a rule* is a variant of the rule where all variables have been substituted by entities (so that all atoms are grounded). If we substitute $x = Elvis$ and $y = Priscilla$ in our example rule, we obtain the following instantiation:

$$marriedTo(Elvis, Priscilla) \Rightarrow marriedTo(Priscilla, Elvis)$$

Thus, an instantiation of a rule is an application of the rule to one concrete case. Let us now see what rules can predict:

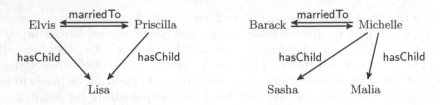

Fig. 1. Example KB

Definition 14 (Prediction of a rule): *The predictions P of a rule $B \Rightarrow h$ in a KB $\mathcal{K}$ are the head atoms of all instantiations of the rule where the body atoms appear in $\mathcal{K}$. We write $\mathcal{K} \wedge (B \Rightarrow h) \models P$. The predictions of a set of rules are the union of the predictions of each rule.*

For example, consider the KB in Fig. 1. The predictions of the rule $marriedTo(x, y) \wedge hasChild(y, z) \Rightarrow hasChild(x, z)$ are *hasChild(Priscilla, Lisa), hasChild(Elvis, Lisa), hasChild(Barack, Sasha), hasChild(Barack, Malia), hasChild(Michelle, Sasha), hasChild(Michelle, Malia).* This is useful, because two of these facts are not yet in the KB.

Logic. From a logical perspective, all variables in a rule are implicitly universally quantified (over every entity defined in the KB). Thus, our example rule is more explicitly written as

$$\forall x, y, z : marriedTo(x, y) \wedge hasChild(y, z) \Rightarrow hasChild(x, z)$$

It can be easily verified that such a rule is equivalent to the following disjunction:

$$\forall x, y, z : \neg marriedTo(x, y) \vee \neg hasChild(y, z) \vee hasChild(x, z)$$

While every Horn rule corresponds to a disjunction with universally quantified variables, not every such disjunction corresponds to a Horn rule. Only those disjunctions with exactly one positive atom correspond to Horn rules. In principle, we could mine arbitrary disjunctions, and not just those that correspond to Horn rules. We could even mine arbitrary first-order expressions, such as $\forall x : person(x) \Rightarrow \neg(underage(x) \wedge adult(x))$. For simplicity, we stay with Horn rules in what follows, and point out when an approach can be generalized to disjunctions or arbitrary formulae.

3.2   Rule Mining

3.2.1   Inductive Logic Programming

We now turn to mining rules automatically from a KB. This endeavor is based on *Inductive Reasoning*. To reason by induction is to expect that events that always appeared together in the past will always appear together in the future. For example, inductive reasoning could tell us: "All life forms we have seen so far need water. Therefore, all life forms in general need water.". This is the fundamental principle of empirical science: the generalization of past experiences to a scientific theory. Of course, inductive reasoning can never deliver the logical certitude of deductive reasoning. This is illustrated by Bertrand Russel's analogy of the turkey [61]: The turkey is fed every day by its owner, and so it comes to believe that the owner will always feed the turkey – which is true only until Christmas day. The validity and limitations of modeling the reality using inductive reasoning are a debated topic in philosophy of science. For more perspectives on the philosophical discussions, we refer the reader to [29] and [31]. In the setting of KBs, inductive reasoning is formalized as *Inductive Logic Programming* [51,57,63]:

Definition 15 (Inductive Logic Programming): *Given a background knowledge $\mathcal{B}$ (in general, any first order logic expression; in our case: a KB), a set of positive example facts E^+, and a set of negative example facts E^-, Inductive Logic Programming (ILP) is the task of finding an hypothesis $\mathfrak{h}$ (in general, a set of first order logic expressions; in our case: a set of rules) such that $\forall e^+ \in E^+ : \mathcal{B} \wedge \mathfrak{h} \models e^+$ and $\forall e^- \in E^- : \mathcal{B} \wedge \mathfrak{h} \not\models e^-$.*

This means that the rules we seek have to predict all positive examples (they have to be *complete*), and they may not predict a negative example (they have to be *correct*). For example, consider again the KB from Fig. 1 as background knowledge, and let the sets of examples be:

$$E^+ = \{isMarriedTo(Elvis, Priscilla), isMarriedTo(Priscilla, Elvis),$$
$$isMarriedTo(Barack, Michelle), isMarriedTo(Michelle, Barack)\}$$
$$E^- = \{isMarriedTo(Elvis, Michelle), isMarriedTo(Lisa, Barack),$$
$$isMarriedTo(Sasha, Malia)\}$$

Now consider the following hypothesis:

$$\mathfrak{h} = \{isMarriedTo(x, y) \Rightarrow isMarriedTo(y, x)\}$$

This hypothesis is complete, as every positive example is a prediction of the rule, and it is correct, as no negative example is predicted.

The attentive reader will notice that the difficulty is now to correctly determine the sets of positive and negative examples. In the ideal case the positive examples should contain any fact that is true in the real world and the negative examples contain any other fact. Thus, in a correct KB, every fact is a positive example.

Definition 16 (Rule Mining): *Given a KB, Rule Mining is the ILP task with the KB as background knowledge, and every single atom of the KB as a positive example.*

This means that the rule mining will find several rules, in order to explain all facts of the KB. Three problems remain: First, we have to define the set of negative examples (Sect. 3.2.2). Second, we have to define what types of rules we are interested in (Sect. 3.2.3). Finally, we have to adapt our mining to cases where the rule does not always hold (Sect. 3.2.4).

3.2.2 The Set of Negative Examples

Rule mining needs negative examples (also called *counter-examples*). The problem is that KBs usually do not contain negative information (Sect. 2.6). We can think of different ways to generate negative examples.

Closed World Assumption. The Closed World Assumption (CWA) says that any statement that is not in the KB is wrong (Sect. 2.4.1). Thus, under the Closed-World Assumption, any fact that is not in the KB can serve as a negative example. The problem is that these may be exactly the facts that we want to predict. In our example KB from Fig. 1, we may want to learn the rule $marriedTo(x, y) \wedge hasChild(y, z) \Rightarrow hasChild(x, z)$. For this rule, the fact *hasChild(Barack, Malia)* is a counter-example. However, this fact is exactly what we want to predict, and so it would be a counter-productive counter-example.

Open World Assumption. Under the Open-World Assumption (OWA), any fact that is not in the KB can be considered either a negative or a positive example (see again Sect. 2.4.1). Thus the OWA does not help in establishing counter-examples. Without counter-examples, we can learn any rule. For example, in our KB, the rule *type(x, person) ⇒ marriedTo(x, Barack)* has a single positive example (for $x = Michelle$), and no counter-examples under the Open World Assumption. Therefore, we could deduce that everyone is married to Barack.

Partial Completeness Assumption. Another strategy to generate negative examples is to assume that entities are complete for the relations they already have. For example, if we know that Michelle has the children Sasha and Malia, then we assume (much like Barack) that Michelle has no other children. If, in contrast, Barack does not have any children in the KB, then we do not conclude anything. This idea is called the Partial-Completeness Assumption (PCA) or the Local Closed World Assumption [25]. It holds trivially for functions (such as *hasBirthDate*), and usually [26] for relations with a high functionality (such as *hasNationality*). The rationale is that if the KB curators took the care to enter some objects for the relation, then they will most likely have entered all of them, if there are few of them. In contrast, the assumption does usually not hold for relations with low functionality (such as *starsInMovie*). Fortunately, relations usually have a higher functionality than their inverses (see Sect. 2.3.3). If that is not the case, we can apply the PCA to the object of the relation instead.

Random Examples. Another strategy to find counter-examples is to generate random statements [50]. Such random statements are unlikely to be correct, and can thus serve as counter-examples. This is one of the methods used by DL-Learner [30]. As we shall see in Sect. 4.3.1, it is not easy to generate helpful random counter-examples. If, e.g., we generate the random negative example *marriedTo(Barack, USA)*, then it is unlikely that a rule will try to predict this example. Thus, the example does not actually help in filtering out any rule. The challenge is hence to choose counter-examples that are false, but still reasonable. The authors of [55] describe a method to sample negative statements about semantically connected entities by help of the PCA. We will also revisit the problem in the context of representation learning (Sect. 4.3.1).

3.2.3   The Language Bias

After solving the problem of negative examples, the next question is what kind of rules we should consider. This choice is called the *language bias*, because it restricts the "language" of the hypothesis. We have already limited ourselves to Horn Rules, and in practice we even restrict ourselves to connected and closed rules.

Definition 17 (Connected rules): *Two atoms are connected if they share a variable, and a rule is connected if every non-ground atom is transitively connected to one another.*

For example, the rule *presidentOf(x, America)* ⇒ *hasChild(Elvis, y)* is not connected. It is an uninteresting and most likely wrong rule, because it makes a prediction about arbitrary *y*.

Definition 18 (Closed rules): *A rule is closed if every variable appears in at least two atoms.*

For example the rule *marriedTo(x, y)* ∧ *worksAt(x, z)* ⇒ *marriedTo(y, x)* is not closed. It has a "dangling edge" that imposes that *x* works somewhere. While such rules are perfectly valid, they are usually less interesting than the more general rule without the dangling edge.

Finally, one usually imposes a limit on the number of atoms in the rule. Rules with too many atoms tend to be very convoluted [26]. That said, mining rules without such restrictions is an interesting field of research, and we will come back to it in Sect. 3.5.

3.2.4   Support and Confidence

One problem with classical ILP approaches is that they will find rules that apply to very few entities, such as *marriedTo(x, Elvis)* ⇒ *hasChild(x, Lisa)*. To avoid this type of rules, we define the *support* of a rule:

Definition 19 (Support): *The support of a rule in a KB is the number of positive examples predicted by the rule.*

Usually, we are interested only in rules that have a support higher than a given threshold (say, 100). Alternatively, we can define a relative version of support, the *head coverage* [25], which is the number of positive examples predicted by the rule divided by the number of all positive examples with the same relation. Another problem with classical ILP approaches is that they will not find rules if there is a single counter-example. To mitigate this problem, we define the *confidence*:

Definition 20 (Confidence): *The confidence of a rule is the number of positive examples predicted by the rule (i.e., the support of the rule), divided by the number of examples predicted by the rule.*

This notion depends on how we choose our negative examples. For instance, under the CWA, the rule $marriedTo(x, y) \wedge hasChild(y, z) \Rightarrow hasChild(x, z)$ has a confidence of 4/6 in Fig. 1. We call this value the *standard confidence*. Under the PCA, in contrast, the confidence for the example rule is 4/4. We call this value the *PCA confidence*. While the standard confidence tends to "punish" rules that predict many unknown statements, the PCA confidence will permit more such rules. We present in Appendix A the exact mathematical formula of these measures.

In general, the support of a rule quantifies its completeness, and the confidence quantifies its correctness. A rule with low support and high confidence indicates a conservative hypothesis and may be overfitting, i.e. it will not generalize to new positive examples. A rule with high support and low confidence, in contrast, indicates a more general hypothesis and may be overgeneralizing, i.e., it does not generalize to new negative examples. In order to avoid these effects we are looking for a trade-off between support and confidence.

Definition 21 (Frequent Rule Mining): *Given a KB $\mathcal{K}$, a set of positive examples (usually $\mathcal{K}$), a set of negative examples (usually according to an assumption above) and a language of rules, Frequent rule mining is the task of finding all rules in the language with a support and a level of confidence superior to given thresholds.*

3.3   Rule Mining Approaches

Using substitutions (see Definition 13), we can define a syntactical order on rules:

Definition 22 (Rule order): *A rule $R \equiv (\boldsymbol{B} \Rightarrow h)$ subsumes a rule $R' \equiv (\boldsymbol{B'} \Rightarrow h')$, or R is "more general than" R', or R' "is more specific than" R, if there is a substitution σ such that $\sigma(\boldsymbol{B}) \subseteq \boldsymbol{B'}$ and $\sigma(h) = h'$. If both rules subsume each other, the rules are called equivalent.*

For example, consider the following rules:

$$
\left\{
\begin{array}{ll}
hasChild(x, y) \Rightarrow hasChild(z, y) & (R_0) \\
hasChild(Elvis, y) \Rightarrow hasChild(Priscilla, y) & (R_1) \\
hasChild(x, y) \Rightarrow hasChild(z, Lisa) & (R_2) \\
hasChild(x, y) \wedge marriedTo(x, z) \Rightarrow hasChild(z, y) & (R_3) \\
marriedTo(v_1, v_2) \wedge hasChild(v_1, v_3) \Rightarrow hasChild(v_2, v_3) & (R_4) \\
hasChild(x, y) \wedge marriedTo(z, x) \Rightarrow hasChild(z, y) & (R_5)
\end{array}
\right.
$$

The rule R_0 is more general than the rule R_1, because we can rewrite the variables x and z to $Elvis$ and $Priscilla$ respectively. However R_0 and R_2 are incomparable as we cannot choose to bind only one y and not the other in R_0. The rules R_3, R_4 and R_5 are more specific than R_0. Finally R_3 is equivalent to R_4 but not to R_5.

Proposition 23 (Prediction inclusion): *If a rule R is more general than a rule R', then the predictions of R' on a KB are a subset of the predictions of R. As a corollary, R' cannot have a higher support than R.*

This observation gives us two families of rule mining algorithms: top-down rule mining starts from very general rules and specializes them until they become too specific (i.e., no longer meet the support threshold). Bottom-up rule mining, in contrast, starts from multiple ground rules and generalizes them until the rules become too general (i.e., too many negative examples are predicted).

3.3.1 Top-Down Rule Mining

The concept of specializing a general rule to more specific rules can be traced back to [63] in the context of an exact ILP task (under the CWA). Such approaches usually employ a *refinement operator*, i.e. a function that takes a rule (or a set of rules) as input and returns a set of more specific rules. For example, a refinement operator could take the rule $hasChild(y, z) \Rightarrow hasChild(x, z)$ and produce the more specific rule $marriedTo(x, y) \wedge hasChild(y, z) \Rightarrow hasChild(x, z)$. This process is iterated, and creates a set of rules that we call the *search space* of the rule mining algorithm. On the one hand, the search space should contain every rule of a given rule mining task, so as to be complete. On the other hand, the smaller the search space is, the more efficient the algorithm is.

Usually, the search space is *pruned*, i.e., less promising areas of the search space are cut away. For example, if a rule does not have enough support, then any refinement of it will have even lower support (Proposition 23). Hence, there is no use refining this rule.

AMIE. AMIE [25] is a top-down rule mining algorithm that aims to mine any connected rule composed of binary atoms for a given support and minimum level of confidence in a KB. AMIE starts with rules composed of only a head atom for all possible head atoms (e.g., $\Rightarrow marriedTo(x, y)$). It uses three refinement operators, each of which adds a new atom to the body of the rule.

The first refinement operator, `addDanglingAtom`, adds an atom composed of a variable already present in the input rule and a new variable.

$$\text{Some refinements of:} \qquad \Rightarrow hasChild(z, y) \ (R_h)$$
$$\text{are:} \quad \begin{cases} hasChild(x, y) \Rightarrow hasChild(z, y) \ (R_0) \\ marriedTo(x, z) \Rightarrow hasChild(z, y) \ (R_a) \\ marriedTo(z, x) \Rightarrow hasChild(z, y) \ (R_b) \end{cases}$$

The second operator, `addInstantiatedAtom`, adds an atom composed of a variable already present in the input rule and an entity of the KB.

$$\text{Some refinements of:} \qquad \Rightarrow hasChild(Priscilla, y) \ (R'_h)$$
$$\text{are:} \quad \begin{cases} hasChild(Elvis, y) \Rightarrow hasChild(Priscilla, y) \ (R_1) \\ hasChild(Priscilla, y) \Rightarrow hasChild(Priscilla, y) \ (R_\top) \\ marriedTo(Barack, y) \Rightarrow hasChild(Priscilla, y) \ (R_\bot) \end{cases}$$

The final refinement operator, `addClosingAtom`, adds an atom composed of two variables already present in the input rule.

$$\text{Some refinements of:}\quad marriedTo(x,z) \Rightarrow hasChild(z,y)\ (R_a)$$

$$\text{are:}\quad \begin{cases} hasChild(x,y) \wedge marriedTo(x,z) \Rightarrow hasChild(z,y)\ (R_3) \\ marriedTo(z,y) \wedge marriedTo(x,z) \Rightarrow hasChild(z,y)\ (R_\alpha) \\ marriedTo(x,z) \wedge marriedTo(x,z) \Rightarrow hasChild(z,y)\ (R_a^2) \end{cases}$$

As every new atom added by an operator contains at least a variable present in the input rule, the generated rules are connected. The last operator is used to close the rules (for example R_3), although it may have to be applied several times to actually produce a closed rule (cf. Rules R_α or R_a^2).

The AMIE algorithm works on a queue of rules. Initially, the queue contains one rule of a single head atom for each relation in the KB. At each step, AMIE dequeues the first rule, and applies all three refinement operators. The resulting rules are then pruned: First, any rule with low support (such as $R_\perp$) is discarded. Second, different refinements may generate equivalent rules (using the closing operator on R_0 or R_a, e.g., generates among others two equivalent "versions" of R_3). AMIE prunes out these equivalent versions. AMIE+ [26] also detects equivalent atoms as in $R_\top$ or R_a^2 and rewrites or removes those rules. There are a number of other, more sophisticated pruning strategies that estimate bounds on the support or confidence. The rules that survive this pruning process are added to the queue. If one of the rules is a closed rule with a high confidence, it is also output as a result. In this way, AMIE enumerates the entire search space.

The top-down rule mining method is generic, but its result depends on the initial rules and on the refinement operators. The operators directly impact the language of rules we can mine (see Sect. 3.2.3) and the performance of the method. We can change the refinement operators to mine a completely different language of rules. For example, if we don't use the `addInstantiatedAtom` operator, we restrict our search to any rule without instantiated atoms, which also drastically reduce the size of the search space[4].

Apriori Algorithm. There is an analogy between top-down rule mining and the Apriori algorithm [1]. The Apriori algorithm considers a set of transactions (sales, products bought in a supermarket), each of which is a set of items (items bought together, in the supermarket analogy). The goal of the Apriori algorithm is to find a set of items that are frequently bought together.

These are frequent patterns of the form $\boldsymbol{P} \equiv I_1(x) \wedge \cdots \wedge I_n(x)$, where $I(t)$ is in our transaction database if the item I has been bought in the transaction t. Written as the set (called an "itemset") $\boldsymbol{P} \equiv \{I_1, \ldots, I_n\}$, any subset of $\boldsymbol{P}$ forms a "more general" itemset than $\boldsymbol{P}$, which is at least as frequent as P. The Apriori algorithm uses the dual view of the support pruning strategy: Necessarily, all

---

[4] Let $|\mathcal{K}|$ be the number of facts and $|r(\mathcal{K})|$ the number of relations in a KB $\mathcal{K}$. Let d be the maximal length of a rule. The size of the search space is reduced from $O(|\mathcal{K}|^d)$ to $O(|r(\mathcal{K})|^d)$ when we remove the `addInstantiatedAtom` operator.

patterns more general than P must be frequent for P to be frequent[5]. The refinement operator of the Apriori algorithm takes as input all frequent itemsets of size n and generate all itemsets of size $n + 1$ such that any subset of size n is a frequent itemset. Thus, Apriori can be seen as a top-down rule mining algorithm over a very specific language where all atoms are unary predicates.

The WARMR algorithm [13], an ancestor of AMIE, was the first to adapt the Apriori algorithm to rule mining over multiple (multidimensional) relations.

3.3.2   Bottom-Up Rule Mining

As the opposite of a refinement operator, one can define a generalization operator that considers several specific rules, and outputs a rule that is more general than the input rules. For this purpose, we will make use of the observation from Sect. 3.1 that a rule $b_1 \wedge \ldots \wedge b_n \Rightarrow h$ is equivalent to the disjunction $\neg b_1 \vee \cdots \vee \neg b_n \vee h$. The disjunction, in turn, can be written as a set $\{\neg b_1, \ldots, \neg b_n, h\}$ – which we call a *clause*. For example, the rule $marriedTo(x, y) \wedge hasChild(y, z) \Rightarrow hasChild(x, z)$ can be written as the clause $\{\neg marriedTo(x, y), \neg hasChild(y, z), hasChild(x, z)\}$. Bottom-up rule mining approaches work on clauses. Thus, they work on universally quantified disjunctions – which are more general than Horn rules. Two clauses can be combined to a more general clause using the "least general generalization" operator [57]:

Definition 24 (Least general generalization): *The least general generalization (lgg) of two clauses is computed in the following recursive manner:*

- *The lgg of two terms (i.e., either entities or variables) t and t' is t if $t = t'$ and a new variable $x_{t/t'}$ otherwise.*
- *The lgg of two negated atoms is the negation of their lgg.*
- *The lgg of $r(t_1, \ldots, t_n)$ and $r(t'_1, \ldots, t'_n)$ is $r(lgg(t_1, t'_1), \ldots, lgg(t_n, t'_n))$.*
- *The lgg of a negated atom with a positive atom is undefined.*
- *Likewise, the lgg of two atoms with different relations is undefined.*
- *The lgg of two clauses R and R' is the set of defined pair-wise generalizations:*

$$lgg(R, R') = \{lgg(l_i, l'_j) \ : \ l_i \in R, \ l'_j \in R', \ and \ lgg(l_i, l'_j) \ is \ defined\}$$

For example, let us consider the following two rules:

$$hasChild(Michelle, Sasha) \wedge marriedTo(Michelle, Barack)$$
$$\Rightarrow hasChild(Barack, Sasha) \ (R)$$
$$hasChild(Michelle, Malia) \wedge marriedTo(Michelle, x)$$
$$\Rightarrow hasChild(x, Malia) \qquad (R')$$

In the form of clauses, these are

$$\{\neg hasChild(Michelle, Sasha), \neg marriedTo(Michelle, Barack),$$
$$hasChild(Barack, Sasha)\} \quad (R)$$
$$\{\neg hasChild(Michelle, Malia), \neg marriedTo(Michelle, x),$$
$$hasChild(x, Malia)\} \qquad (R')$$

---

[5] Instead of: if a rule is not frequent, none of its refinements can be frequent.

Now, we have to compute the lgg of every atom of the first clause with every atom of the second clause. As it turns out, there are only 3 pairs where the lgg is defined:

$$lgg(\neg hasChild(Michelle, Sasha), \neg hasChild(Michelle, Malia))$$
$$= \neg lgg(hasChild(Michelle, Sasha), hasChild(Michelle, Malia))$$
$$= \neg hasChild(lgg(Michelle, Michelle), lgg(Sasha, Malia))$$
$$= \neg hasChild(Michelle, x_{Sasha/Malia})$$

$$lgg(\neg marriedTo(Michelle, Barack), \neg marriedTo(Michelle, x))$$
$$= \neg marriedTo(Michelle, x_{Barack/x})$$

$$lgg(hasChild(Barack, Sasha), hasChild(x, Malia))$$
$$= hasChild(x_{Barack/x}, x_{Sasha/Malia})$$

This yields the clause

$$\{\neg hasChild(Michelle, x_{Sasha/Malia}), \neg marriedTo(Michelle, x_{Barack/x}),$$
$$hasChild(x_{Barack/x}, x_{Sasha/Malia})\}$$

This clause is equivalent to the rule

$$hasChild(Michelle, x) \wedge marriedTo(Michelle, y) \Rightarrow hasChild(x, y)$$

Note that the generalization of two different terms in an atom should result in the same variable as the generalization of these terms in another atom. In our example, we obtain only two new variables $x_{Sasha/Malia}$ and $x_{Barack/x}$. In this way, we have generalized the two initial rules to a more general rule. This can be done systematically with an algorithm called *GOLEM*.

GOLEM. The GOLEM/RLGG algorithm [51] creates, for each positive example $e \in E^+$, the rule $\mathcal{B} \Rightarrow e$, where $\mathcal{B}$ is the background knowledge. In our case, $\mathcal{B}$ is the entire KB, and so a very long conjunction of facts. The algorithm will then generalize these rules to shorter rules. More precisely, the *relative lgg* (rlgg) of a tuple of ground atoms $(e_1, \ldots, e_n)$ is the rule obtained by computing the lgg of the rules $\mathcal{B} \Rightarrow e_1$, ..., $\mathcal{B} \Rightarrow e_n$. We will call a rlgg *valid* if it is defined and does not predict any negative example.

The algorithm starts with a randomly sampled pair of positive examples (e_1, e_2) and selects the pair for which the rlgg is valid and predicts ("covers") the most positive examples. It will then greedily add positive examples, chosen among a sample of "not yet covered positive examples", to the tuple – as long as the corresponding rlgg is valid and covers more positive examples. The resulting rule will still contain ground atoms from $\mathcal{B}$. These are removed, and the rule is output. Then the process starts over to find other rules for uncovered positive examples.

Progol and Others. More recent ILP algorithms such as Progol [49], HAIL [59], Imparo [36] and others [34,88] use inverse entailment to compute the hypothesis more efficiently. This idea is based on the observation that a hypothesis $\mathfrak{h}$ that satisfies $\mathcal{B} \wedge \mathfrak{h} \models E^+$ should equivalently satisfy $\mathcal{B} \wedge \neg E^+ \models \neg \mathfrak{h}$ (by logical contraposition). The algorithms work in two steps: they will first construct an intermediate theory F such that $\mathcal{B} \wedge \neg E^+ \models F$ and then generalize its negation $\neg F$ to the hypothesis $\mathfrak{h}$ using inverse entailment.

3.4 Related Approaches

This article cannot give a full review of the field of rule mining. However, it is interesting to point out some other approaches in other domains that deal with similar problems:

OWL. OWL is a Description logic language designed to define rules and constraints on the KB. For example, an OWL rule can say that every person must have a single birth date. Such constraints are usually defined upfront by domain experts and KB architects when they design the KB. They are then used for automatic reasoning and consistency checks. Thus, constraints *prescribe* the shape of the data, while the rules we mine *describe* the shape of the data. In other words, constraints are used deductively – instead of being found inductively. As such, they should suffer no exception. However, rule mining can provide candidate constraints to experts when they want to augment their theory [30].

Probabilistic ILP. As an extension of the classic ILP problem, Probabilistic ILP [12] aims to find the logical hypothesis $\mathfrak{h}$ that, given probabilistic background knowledge, maximizes the probability to observe a positive example, and minimizes the probability to observe a negative example. In our case, it would require a probabilistic model of the real world. Such models have been proposed for some specific use cases [38,90], but they remain an ongoing subject of research.

Graph Mining and Subgraph Discovery. Subgraph discovery is a well studied problem in the graph database community (see [27] Part 8 for a quick overview). Given a set of graphs, the task is to mine a subgraph that appears in most of them. Rule mining, in contrast, is looking for patterns that are frequent in the same graph. This difference may look marginal, but the state-of-the-art algorithms are very different and further work would be needed to determine how to translate one problem to the other.

Link Prediction. Rules can be used for link prediction, i.e., to predict whether a relation links two entities. This task can also be seen as a classification problem ([27] Part 7): given two entities, predict whether there is a relation between them. A notable work that unites both views [39] uses every conjunction of atoms (a possible body for a rule, which they call a "path") as a feature dimension for this classification problem. We will extensively present a more recent approach to this problem in Sect. 4.

3.5 Challenges in Rule Mining

Today, Horn rules can be mined efficiently on large KBs [68]. However, many challenges remain.

Negation. KBs usually do not contain negative information. Therefore, it is difficult to mine rules that have a negated atom in the body or in the head, such as $marriedTo(x, y) \land y \neq z \Rightarrow \neg marriedTo(x, z)$. Newer approaches use a variant of the PCA [55], class information [22], or new types of confidence measures [16].

External Information. Since KBs are both incomplete and lacking negative information, it is tempting to add in data from other sources to guide the rule mining. One can e.g., add in information about cardinalities [56], or embeddings computed on text [32].

Numerical Rules. We can imagine rules that detect numerical correlations (say, between the population of a city and the size of its area), bounds on numerical values (say, on the death year of Ancient Romans), or even complex numerical formulae (say, that the ratio of inhabitants of the capital is larger in city states) [18, 24, 48].

Scaling. While today's algorithms work well on large KBs, they do less well once we consider rules that do not just contain variables, but also entities. Furthermore, KBs grow larger and larger. Thus, scalability remains a permanent problem. It can be addressed, e.g., by smarter pruning strategies, parallelization, or by precomputing cliques in the graph of the KB.

4   Representation Learning

After having discussed symbolic representations of entities and rules, we now turn to subsymbolic representations. In this setting, entities are represented not as identifiers with relations, but as numerical vectors. Facts are predicted not by logical rules, but by computing a score for fact candidates.

4.1   Embedding

The simplest way to represent an entity as a vector is by a *one-hot encoding*:

Definition 25 (One-hot encoding): *Given an ordered set of objects $S = \{o_1, ... o_n\}$, the one-hot encoding of the object o_i is the vector $h(o_i) \in \mathbb{R}^n$ that contains only zeros, and a single one at position i.*

For example, in our KB in Fig. 1, we have 7 entities. We can easily order them, say alphabetically. Then, *Barack* is the first entity, and hence his one-hot encoding is $\begin{pmatrix} 1 & 0 & 0 & 0 & 0 & 0 & 0 \end{pmatrix}^T$ (where the T just means that we wrote the vector horizontally instead of vertically). Such representations are not particularly useful, because they do not reflect any semantic similarity: The vector of Barack has the same distance to the vector of Michelle as to the vector of Lisa.

Definition 26 (Embedding): *An n-dimensional embedding for a group of objects (e.g. words, entities) is an injective function that maps each object to a vector in $\mathbb{R}^n$, so that the intrinsic relations between the objects are maintained.*

For example, we want to embed the entity *Barack* in such a way that his vector is close to the vector of Michelle, or maybe to the vectors of other politicians. Embeddings can also be used for words of natural language. In that case, the goal is to find an embedding where the vectors of related words are close. For example, the vector of the word "queen" and the vector of "king" should be close to each other. An ideal word embedding would even permit arithmetic relations such as $v(king) - v(man) + v(woman) = v(queen)$ (where $v(\cdot)$ is the embedding function). This means that removing the vector for "man" from "king", and adding "woman" should yield the vector for "queen". Vectors are usually denoted with bold letters.

Embeddings are interesting mainly for two reasons: first they lower the dimensions of object representations. For example, there may be millions of entities in a KB, but they can be embedded in vectors of a few hundred dimensions. It is typically easier for down-stream tasks to deal with vectors than with sets of this size. Second, the structure of the embedding space makes it possible to compare objects that were incomparable in their original forms (e.g. it is now easy to define a distance between entities or between words by measuring the euclidean distance between their embeddings).

There are many ways to compute embeddings. A very common one is to use neural networks, as we shall discuss next.

4.2   Neural Networks

4.2.1   Architecture

We start our introduction to neural networks with the notion of an *activation function*:

Definition 27 (Activation Function): *An activation function is a non-linear real function.*

Typical examples of activation functions are the hyperbolic tangent, the sigmoid function $\sigma : x \mapsto (1 + e^{-x})^{-1}$ and the rectified linear unit function ReLU: $x \mapsto \max(0, x)$. Although these functions are defined on a single real value, they are usually applied point-wise on a vector of real values. For example, we write $\sigma(\langle x_1, ..., x_n \rangle)$ to mean $\langle \sigma(x_1), ..., \sigma(x_n) \rangle$. Neural networks consist of several layers with such activation functions:

Definition 28 (Layer): *In the context of neural networks, a layer is a function $\ell : \mathbb{R}^i \to \mathbb{R}^j$ that is either linear or the composition of a linear function and an activation function (i and j are non-zero naturals).*

Thus, a layer is a function that takes as input a vector $v \in \mathbb{R}^i$, and does two things with it. First, it applies a linear function to v, i.e., it multiplies v with a matrix $W \in \mathbb{R}^i \times \mathbb{R}^j$ (the *weight matrix*). This yields $W \cdot v \in \mathbb{R}^j$. Then, it applies the activation function to this vector, which yields again a vector of size j. We can now compose the layers to neural networks:

Definition 29 (Neural Network): *In its simplest form (that of a Multilayer perceptron), a neural network is a function $g\colon \mathbb{R}^n \to \mathbb{R}^p$, such that g is a composition of layers. The parameters p, n and the intermediate dimensions of the layers are non-zero naturals.*

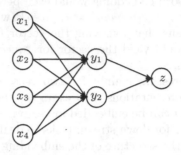

Fig. 2. Example of a one-hidden-layer network.

Figure 2 shows a one-hidden-layer network that takes as input vectors $x \in \mathbb{R}^4$ and outputs real values $z \in \mathbb{R}^1$. The function g of the network can be decomposed as $g = \ell_2 \circ \ell_1$ where $\ell_1\colon \mathbb{R}^4 \to \mathbb{R}^2$ is the hidden layer and $\ell_2\colon \mathbb{R}^2 \to \mathbb{R}^1$ is the output layer. Let us now see how such a network computes its output. Let us assume that the weight matrix of the first layer is A, that the weight matrix of the second layer is B and that the input is x:

$$x = \begin{pmatrix} 0 \\ 1 \\ 0 \\ 0 \end{pmatrix}, A = \begin{pmatrix} 0.4 & 0.5 & -0.3 & 0.1 \\ 0.8 & -0.6 & 0.4 & 0.2 \end{pmatrix}, B = \begin{pmatrix} 0.5 & -0.6 \end{pmatrix} \tag{1}$$

If both layers use the sigmoid activation function σ, we can compute the result of the first layer as $y = \ell_1(x) = \sigma(A \cdot x)$, and the result of the second layer (and thus of the entire network) as $z = \ell_2(y) = \sigma(B \cdot y)$:

$$y = \sigma \left(\begin{pmatrix} 0.2 & 0.5 & -0.3 & 0.1 \\ 0.8 & -0.6 & 0.4 & 0.2 \end{pmatrix} \cdot \begin{pmatrix} 0 \\ 1 \\ 0 \\ 0 \end{pmatrix} \right) = \sigma \left(\begin{pmatrix} 0.5 \\ -0.6 \end{pmatrix} \right) = \begin{pmatrix} \sigma(0.5) \\ \sigma(-0.6) \end{pmatrix} \approx \begin{pmatrix} 0.62 \\ 0.35 \end{pmatrix}$$

$$\tag{2}$$

$$z = \sigma \left(\begin{pmatrix} 0.5 & -0.7 \end{pmatrix} \cdot \begin{pmatrix} 0.62 \\ 0.35 \end{pmatrix} \right) = \sigma \left(\begin{pmatrix} 0.063 \end{pmatrix} \right) = \begin{pmatrix} \sigma(0.063) \end{pmatrix} \approx \begin{pmatrix} 0.51 \end{pmatrix} \tag{3}$$

Figure 3 shows this computation graphically. The function computed by the network is $g = \ell_2 \circ \ell_1 = \sigma \circ b \circ \sigma \circ a$, where a and b are linear functions defined by the matrices A and B.

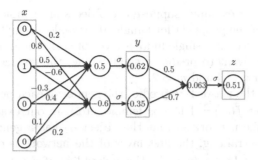

Fig. 3. Example of a one-hidden-layer network with computations.

4.2.2   Training

We want the neural network to compute an embedding of KB entities and relations. For this, we first have to understand how neural networks can perform supervised learning tasks. In supervised learning, we are interested in approximating some function $f : \mathbb{R}^p \mapsto \mathbb{R}^k$. We do not know f. We only know some datapoints of f: $\{(x_i, \alpha_i) \in \mathbb{R}^p \times \mathbb{R}^k | i = 0, \dots, n\}$, with $f(x_i) = \alpha_i$ for any i. The goal is to find the best estimation of f. With our notations, we would like to find the neural network whose function $\hat{f}$ approximates f.

We first decide on an architecture of the neural network. We already know how many input nodes it has (namely p), and how many output nodes it has (namely k). We just have to decide how many hidden layers it has (typically a single one for simple tasks), and what the activation functions are (the sigmoid function is a popular choice). Then, we initialize the weight matrices of the layers randomly. Training the network model now means adapting the weight matrices for each layer so that $\hat{f}(x_i) = \alpha_i$ for all datapoints (or at least for as many as possible). This is achieved using gradient descent: for each sample x_i of the dataset, a loss is computed (comparing the output $\hat{f}(x_i)$ to the true value α_i) and the weights are updated in the opposite direction of the gradient of the sum of the losses with respect to the weights of the network.

Interestingly, neural networks can be trained for many functions between vectors. It has been proven that the range of functions neural networks can approximate is very large and grows very rapidly with the depth (number of layers) of the network [74]. This is a big strength of these models.

4.2.3   Embeddings

Let us now see how we can use neural networks to compute an m-dimensional embedding function for a set of objects $S = \{o_1, ..., o_n\}$. The input to the network will be the one-hot encoding of the object, i.e., we need n input nodes. In the ideal case, the neural network would directly output a vector of size m (the embedding). Then, however, we would not know how to train the network, because we have no given embeddings to compare to. Therefore, we use a trick: We do not let the network compute the embedding directly, but a function whose output we know. For example, suppose the objects are people, and suppose we know the gender of the people (1 for female, 0 for male, or anything in between). We build a network with a single hidden layer of size m and an output layer of size 1 (because we want to predict a single value, the gender). Figure 2 shows such a network for $n = 4$ and $m = 2$.

Then we train the network to predict the gender of each person (i.e., we find the weights so that $\hat{f}(o_i) = 1$ if o_i is female, etc.). In our example from Fig. 3, we have trained the network so that the object o_2 has a gender value of 0.51. Interestingly, after training, the first layer of the network is often a very good embedding function. In our example, the embedding of $x = \begin{pmatrix} 0 & 1 & 0 & 0 \end{pmatrix}^T$ would be $y = \begin{pmatrix} 0.62 & 0.35 \end{pmatrix}^T$.

Why is that a good choice? We first observe that the embedding function has the right dimensions: it maps a one-hot encoded vector of dimension n to a vector of dimension m, as desired ($v : \mathbb{R}^n \mapsto \mathbb{R}^m$). Then, we observe that the second layer (which computes the gender) bases its computation purely on the outputs of the first layer (the sigmoid of the embedding). Therefore, the output of the hidden layer provided enough information to reconstitute the gender, i.e., our embedding maintains the crucial information. Selecting a hidden layer as embedding comes down to dividing the network in two parts: the first layer computes features (the components of y) that should capture the information relevant for the application it is trained on; the second layer computes the value $\hat{f}(x)$ using only those extracted features contained in y. This division makes it intuitive that if the training task is well-chosen, the computed features should capture interesting aspects of the data and constitute a good embedding candidate. Note that even if we are interested only in y it is still necessary to train the entire network as we can only evaluate the performance of the embedding by comparing $\hat{f}$ to f.

The method that creates an embedding (in our case: a neural network) is often called a *model*. We will now see how to create models for facts in KBs.

4.3   Knowledge Base Embeddings

If we want to embed a KB, we can either embed entities, relations, or facts. Most models in the literature embed entities and relations together. These models take as input a fact of a subject s, a relation r and an object o as one-hot encoded vectors, which are concatenated together to one long vector with three 1s. Let's take as example the knowledge base from Fig. 1. This KB has 7 entities (*Barack, Michelle, Sasha, Malia, Elvis, Priscilla Lisa*) and 2 relations (*marriedTo, hasChild*). To feed the fact *marriedTo(Barack, Michelle)* into the model, we create the one-hot encoded vectors and concatenate them to one long vector: $((1\,0\,0\,0\,0\,0\,0),(1\,0),(0\,1\,0\,0\,0\,0\,0))^T$. The output of the model will be a scoring function:

Definition 30 (Scoring Function): *In the context of knowledge base embeddings, a scoring function maps a fact $r(s, o)$ to a real-valued score.*

The score of a fact is an estimate of the true theoretical and unknown function deciding whether the fact is true. Obviously, the score should be high for the facts in a correct KB (Sect. 2.4.1). In certain probabilistic contexts, the score can be interpreted as the likelihood of the fact to be true. We denote the scoring function of the fact $r(s, o)$ by $f_r(s, o)$.

As for the embeddings we already saw, models are divided in two parts: the first one which links the one-hot encoded vectors to the embeddings r, s and o and the second part which computes $f_r(s, o)$. We now have to train the model to predict whether an input fact is true (has a high score) or not (has a low score). Once the model is trained, we will be able to read off the embeddings from the weight matrix of one of the hidden layers.

Let us now see where we can find training data. For the true facts, the KB obviously provides lots of datapoints. However, if we just train the model on positive facts, it will just learn to always predict a high score. This is the same problem we already saw in Sect. 3.2.2. Therefore, we also need to provide negative facts:

Definition 31 (Negative fact): *Given a fact $r(s, o)$ from a KB, a negative fact is a statement $r(s', o')$ that is not in the KB.*

The process of generating negative fact is called *negative sampling* and is detailed in Sect. 4.3.1. For now, let us just assume that we have such negative facts at our disposal. To train the network, we use a loss function:

Definition 32 (Loss function): *A loss function ℓ is a function from $\mathbb{R}^2$ to $\mathbb{R}$.*

We will apply the loss function to the score $f_r(s, o)$ that the network computed for a true fact $r(s, o)$ and the score $f_r(s', o')$ that the network computed for a negative fact $r(s', o')$. Naturally, the two scores should be very different: the first score should be high, and the second one should be low. If the two scores are close, the network is not trained well. Therefore, the loss function $\ell(x_1, x_2)$ should be larger the closer x_1 and x_2 are. The logistic loss or the margin loss are usual examples. They are defined respectively in Eqs. 4 and 5, where γ is a parameter, and $\eta_z = 1$ if z is the score for a positive example and $\eta_z = -1$ if z is the score for a negative example:

$$(x, y) \mapsto \log(1 + \exp(-\eta_x \times x)) + \log(1 + \exp(-\eta_y \times y)) \qquad (4)$$
$$(x, y) \mapsto \max(0, \gamma + \eta_x \times x + \eta_y \times y) \qquad (5)$$

Definition 33 (Training): *Training a knowledge base embedding model is finding the best parameters of the model (and then the best embeddings) so that the scoring function $f_r(s, o)$ is maximized for true facts and minimized for negative ones.*

Training is done by minimizing the sum of loss functions by gradient descent over a training set of facts. The sum is usually computed as follows, where $r(s, o)$

is a fact, $r(s', o')$ is a negative fact generated from $r(s, o)$ (c.f. Sect. 4.3.1), and ℓ a loss function:

$$\mathcal{L} = \sum_{(s,r,o) \in \mathcal{K}} \ell\left(f_r(s, o), f_r(s', o')\right) \tag{6}$$

4.3.1 Negative Sampling

Let us now see how we can generate the negative facts for our model. Feeding negative samples to the model is vital during training. If the model was only trained on true samples, then it could minimize any loss by trivially returning a large score for any fact it is fed with. This is the same problem that we already saw in Sect. 3.2.2, and in principle the same considerations and methods apply here as well. In the context of knowledge base embeddings, the generation of negative facts is usually done by *negative sampling*. Negative sampling is the process of corrupting a true fact's subject or object in order to create a wrong statement. This is very related to the Partial Completeness Assumption that we already saw in Sect. 3.2.2: If we have a fact *hasChild(Michelle, Sasha)*, then any variant of this fact with a different object is assumed to be a wrong statement – unless it is already in the KB. For example, we would generate the negative facts *hasChild(Michelle, Elvis)* and *hasChild(Michelle, Barack)*.

It has been observed that the quality of the resulting embedding highly depends on the quality of the negative sampling. Thus, we have to choose wisely which facts to generate. Intuitively, negative samples introduce repulsive forces in the embedding space so that entities that are not interchangeable in a fact should have embeddings far away from each other. It is of course easy to generate negative facts, simply by violating type constraints. For example, we can generate *hasChild(Michelle, USA)*, which is certain to be a false statement (due to the domain and range constraints, see Sect. 2.3.2). Then, however, we run into the *zero loss problem* [78]: The model learns to compute a low score for statements that are so unrealistic that they are not of interest anyway. It will not learn to compute a low score for statements such as *hasChild(Michelle, Lisa)*, which is the type of statements that we are interested in.

We thus have to choose negative facts that are as realistic as possible. As training goes on, we will provide the model with negative samples that are closer and closer to true facts, in order to adjust in a finer way the embeddings. To this end, various methods have been presented. One of them uses rules on the types of entities in order to avoid impossible negative facts such as *hasChild(Michelle, USA)* [42]. Another more complex method is adversarial negative sampling [78].

4.3.2 Shallow Models

Shallow models rely on the intuition that for a given fact $r(s, o)$, we would like the vectors $s+r$ and o to be close in the embedding space. For example, we would like the vectors $\boldsymbol{Barack} + \boldsymbol{marriedTo}$ and $\boldsymbol{Michelle}$ to be close. There are two ways to define "close": The vectors can have a small vector difference (which is what translational models aim at) or they can have a small angle between them (which is what semantic-matching models do). The simplest translational model is TransE [8]. Its scoring function is simply the opposite of the distance between $s + r$ and o: $f_r(s, o) = -||s + r - o||$ (where $|| \cdot ||$ is either the 1-norm or the 2-norm). Maximizing this scoring function for true facts and minimizing it for negative facts leads to embeddings that should verify the simple arithmetic equation $s + r \approx o$.

Let us now see how a network can be made to compute this score. As an example, let us embed in $\mathbb{R}^3$ the KB of Fig. 1 with TransE and the 2-norm. The network to this end is shown in Fig. 4.

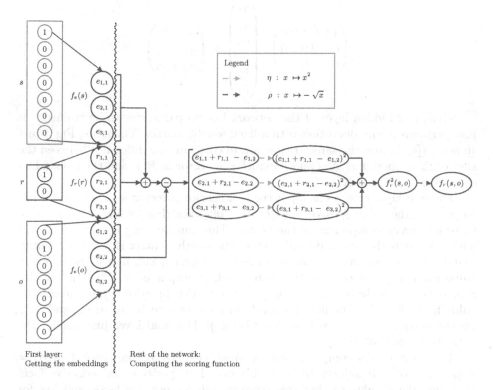

First layer: Rest of the network:
Getting the embeddings   Computing the scoring function

Fig. 4. Graphical representation of the TransE Model (with 2-norm) applied to the fact *marriedTo(Barack,Michelle)* for Fig. 1, with $s = (1\ 0\ 0\ 0\ 0\ 0\ 0)^T$, $o = (0\ 1\ 0\ 0\ 0\ 0\ 0)^T$ and $r = (1\ 0)^T$.

The network takes as input a fact $r(s, o)$, i.e., the concatenation of the one-hot encodings of s, r, and o. The first layer computes the embeddings of these items. The trick is that we will use not a single weight matrix for the first layer, but two: One weight matrix $E \in \mathbb{R}^{3 \times 7}$ to compute the embedding of entities s and o and one weight matrix $R \in \mathbb{R}^{3 \times 2}$ to compute the embedding of relations r. Thus, when we train the network, we will learn the same matrix (and thus the same embeddings) for entities independently of their roles (as subject or object of the facts). We use no activation function in the first layer (or the identity function but it is not really an activation function as it is linear). Thus, the first layer computes simply, for a given fact $r(s, o)$, the embeddings $E \cdot h(s)$, $R \cdot h(r)$ and $E \cdot h(o)$ where h is the one-hot encoding function.

The next layer of the network will add the embeddings of s and r, i.e., it will take as input two 3-dimensional vectors, and produce as output a single 3-dimensional vector. This can be done by a simple matrix multiplication of the concatenated vectors with a fixed matrix, as shown here:

$$
\begin{pmatrix} 1\,0\,0\,1\,0\,0 \\ 0\,1\,0\,0\,1\,0 \\ 0\,0\,1\,0\,0\,1 \end{pmatrix} \cdot \begin{pmatrix} a_1 \\ a_2 \\ a_3 \\ b_1 \\ b_2 \\ b_3 \end{pmatrix} = \begin{pmatrix} a_1 + b_1 \\ a_2 + b_2 \\ a_3 + b_3 \end{pmatrix} \tag{7}
$$

Thus, this hidden layer of the network has no parameter to be trained – it just performs a sum operation with a fixed weight matrix. Therefore, Fig. 4 just shows a $\oplus$. We use the same trick again to compute the difference between the sum of the embeddings of s and r (which is a vector in $\mathbb{R}^3$), and the embedding of o (also a vector in $\mathbb{R}^3$).

The next layer receives as input the 3-dimensional vector $s + r - o$, and it has to produce the value $f_r(s, o) = -\|s + r - o\|_2$. For this purpose, we first have to square every component of the vector. This can be done by a hidden layer with 3 input nodes and 3 output nodes. The weight matrix is just the identity matrix: It has only zeroes, and one's in the diagonal, i.e., it just passes every value from an input node to the corresponding output node. This matrix also remains constant during the training. The activation function η just squares the value for each node. The next hidden layer receives a vector of size 3, and adds up the components (as we have done before). The final layer just applies the activation function $\rho : x \mapsto -\sqrt{x}$.

Thus, the model computes, for an input fact $r(s, o)$, the score $f_r(s, o) = -\|s + r - o\|_2$. It suffices to train this model on positive and negative facts to adjust the weights so that this score is high for positive facts, and low for negative ones.

Limitations of the TransE Model. Although TransE yields promising results, it does not work well for one-to-many, many-to-one or many-to-many relations. As an example, consider the relation *wrote*. This is a one-to-many relation because one author can write several books: the facts *wrote(Albert Camus, La Peste)* and (*wrote(Albert Camus, L'étranger)* are both true. The TransE approach tends to return the same embeddings for the books of an author (i.e., *La Peste* and *L'étranger*), because they are almost interchangeable for the loss function, they have very few facts to distinguish them.

This limitation has been overcome by adding a relation-specific projection step. Entities should be projected into a relation-specific subspace before the translation happens. Let p_r be a relation-specific projection, i.e., a linear function from entity embeddings to vectors of possibly smaller size (projecting into a subspace). Then we want $p_r(s)+r$ and $p_r(o)$ to be close (in distance or in similarity). Since a projection is a linear operation, it just adds another layer in the network after the first one (which returns the embeddings) and before the computation of the scoring function starts.

If, in our previous example, all entities are projected into a subspace that is specific to *wrote*, the projections of the books could all be the same. We can thus arrive at the desired arithmetic relations between the projected entities without forcing all books to have the same embedding. This idea has given rise to various refinements of the TransE model, depending on the projections considered. Here are a couple of the methods that followed TransE: TransH (projections on hyperplanes), TransR (projections on any type of subspace), TransD (projections with dynamic mapping matrices built both from entities and relations) [35, 43, 79].

We mainly presented translational models because they are the most intuitive ones, but semantic models achieve good results as well. The first such model (RESCAL [54]) lacks some representation capacity (as TransE) and was later refined to more complex models such as DistMul, HolE, ComplEx, ANALOGY [45, 53, 75, 89].

4.3.3   Deep Models
Deeper architectures have also been introduced with the hope that hidden layers can capture more complex interaction patterns between entities and relations (and then estimate more complex scoring functions). In such models, the first part of the network (which, in shallow networks, just maps facts to their embeddings or their projections) now adds additional layers (possibly numerous) that receive as inputs the embeddings, and produce as outputs some extracted features. The second part of the network now computes the scoring function from the features extracted by the first part of the network and not directly from the embedding (or its projection) as in shallow models. The scoring function also becomes a parameter of the model (to be trained) and is not defined a priori anymore (in TransE for example it was only a distance between $s + r$ and o). Note that we often loose the interpretability of the scoring function in this process. Examples of such methods are SME, NTN, and MLP [7, 15, 64] and more recent ones that include convolutional structures ConvE, ConvKB [14, 52].

4.3.4   Fact Prediction
Commonly, authors compare the performance of their embedding methods on two tasks: fact checking and link prediction. Fact checking is simply deciding whether a given fact is true or false. To see how the embedding performs on this task, we train the network on a portion of the KB (i.e., on a subset of the facts) in which all entities and relations appear at least once. Then, we check for each fact from the KB that the network has not seen during training whether the network computes a score that is higher than a threshold (i.e., whether the

network correctly assumes the fact to be true). The thresholds are determined on a validation set extracted from the training set.

Link prediction is a bit more complex. Again, we train the network on a portion of the KB. For a given fact $r(s, o)$ that the network did not see during training, the value of the scoring function $f_r(s, e)$ (resp. $f_r(e, o)$) is computed with the model for all entities e. This allows ranking candidate entities by decreasing order of scoring function. Then we count the share of unseen facts that the model manages to recover when the object (resp. subject) is hidden. Such evaluations are usually done under the Closed World Assumption (Sect. 2.4.1): If the network predicts a fact that is not in the KB, this is counted against the network – although the fact may be true but just unknown to the KB.

Thus, link prediction amounts to predicting facts – much like rules predict facts (Sect. 3.1). The difference is that rules are explicit: they tell us which circumstances lead to a prediction. Networks are not: they deduce new facts from the overall similarity of the facts. Another difference is that a rule does not know about the other rules: The confidence of a prediction does not increase if another rule makes the same prediction. Networks, in contrast, combine evidence from all types of correlations, and may thus assign a higher score to a fact for which is has more evidence.

4.4   Challenges in Representation Learning

While representation learning for knowledge graphs has made big advances these recent years, some challenges remain to be tackled:

Generalization of Performances. Current models tend to have performances that do not generalize well from one dataset to the other. Most methods are heuristics executing a more or less intuitive approach. A theory that could explain the variation of performance is missing.

Negative Sampling. Finding realistic negative facts remains a challenge in knowledge base embedding – much like in rule mining (Sect. 3.5). Here, we could use logical constraints. For example, if we know that Lisa cannot have more than two parents, then we could use *hasChild(Michelle, Lisa)* as a negative fact.

Dealing with Literals. Most current methods consider literals as monolithic entities. Thus, they are unable to see, e.g., that the date "2019-01-01" is close to the date "2018-12-31", or that the number "99" is close to the number "100". Such knowledge could lead to more accurate fact scoring functions.

Scalability. The development of massive KBs such as Wikidata requires algorithms to be able to scale. There is still room for improvement here: Embedding methods are usually tested on the FB15k dataset, which counts only 500 thousand facts – while Wikidata counts more than 700 million.

5   Conclusion

In this article, we have investigated how entities, relations, and facts in a knowledge base can be represented. We have seen the standard knowledge representation model of instances and classes. We have also seen an alternative representation of entities, as embeddings in a vector space. We have then used these representations to predict new facts – either through logical rules (by the help of rule mining), or through link prediction (with the help of neural networks).

Many challenges remain: the knowledge representation of today's KBs remains limited to subject-relation-object triples (Sect. 2.6). In Rule Mining, we have only just started looking beyond Horn Rules (Sect. 3.5). In Knowledge Base Embeddings, we have to learn how to generate more realistic negative examples (Sect. 4.4). This is one of the areas where the Semantic Web community and the Machine Learning community can have fruitful interchanges.

A   Computation of Support and Confidence

Notation. Given a logical formula ϕ with some free variables $x_1, \ldots, x_n$, all other variables being by default *existentially* quantified, we define:

$$\#(x_1, \ldots, x_n) : \phi \quad := \quad |\{(x_1, \ldots, x_n) : \phi(x_1, \ldots, x_n) \text{ is } true\}|$$

We remind the reader of the two following definitions:

Definition 14 (Prediction of a rule): *The predictions P of a rule $B \Rightarrow h$ in a KB $\mathcal{K}$ are the head atoms of all instantiations of the rule where the body atoms appear in $\mathcal{K}$. We write $\mathcal{K} \wedge (B \Rightarrow h) \models P$.*

Definition 19 (Support): *The support of a rule in a KB is the number of positive examples predicted by the rule.*

A prediction of a rule is a positive example if and only if it is in the KB. This observation gives rise to the following property:

Proposition 34 (Support in practice): *The support of a rule $B \Rightarrow h$ is the number of instantiations of the head variables that satisfy the query $B \wedge h$. This value can be written as:*

$$support(B \Rightarrow h(x,y)) = \#(x,y) : B \wedge h(x,y)$$

Definition 20 (Confidence): *The confidence of a rule is the number of positive examples predicted by the rule (the support of the rule), divided by the number of examples predicted by the rule.*

Under the CWA, all the predicted examples are either positive examples or negative examples. Thus, the standard confidence of a rule is the support of the rule divided by the number of prediction of the rule, written:

$$std\text{-}conf(B \Rightarrow h(x,y)) = \frac{\#(x,y) : B \wedge h(x,y)}{\#(x,y) : B}$$

Assume h is more functional than inverse functional. Under the PCA, a predicted negative example is a prediction $h(x, y)$ that is not in the KB, such that, for this x there exists another entity y' such that $h(x, y')$ is in the KB. When we add the predicted positive examples, the denominator of the PCA confidence becomes:

$$\#(x, y) : (\boldsymbol{B} \wedge h(x, y)) \vee (\boldsymbol{B} \wedge \neg h(x, y) \wedge \exists y'.h(x, y'))$$

We can simplify this logical formula to deduce the following formula for computing the PCA confidence:

$$pca\text{-}conf(\boldsymbol{B} \Rightarrow h(x, y)) = \frac{\#(x, y) : \boldsymbol{B} \wedge h(x, y)}{\#(x, y) : \boldsymbol{B} \wedge \exists y'.h(x, y')}$$

References

1. Agrawal, R., Srikant, R., et al.: Fast algorithms for mining association rules. In: Proceedings of the 20th International Conference on Very Large Data Bases, VLDB, vol. 1215, pp. 487–499 (1994)
2. Aho, A.V., Garey, M.R., Ullman, J.D.: The transitive reduction of a directed graph. SIAM J. Comput. **1**(2), 131–137 (1972)
3. Baader, F., Calvanese, D., McGuinness, D.L., Nardi, D., Patel-Schneider, P.F. (eds.): The Description Logic Handbook. Cambridge University Press, Cambridge (2003)
4. Bienvenu, M., Deutch, D., Suchanek, F.M.: Provenance for web 2.0 data. In: Jonker, W., Petković, M. (eds.) SDM 2012. LNCS, vol. 7482, pp. 148–155. Springer, Heidelberg (2012). https://doi.org/10.1007/978-3-642-32873-2_10
5. Bizer, C., Heath, T., Idehen, K., Berners-Lee, T.: Linked data on the web. In: WWW (2008)
6. Bollacker, K., Evans, C., Paritosh, P., Sturge, T., Taylor, J.: Freebase: a collaboratively created graph database for structuring human knowledge. In: SIGMOD (2008)
7. Bordes, A., Glorot, X., Weston, J., Bengio, Y.: A semantic matching energy function for learning with multi-relational data. Mach. Learn. **94**(2), 233–259 (2014)
8. Bordes, A., Usunier, N., Garcia-Duran, A., Weston, J., Yakhnenko, O.: Translating embeddings for modeling multi-relational data. In: Burges, C.J.C., Bottou, L., Welling, M., Ghahramani, Z., Weinberger, K.Q. (eds.) Advances in Neural Information Processing Systems, vol. 26, pp. 2787–2795. Curran Associates Inc. (2013)
9. Carlson, A., Betteridge, J., Kisiel, B., Settles, B., Hruschka Jr., E., Mitchell, T.: Toward an architecture for never-ending language learning. In: AAAI (2010)
10. Carroll, J.J., Bizer, C., Hayes, P., Stickler, P.: Named graphs, provenance and trust. In: WWW (2005)
11. Chen, Y., Wang, D.Z., Goldberg, S.: Scalekb: scalable learning and inference over large knowledge bases. In: VLDBJ (2016)
12. De Raedt, L., Kersting, K.: Probabilistic inductive logic programming. In: De Raedt, L., Frasconi, P., Kersting, K., Muggleton, S. (eds.) Probabilistic Inductive Logic Programming. LNCS (LNAI), vol. 4911, pp. 1–27. Springer, Heidelberg (2008). https://doi.org/10.1007/978-3-540-78652-8_1

13. Dehaspe, L., De Raedt, L.: Mining association rules in multiple relations. In: Lavrač, N., Džeroski, S. (eds.) ILP 1997. LNCS, vol. 1297, pp. 125–132. Springer, Heidelberg (1997). https://doi.org/10.1007/3540635149_40
14. Dettmers, T., Minervini, P., Stenetorp, P., Riedel, S.: Convolutional 2D knowledge graph embeddings. In: Proceedings of the 32nd AAAI Conference on Artificial Intelligence (AAAI 2018), New Orleans, LA, USA, vol. 32, February 2018. arXiv: 1707.01476
15. Dong, X.L., et al.: Knowledge vault: a web-scale approach to probabilistic knowledge fusion. In: The 20th ACM SIGKDD International Conference on Knowledge Discovery and Data Mining, KDD 2014, New York, NY, USA, 24–27 August 2014, pp. 601–610 (2014)
16. Duc Tran, M., d'Amato, C., Nguyen, B.T., Tettamanzi, A.G.B.: Comparing rule evaluation metrics for the evolutionary discovery of multi-relational association rules in the semantic web. In: Castelli, M., Sekanina, L., Zhang, M., Cagnoni, S., García-Sánchez, P. (eds.) EuroGP 2018. LNCS, vol. 10781, pp. 289–305. Springer, Cham (2018). https://doi.org/10.1007/978-3-319-77553-1_18
17. Etzioni, O., et al.: Web-scale information extraction in knowitall. In: WWW (2004)
18. Fanizzi, N., d'Amato, C., Esposito, F., Minervini, P.: Numeric prediction on owl knowledge bases through terminological regression trees. Int. J. Semant. Comput. 6(04), 429–446 (2012)
19. Fellbaum, C. (ed.): WordNet: An Electronic Lexical Database. MIT Press, Cambridge (1998)
20. Ferrucci, D., et al.: Building Watson: an overview of the DeepQA project. AI Mag. 31(3), 59–79 (2010)
21. Fisher, M.D., Gabbay, D.M., Vila, L.: Handbook of Temporal Reasoning in Artificial Intelligence. Elsevier, Amsterdam (2005)
22. Gad-Elrab, M.H., Stepanova, D., Urbani, J., Weikum, G.: Exception-enriched rule learning from knowledge graphs. In: Groth, P., et al. (eds.) ISWC 2016. LNCS, vol. 9981, pp. 234–251. Springer, Cham (2016). https://doi.org/10.1007/978-3-319-46523-4_15
23. Galárraga, L., Razniewski, S., Amarilli, A., Suchanek, F.M.: Predicting completeness in knowledge bases. In: WSDM (2017)
24. Galárraga, L., Suchanek, F.M.: Towards a numerical rule mining language. In: AKBC Workshop (2014)
25. Galárraga, L., Teflioudi, C., Hose, K., Suchanek, F.M.: AMIE: association rule mining under incomplete evidence in ontological knowledge bases. In: WWW (2013)
26. Galárraga, L., Teflioudi, C., Hose, K., Suchanek, F.M.: Fast rule mining in ontological knowledge bases with AMIE+. In: VLDBJ (2015)
27. Getoor, L., Diehl, C.P.: Link mining: a survey. ACM SIGKDD Explor. Newsl. 7(2), 3–12 (2005)
28. Gutierrez, C., Hurtado, C.A., Vaisman, A.: Introducing time into RDF. IEEE Trans. Knowl. Data Eng. 19(2), 207–218 (2007)
29. Hawthorne, J.: Inductive logic. In: Zalta, E.N. (ed.) The Stanford Encyclopedia of Philosophy. Metaphysics Research Lab, Stanford University, spring 2018 edition (2018)
30. Hellmann, S., Lehmann, J., Auer, S.: Learning of owl class descriptions on very large knowledge bases. Int. J. Semant. Web Inf. Syst. (IJSWIS) 5(2), 25–48 (2009)
31. Henderson, L.: The problem of induction. In: Zalta, E.N. (ed.) The Stanford Encyclopedia of Philosophy. Metaphysics Research Lab, Stanford University, spring 2019 edition (2019)

32. Ho, V.T., Stepanova, D., Gad-Elrab, M.H., Kharlamov, E., Weikum, G.: Rule learning from knowledge graphs guided by embedding models. In: Vrandečić, D., et al. (eds.) ISWC 2018. LNCS, vol. 11136, pp. 72–90. Springer, Cham (2018). https://doi.org/10.1007/978-3-030-00671-6_5

33. Hoffart, J., Suchanek, F.M., Berberich, K., Weikum, G.: Yago2: a spatially and temporally enhanced knowledge base from wikipedia. Artif. Intell. **194**, 28–61 (2013)

34. Inoue, K.: Induction as consequence finding. Mach. Learn. **55**(2), 109–135 (2004)

35. Ji, G., He, S., Xu, L., Liu, K., Zhao, J.: Knowledge graph embedding via dynamic mapping matrix. In: Proceedings of the 53rd Annual Meeting of the Association for Computational Linguistics and the 7th International Joint Conference on Natural Language Processing (Vol. 1: Long Papers), pp. 687–696, Beijing, China. Association for Computational Linguistics, July 2015

36. Kimber, T., Broda, K., Russo, A.: Induction on failure: learning connected horn theories. In: Erdem, E., Lin, F., Schaub, T. (eds.) LPNMR 2009. LNCS (LNAI), vol. 5753, pp. 169–181. Springer, Heidelberg (2009). https://doi.org/10.1007/978-3-642-04238-6_16

37. Krötzsch, M., Marx, M., Ozaki, A., Thost, V.: Attributed description logics: reasoning on knowledge graphs. In: IJCAI (2018)

38. Lajus, J., Suchanek, F.M.: Are all people married? Determining obligatory attributes in knowledge bases. In: WWW (2018)

39. Lao, N., Mitchell, T., Cohen, W.W.: Random walk inference and learning in a large scale knowledge base. In: Proceedings of the Conference on Empirical Methods in Natural Language Processing, pp. 529–539. Association for Computational Linguistics (2011)

40. Lehmann, J., et al.: DBpedia - a large-scale, multilingual knowledge base extracted from wikipedia. Semant. Web J. **6**(2), 167–195 (2015)

41. Lenat, D.B., Guha, R.V.: Building Large Knowledge-Based Systems; Representation and inference in the Cyc Project. Addison-Wesley Longman Publishing Co. Inc., Boston (1989)

42. Lerer, A., et al.: PyTorch-BigGraph: a large-scale graph embedding system. In: Proceedings of The Conference on Systems and Machine Learning, March 2019. arXiv: 1903.12287

43. Lin, Y., Liu, Z., Sun, M., Liu, Y., Zhu, X.: Learning entity and relation embeddings for knowledge graph completion. In: Twenty-Ninth AAAI Conference on Artificial Intelligence, February 2015

44. Liu, H., Singh, P.: Conceptnet. BT Tech. J. **22**(4), 211–226 (2004)

45. Liu, H., Wu, Y., Yang, Y.: Analogical inference for multi-relational embeddings. In: Precup, D., Teh, Y.W. (eds.) Proceedings of the 34th International Conference on Machine Learning, International Convention Centre, Sydney, Australia, 06–11 Aug 2017, vol. 70, pp. 2168–2178. PMLR (2017)

46. Margolis, E., Laurence, S.: Concepts. In: Zalta, E.N. (ed.) The Stanford Encyclopedia of Philosophy. Stanford (2014)

47. Marx, M., Krötzsch, M., Thost, V.: Logic on mars: ontologies for generalised property graphs. In: IJCAI (2017)

48. Melo, A., Theobald, M., Völker, J.: Correlation-based refinement of rules with numerical attributes. In: FLAIRS (2014)

49. Muggleton, S.: Inverse entailment and progol. New Gener. Comput. **13**(3–4), 245–286 (1995)

50. Muggleton, S., De Raedt, L.: Inductive logic programming: theory and methods. J. Log. Program. **19**, 629–679 (1994)

51. Muggleton, S., Feng, C.: Efficient induction of logic programs. Citeseer (1990)
52. Nguyen, D.Q., Nguyen, T.D., Nguyen, D.Q., Phung, D.: A novel embedding model for knowledge base completion based on convolutional neural network. In: Proceedings of the 2018 Conference of the North American Chapter of the Association for Computational Linguistics: Human Language Technologies, vol. 2 (Short Papers), pp. 327–333 (2018). arXiv: 1712.02121
53. Nickel, M., Rosasco, L., Poggio, T.: Holographic embeddings of knowledge graphs. In: Proceedings of the Thirtieth AAAI Conference on Artificial Intelligence, pp. 1955–1961, February 2016
54. Nickel, M., Tresp, V., Kriegel, H.-P.: A three-way model for collective learning on multi-relational data. In: Proceedings of the 28th International Conference on International Conference on Machine Learning, ICML 2011, pp. 809–816. Omnipress, Bellevue 92011)
55. Ortona, S., Meduri, V.V., Papotti, P.: Robust discovery of positive and negative rules in knowledge bases. In: ICDE (2018)
56. Pellissier Tanon, T., Stepanova, D., Razniewski, S., Mirza, P., Weikum, G.: Completeness-aware rule learning from knowledge graphs. In: d'Amato, C., et al. (eds.) ISWC 2017. LNCS, vol. 10587, pp. 507–525. Springer, Cham (2017). https:// doi.org/10.1007/978-3-319-68288-4_30
57. Plotkin, G.: Automatic methods of inductive inference (1972)
58. Ponzetto, S., Navigli, R.: BabelNet: the automatic construction, evaluation and application of a wide-coverage multilingual semantic network. Artif. Intell. **193**, 217–250 (2012)
59. Ray, O., Broda, K., Russo, A.: Hybrid abductive inductive learning: a generalisation of progol. In: Horváth, T., Yamamoto, A. (eds.) ILP 2003. LNCS (LNAI), vol. 2835, pp. 311–328. Springer, Heidelberg (2003). https://doi.org/10.1007/978-3-540-39917-9_21
60. Razniewski, S., Suchanek, F.M., Nutt, W.: But what do we actually know? In: AKBC Workshop (2016)
61. Russell, B.: The Problems of Philosophy. Barnes & Noble, New York City (1912)
62. Russell, S., Norvig, P.: Artificial Intelligence: A Modern Approach. Prentice Hall, Upper Saddle River (2002)
63. Shapiro, E.Y.: Inductive inference of theories from facts. Yale University, Department of Computer Science (1981)
64. Socher, R., Chen, D., Manning, C.D., Ng, A.: Reasoning with neural tensor networks for knowledge base completion. In: Burges, C.J.C., Bottou, L., Welling, M., Ghahramani, Z., Weinberger, K.Q. (eds.) Advances in Neural Information Processing Systems, vol. 26, pp. 926–934. Curran Associates Inc. (2013)
65. Soulet, A., Giacometti, A., Markhoff, B., Suchanek, F.M.: Representativeness of knowledge bases with the generalized Benford's law. In: Vrandečić, D., et al. (eds.) ISWC 2018. LNCS, vol. 11136, pp. 374–390. Springer, Cham (2018). https://doi. org/10.1007/978-3-030-00671-6_22
66. Sowa, J.F.: Knowledge Representation: Logical, Philosophical, and Computational Foundations. Brooks/Cole, Boston (2000)
67. Staab, S., Studer, R. (eds.): Handbook on Ontologies. International Handbooks on Information Systems. Springer, Heidelberg (2004). https://doi.org/10.1007/978-3-540-92673-3
68. Stepanova, D., Gad-Elrab, M.H., Ho, V.T.: Rule induction and reasoning over knowledge graphs. In: d'Amato, C., Theobald, M. (eds.) Reasoning Web 2018. LNCS, vol. 11078, pp. 142–172. Springer, Cham (2018). https://doi.org/10.1007/978-3-030-00338-8_6

69. Suchanek, F.M., Abiteboul, S., Senellart, P.: Paris: probabilistic alignment of relations, instances, and schema. In: VLDB (2012)
70. Suchanek, F.M., Kasneci, G., Weikum, G.: Yago - a core of semantic knowledge. In: WWW (2007)
71. Suchanek, F.M., Preda, N.: Semantic culturomics. In: VLDB Short Paper Track (2014)
72. Tandon, N., de Melo, G., De, A., Weikum, G.: Knowlywood: mining activity knowledge from hollywood narratives. In: CIKM (2015)
73. Tandon, N., de Melo, G., Suchanek, F.M., Weikum, G.: WebChild: harvesting and organizing commonsense knowledge from the web. In: WSDM (2014)
74. Telgarsky, M.: Representation benefits of deep feedforward networks. arXiv [cs], September 2015. arXiv: 1509.08101
75. Trouillon, T., Nickel, M.: Complex and holographic embeddings of knowledge graphs: a comparison. arXiv [cs, stat], July 2017. arXiv: 1707.01475
76. Udrea, O., Recupero, D.R., Subrahmanian, V.S.: Annotated rdf. ACM Trans. Comput. Logic **11**(2), 10 (2010)
77. Vrandečić, D., Krötzsch, M.: Wikidata: a free collaborative knowledgebase. Commun. ACM **57**(10), 78–85 (2014)
78. Wang, P., Li, S., Pan, R.: Incorporating GAN for negative sampling in knowledge representation learning. In: Thirty-Second AAAI Conference on Artificial Intelligence, April 2018
79. Wang, Z., Zhang, J., Feng, J., Chen, Z.: Knowledge graph embedding by translating on hyperplanes. In: Twenty-Eighth AAAI Conference on Artificial Intelligence, June 2014
80. Welty, C., Fikes, R., Makarios, S.: A reusable ontology for fluents in owl. In: FOIS (2006)
81. Whitehead, A.N., Russell, B.: Principia mathematica (1913)
82. Word Wide Web Consortium. RDF Primer (2004)
83. Word Wide Web Consortium. RDF Vocabulary Description Language 1.0: RDF Schema (2004)
84. Word Wide Web Consortium. SKOS Simple Knowledge Organization System (2009)
85. Word Wide Web Consortium. OWL 2 Web Ontology Language (2012)
86. Word Wide Web Consortium. SPARQL 1.1 Query Language (2013)
87. Yahya, M., Barbosa, D., Berberich, K., Wang, Q., Weikum, G.: Relationship queries on extended knowledge graphs. In: WSDM (2016)
88. Yamamoto, A.: Hypothesis finding based on upward refinement of residue hypotheses. Theoret. Comput. Sci. **298**(1), 5–19 (2003)
89. Yang, B., Yih, W., He, X., Gao, J., Deng, L.: Embedding entities and relations for learning and inference in knowledge bases. In: Proceedings of the International Conference on Learning Representation (ICLR), December 2014. arXiv: 1412.6575
90. Zupanc, K., Davis, J.: Estimating rule quality for knowledge base completion with the relationship between coverage assumption. In: Proceedings of the 2018 World Wide Web Conference, pp. 1073–1081. International World Wide Web Conferences Steering Committee (2018)

Explaining Data with Formal Concept Analysis

Bernhard Ganter[1], Sebastian Rudolph[1(✉)], and Gerd Stumme[2]

[1] TU Dresden, Dresden, Germany
{bernhard.ganter,sebastian.rudolph}@tu-dresden.de
[2] Uni Kassel, Kassel, Germany
stumme@cs.uni-kassel.de

Abstract. We give a brief introduction into Formal Concept Analysis, an approach to explaining data by means of lattice theory.

Keywords: Formal Concept Analysis · Data visualization · Attribute logic

1 Introduction

Formal Concept Analysis (FCA) is a mathematical discipline which attempts to formalize aspects of human conceptual thinking. For cognitive reasons, humans tend to form categories for objects and situations they encounter in the real world. These groups, defined based on commonalities between their elements, can then be given a name, referred to, and reasoned about in their entirety. They can be ordered by the level of generality or specificity giving rise to what is called "conceptual hierarchies" or "taxonomies". FCA provides a very simplified, yet powerful and elegant formalization of the notion of "concept" by means of lattice theory.

Over the last four decades, FCA has developed in a versatile scientific field, yielding novel approaches to data visualization and data mining. It greatly contributed to the development of data science and can be seen as a bottom-up approach to explain data by means of hierarchical clustering techniques.

Here, we provide a gentle introduction into the basics of FCA. Thereby, we will omit mathematical proofs of the presented theorems and lemmas; the interested reader may consult [4] for more details.

2 TL;DR – Formal Concept Analysis in a Nutshell

This section is meant to be an 'appetizer'. It provides a brief overview over Formal Concept Analysis, in order to allow for a better understanding of the overall picture. To this end, this section introduces the most basic notions of Formal Concept Analysis, namely formal contexts, formal concepts, and concept lattices. These definitions will be repeated and discussed in more detail later on.

© Springer Nature Switzerland AG 2019
M. Krötzsch and D. Stepanova (Eds.): Reasoning Web 2019, LNCS 11810, pp. 153–195, 2019.
https://doi.org/10.1007/978-3-030-31423-1_5

Formal Concept Analysis (FCA) was introduced as a mathematical theory modeling the notion of 'concepts' in terms of lattice theory. To come up with a formal description of concepts and their constituents, extensions and intensions, FCA starts by defining *(formal) contexts.*

Definition 1. A *(formal) context* is a triple $\mathbb{K} := (G, M, I)$, where G is a set whose elements are called *objects*, M is a set whose elements are called *attributes*, and I is a binary relation between G and M (i. e., $I \subseteq G \times M$), where $(g, m) \in I$ is read "object g *has* attribute m". $\qquad\qquad \Diamond$

This definition captures the basic and immediately graspable idea of a collection of entities, each of which might or might not have certain properties. At the same time, this notion is generic enough to be applicable to a vast variety of situations.

On another note, the interested reader might notice that formal contexts are closely related to *bipartite graphs* (where both objects and attributes are nodes in the graph and edges are connecting each object with its attributes). This link enables the study of bipartite graphs using FCA and, likewise, FCA can profit from known results developed for bipartite graphs.

Figure 1 shows a formal context where the object set G comprises all airlines of the Star Alliance group and the attribute set M lists their destinations.[1] The binary relation I is given by the cross table and describes which destinations are served by which Star Alliance member.

Definition 2. For an object set $A \subseteq G$, let

$$A' := \{ m \in M \mid \forall g \in A \colon (g, m) \in I \}$$

and, for an attribute set $B \subseteq M$, let

$$B' := \{ g \in G \mid \forall m \in B \colon (g, m) \in I \}.$$

A *(formal) concept* of a formal context (G, M, I) is a pair (A, B) with $A \subseteq G$, $B \subseteq M$, $A' = B$ and $B' = A$. The sets A and B are called the *extent* and the *intent* of the formal concept (A, B), respectively. The *subconcept–superconcept relation* $\leq$ is formalized by

$$(A_1, B_1) \leq (A_2, B_2) :\Longleftrightarrow A_1 \subseteq A_2 \quad (\Longleftrightarrow B_1 \supseteq B_2).$$

The set of all formal concepts of a formal context $\mathbb{K}$ together with the order relation $\leq$ always constitutes a complete lattice,[2] called the *concept lattice* of $\mathbb{K}$ and denoted by $\underline{\mathfrak{B}}(\mathbb{K})$. $\qquad\qquad \Diamond$

Figure 2 visualizes the concept lattice of the context in Fig. 1 by means of a *line diagram.* In a line diagram, each node represents a formal concept. A concept

---

[1] Note that the underlying data is somewhat outdated, if not to say antiquated.

[2] I. e., for each subset of concepts, there is always a unique greatest common subconcept and a unique least common superconcept.

	Latin America	Europe	Canada	Asia Pacific	Middle East	Africa	Mexico	Caribbean	United States
Air Canada	X	X	X	X	X		X	X	X
Air New Zealand		X		X					X
All Nippon Airways		X		X					X
Ansett Australia				X					
The Austrian Airlines Group		X	X	X	X	X			X
British Midland		X							
Lufthansa	X	X	X	X	X	X	X		X
Mexicana	X	X		X			X	X	X
Scandinavian Airlines	X	X	X		X				X
Singapore Airlines		X	X	X	X	X			X
Thai Airways International	X	X		X				X	X
United Airlines	X	X	X	X			X	X	X
VARIG	X	X		X		X	X		X

Fig. 1. A formal context about the destinations of the Star Alliance members

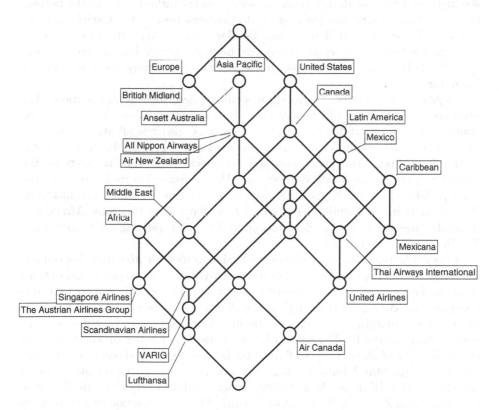

Fig. 2. The concept lattice of the context in Fig. 1

c_1 is a subconcept of a concept c_2 if and only if there is a path of descending edges from the node representing c_2 to the node representing c_1. The name of an object g is always attached to the node representing the smallest concept with g in its extent; dually, the name of an attribute m is always attached to the node representing the largest concept with m in its intent. We can read the context relation from the diagram because an object g has an attribute m if and only if the concept labeled by g is a subconcept of the one labeled by m. The extent of a concept consists of all objects whose labels are attached to subconcepts, and, dually, the intent consists of all attributes attached to superconcepts. For example, the concept labeled by 'Middle East' has {Singapore Airlines, The Austrian Airlines Group, Lufthansa, Air Canada} as extent, and {Middle East, Canada, United States, Europe, Asia Pacific} as intent.

High up in the diagram, we find the destinations which are served by most of the members: Europe, Asia Pacific, and the United States. For instance, besides British Midland and Ansett Australia, all airlines are serving the United States. Those two airlines are located at the top of the diagram, as they serve the fewest destinations—they operate only in Europe and Asia Pacific, respectively.

The further we go down in the concept lattice, the more globally operating are the airlines. The most destinations are served by the airlines close to the bottom of the diagram: Lufthansa (serving all destinations besides the Caribbean) and Air Canada (serving all destinations besides Africa). Also, the further we go down in the lattice, the lesser served are the destinations. For instance, Africa, the Middle East, and the Caribbean are served by relatively few Star Alliance members.

Dependencies between the attributes can be described by implications. For attribute sets $X, Y \subseteq M$, we say that the *implication* $X \to Y$ *holds* in the context, if each object having all attributes in X also has all attributes in Y. For instance, the implication {Europe, United States} $\to$ {Asia Pacific} holds in the Star Alliance context. It can be read directly from the line diagram: the largest concept having both 'Europe' and 'United States' in its intent (i. e., the concept labeled by 'All Nippon Airways' and 'Air New Zealand') also has 'Asia Pacific' in its intent. Similarly, one can detect that the destinations 'Africa' and 'Canada' together imply the destination 'Middle East' (and also 'Europe', 'Asia Pacific', and 'United States').

Concept lattices can also be visualized using *nested line diagrams*. For obtaining a nested line diagram, one splits the set of attributes in two parts, and obtains thus two formal contexts with identical object sets. For each formal context, one computes its concept lattice and a line diagram. The nested line diagram is obtained by enlarging the nodes of the first line diagram and by drawing the second diagram inside. The second lattice is used to further differentiate each of the extents of the concepts of the first lattice. Figure 3 shows a nested line diagram for the Star Alliance context. It is obtained by splitting the attribute set as follows: $M = \{$Europe, Asia Pacific, Africa, Middle East$\} \cup \{$United States, Canada, Latin America, Mexico, Caribbean$\}$. The order relation can be read by replacing each of the lines of the large diagram by eight parallel lines linking

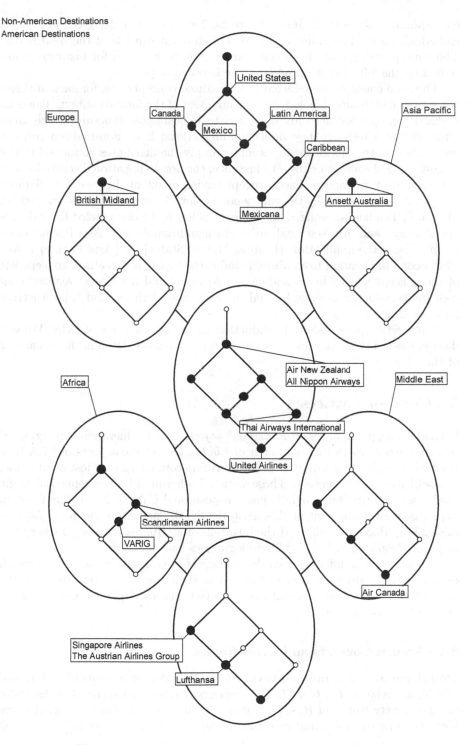

Non-American Destinations
American Destinations

United States

Canada   Latin America

Europe   Asia Pacific

Mexico

Caribbean

British Midland

Mexicana   Ansett Australia

Air New Zealand
All Nippon Airways

Africa   Middle East

Thai Airways International

United Airlines

Scandinavian Airlines

VARIG

Air Canada

Singapore Airlines
The Austrian Airlines Group

Lufthansa

Fig. 3. A nested diagram of the concept lattice in Fig. 2

corresponding nodes in the inner diagrams. The concept lattice given in Fig. 2 is embedded (as a join–semilattice) in this diagram, it consists of the solid nodes. The concept mentioned above (labeled by 'Middle East') is for instance represented by the left-most solid node in the lower right part.

The solid concepts are referred to as 'realized concepts', as, for each of them, the set of all attributes labeled above is an intent of the formal context. The non-realized concepts are not only displayed to indicate the structure of the inner scale, but also because they indicate implications: Each non-realized concept indicates that the attributes in its intent imply the attributes contained in the largest realized concept below. For instance, the first implication discussed above is indicated by the non-realized concept having as intent 'Europe' and 'United States', it is represented by the empty node below the concept labeled by 'British Midland'. The largest realized sub-concept below is the one labeled by 'All Nippon Airways' and 'Air New Zealand'—which additionally has 'Asia Pacific' in its intent. Hence the implication {Europe, United States} → {Asia Pacific} holds. The second implication from above is indicated by the non-realized concept left of the concept labeled by 'Scandinavian Airlines', and the largest realized concept below, which is the one labeled by 'Singapore Airlines' and 'The Austrian Airlines Group'.

This section gave a short introduction to the core notions of FCA. We will discuss most of them (and more advanced topics) in more detail in the remainder of this chapter.

3   Concept Lattices

Formal Concept Analysis studies how *objects* can be hierarchically grouped together according to their common *attributes*. One of the aspects of FCA thus is *attribute logic*, the study of possible attribute combinations. Most of the time, this will be very elementary. Those with a background in Mathematical Logic might say that attribute logic is just Propositional Calculus, and thus Boolean Logic, or even a fragment of this. Historically, the name *Propositional Logic* is misleading: Boole himself used the intuition of attributes ("signs") rather than of propositions. So in fact, attribute logic goes back to Boole.

But our style is different from that of logicians. Our logic is *contextual*, which means that we are interested in the logical structure of concrete data (of the *context*). Of course, the general rules of mathematical logic are important for this and will be utilized.

3.1   Formal Contexts and Cross Tables

Definition 3. A **Formal Context** (G, M, I) consists of two sets G and M and of a binary relation $I \subseteq G \times M$. The elements of G are called the **objects**, those of M the **attributes** of (G, M, I). If $g \in G$ and $m \in M$ are in relation I, we write $(g, m) \in I$ or $g\,I\,m$ and read this as *"object g* **has** *attribute m"*.   ◇

The simplest format for writing down a formal context is a **cross table**: we write a rectangular table with one row for each object and one column for each attribute, having a cross in the intersection of row g with column m iff $(g, m) \in I$. The simplest data type for computer storage is that of a bit matrix.[3]

Note that the definition of a formal context is very general. There are no restrictions about the nature of objects and attributes. We may consider physical objects, or persons, numbers, processes, structures, etc. – virtually everything. Anything that is a *set* in the mathematical sense may be taken as the set of objects or of attributes of some formal context. We may interchange the rôle of objects and attributes: if (G, M, I) is a formal context, then so is the **dual context** (M, G, I^{-1}) (with $(m, g) \in I^{-1} : \iff (g, m) \in I$). It is also not necessary that G and M are disjoint, they need not even be different.

On the other hand, the definition is rather restrictive when applied to real world phenomena. Language phrases like "all human beings" or "all chairs" do not denote sets in our sense. There is no "set of all chairs", because the decision if something is a chair is not a matter of fact but a matter of subjective inter-pretation. The notion of "formal concept" which we shall base on the definition of "formal context" is much, much narrower than what is commonly understood as a concept of human cognition. The step from "context" to "formal context" is quite an incisive one. It is the step from "real world" to "data". Later on, when we get tired of saying "formal concepts of a formal context", we will some-times omit the word "formal". But we should keep in mind that it makes a big difference.

3.2   The Derivation Operators

Given a selection $A \subseteq G$ of objects from a formal context (G, M, I), we may ask which attributes from M are common to all these objects. This defines an operator that produces for every set $A \subseteq G$ of objects the set $A^\uparrow$ of their common attributes.

Definition 4. For $A \subseteq G$, we let

$$A^\uparrow := \{m \in M \mid g \ I \ m \text{ for all } g \in A\}.$$

Dually, we introduce for a set $B \subseteq M$ of attributes

$$B^\downarrow := \{g \in G \mid g \ I \ m \text{ for all } m \in B\}.$$

These two operators are the **derivation operators** for (G, M, I).   ◇

The set $B^\downarrow$ denotes thus the set consisting of those objects in G that have (at least) all the attributes from B.

---

[3] It is not easy to say which is the *most efficient* data type for formal contexts. This depends, of course, on the operations we want to perform with formal contexts. The most important ones are the derivation operators, to be defined in the next subsection.

Usually, we do not distinguish the derivation operators in writing and use the notation A', B' instead. This is convenient, as long as the distinction is not explicitly needed.

If A is a set of objects, then A' is a set of attributes, to which we can apply the second derivation operator to obtain A'' (more precisely: $(A^{\uparrow})^{\downarrow}$), a set of objects. Dually, starting with a set B of attributes, we may form the set B'', which is again a set of attributes. We have the following simple facts:

Proposition 1. *For subsets $A, A_1, A_2 \subseteq G$ we have*

1. $A_1 \subseteq A_2 \Rightarrow A_2' \subseteq A_1'$,
2. $A \subseteq A''$,
3. $A' = A'''$.

Dually, for subsets $B, B_1, B_2 \subseteq M$ we have

1'. $B_1 \subseteq B_2 \Rightarrow B_2' \subseteq B_1'$,
2'. $B \subseteq B''$,
3'. $B' = B'''$.

The reader may confer to [4] for details and proofs. The mathematically interested reader may notice that the derivation operators constitute an (antitone) **Galois connection** between the (power sets of the) sets G and M.

The not so mathematically oriented reader should try to express the statements of the Proposition in common language. We give an example: Statement 1. says that *if a selection of objects is enlarged, then the attributes which are common to all objects of the larger selection are among the common attributes of the smaller selection.* Try to formulate 2 and 3 in a similar manner!

3.3   Formal Concepts, Extent and Intent

In what follows, (G, M, I) always denotes a formal context.

Definition 5. (A, B) is a **formal concept** of (G, M, I) iff

$$A \subseteq G, \quad B \subseteq M, \quad A' = B, \text{ and } \quad A = B'.$$

The set A is called the **extent** while the set B is called the **intent** of the formal concept (A, B). $\Diamond$

According to this definition, a formal concept has two parts: its extent and its intent. This follows an old tradition in philosophical concept logic, as expressed in the *Logic of Port Royal, 1654* [2], and in the International Standard ISO 704 (*Terminology work – Principles and methods*, translation of the German Standard DIN 2330).

The description of a concept by extent and intent is redundant, because each of the two parts determines the other (since $B = A'$ and $A = B'$). But for many reasons this redundant description is very convenient.

When a formal context is written as a cross table, then every formal concept (A, B) corresponds to a (filled) rectangular subtable, with row set A and column set B. To make this more precise, note that in the definition of a formal context, there is no order on the sets G or M. Permuting the rows or the columns of a cross table therefore does not change the formal context it represents. A *rectangular subtable* may, in this sense, omit some rows or columns; it must be rectangular after an appropriate rearrangement of the rows and the columns. It is then easy to characterize the rectangular subtables that correspond to formal concepts: they are full of crosses and maximal with respect to this property.

Lemma 2. (A, B) *is a formal concept of* (G,M,I) *iff* $A \subseteq G$, $B \subseteq M$, *and A and B are each maximal (with respect to set inclusion) for the property* $A \times B \subseteq I$.

A formal context may have many formal concepts. In fact, it is not difficult to come up with examples where the number of formal concepts is exponential in the size of the formal context. The set of all formal concepts of (G, M, I) is denoted

$$\mathfrak{B}(G, M, I),$$

or just $\mathfrak{B}$ if the context is known and fixed. Later on we shall discuss an algorithm to compute all formal concepts of a given formal context.

3.4 Conceptual Hierarchy

Formal concepts can be (partially) ordered in a natural way. Again, the definition is inspired by the way we usually order concepts in a *subconcept–superconcept hierarchy*: "Dog" is a subconcept of "mammal", because every dog is a mammal. Transferring this to formal concepts, the natural definition is as follows:

Definition 6. Let (A_1, B_1) and (A_2, B_2) be formal concepts of (G, M, I). We say that (A_1, B_1) is a **subconcept** of (A_2, B_2) (and, equivalently, that (A_2, B_2) is a **superconcept** of (A_1, B_1)) iff $A_1 \subseteq A_2$. We use the $\leq$-sign to express this relation and thus have

$$(A_1, B_1) \leq (A_2, B_2) : \iff A_1 \subseteq A_2.$$

The set $\mathfrak{B}$ of all formal concepts of (G, M, I), ordered by the relation $\leq$ – that is, the structure $(\mathfrak{B}, \leq)$ – is denoted

$$\underline{\mathfrak{B}}(G, M, I)$$

and is called the **concept lattice** of the formal context (G, M, I).   ◇

We will see in a bit, why the structure is called *lattice*. Arguably, this definition is natural, but irritatingly asymmetric. What about the intents? Well, a look at Proposition 1 shows that for concepts (A_1, B_1) and (A_2, B_2)

$$A_1 \subseteq A_2 \quad \text{is equivalent to} \quad B_2 \subseteq B_1.$$

Therefore

$$(A_1, B_1) \leq (A_2, B_2) : \iff A_1 \subseteq A_2 \quad (\iff B_2 \subseteq B_1).$$

The concept lattice of a formal context is a partially ordered set. We recall the formal definition of such a partial ordered set in the following.

Definition 7. A **partially ordered set** is a pair $(P, \leq)$ where P is a set, and $\leq$ is a binary relation on P (i.e., $\leq$ is a subset of $P \times P$) which is

1. reflexive ($x \leq x$ for all $x \in P$),
2. anti-symmetric ($x \leq y$ and $y \leq x$ imply $x = y$ for all $x, y \in P$), and
3. transitive ($x \leq y$ and $y \leq z$ imply $x \leq z$ for all $x, y, z \in P$).

We write $x \geq y$ for $y \leq x$, and $x < y$ for $x \leq y$ with $x \neq y$. $\qquad \Diamond$

Partially ordered sets appear frequently in mathematics and computer science. Observe that we do not assume a **total order**, which would require the additional condition $x \leq y$ or $y \leq x$ for all $x, y \in P$. Concept lattices have additional properties beyond being partially ordered sets, that is why we call them 'lattices'. This will be the topic of the next section.

3.5 Concept Lattice Diagrams

The concept lattice of (G, M, I) is the set of all formal concepts of (G, M, I), ordered by the subconcept–superconcept order. Ordered sets of moderate size can conveniently be displayed as **order diagram**s, sometimes also referred to as line diagrams. We explain how to *read* such a concept lattice line diagram by means of an example given in Fig. 4. Later on, we will discuss how to *draw* such diagrams.

Figure 4 refers to the following situation: Think of two squares of equal size that are drawn on paper. There are different ways to arrange the two squares: they may be disjoint (i.e., have no point in common), may overlap (i.e., have a common interior point), may share a vertex, an edge or a line segment of the boundary (of length > 0), they may be parallel or not.

Figure 4 shows a concept lattice unfolding these possibilities. It consists of twelve formal concepts, represented by the twelve small circles in the diagram. The names of the six attributes are given. Each name is attached to one of the formal concepts and is written slightly above the respective circle. The ten objects are represented by little pictures; each showing a pair of unit squares. Again, each object is attached to exactly one formal concept; the picture representing the object is drawn slightly below the circle representing the object concept.

Some of the circles are connected by edges. These express the concept order. With the help of the edges, we can read from the diagram which concepts are subconcepts of which other concepts, and which objects have which attributes. To do so, one has to follow *ascending paths* in the diagram.

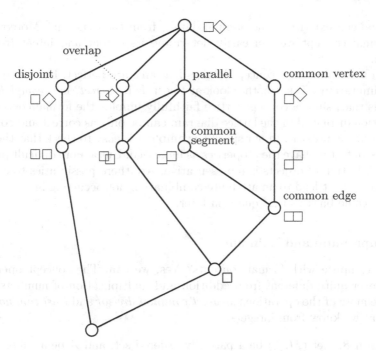

Fig. 4. A concept lattice diagram. The objects are pairs of unit squares. The attributes describe their mutual position.

For example, consider the object ⊡⊡. From the corresponding circle we can reach, via ascending paths, four attributes: "common edge", "common segment", "common vertex", and "parallel". ⊡⊡ does in fact have these properties, and does not have any of the others: the two squares are neither "disjoint" nor do they "overlap".

Similarly, we can find those objects that have a given attribute by following all descending paths starting at the attribute concept. For example, to find all objects which "overlap", we start at the attribute concept labeled "overlap" and follow the edges downward. We can reach three objects (namely ⟐, ⊞, and □, the latter symbolizing two squares at the same position). Note that we cannot reach ⊡⊡, because only at concept nodes it is allowed to make a turn.

With the same method, we can read the intent and the extent of every formal concept in the diagram. For example, consider the concept circle labeled ⊞. Its extent consists of all objects that can be reached from that circle on an descending path. The extent therefore is $\{⊞, □\}$. Similarly, we find by an inspection of the ascending paths that the intent of this formal concept is {overlap, parallel}.

The diagram contains all necessary information. We can read off the objects, the attributes, and the incidence relation I. Thus we can perfectly reconstruct the formal context (i.e., "the original data") from the diagram.[4] Moreover, for each formal concept we can easily determine its extent and intent from the diagram.

So in a certain sense, concept lattice diagrams are perfect. But there are, of course, limitations. Take another look at Fig. 4. Is it *correct*? Is it *complete*? The answer is that, since a concept lattice faithfully unfolds the formal context, the information displayed in the lattice diagram can be only as correct and complete as the formal context is. In our specific example it is easy to check that the given examples in fact do have the properties as indicated. But a more difficult problem is if our selection of objects is representative. Are there possibilities to combine two squares, that lead to an attribute combination not occurring in our sample? We shall come back to that question later.

3.6   Supremum and Infimum

Can we compute with formal concepts? Yes, we can. The concept operations are however quite different from addition and multiplication of numbers. They resemble more of the operations *greatest common divisor* and *least common multiple*, that we know from integers.

Definition 8. Let $(M, \leq)$ be a partially ordered set, and A be a subset of M. A **lower bound** of A is an element s of M with $s \leq a$, for all $a \in A$. An **upper bound** of A is defined dually. If there exists a largest element in the set of all lower bounds of A, then it is called the **infimum** (or **meet**) of A. It is denoted $inf\ A$ or $\bigwedge A$. The **supremum** (or **join**) of A ($sup\ A$, $\bigvee A$) is defined dually. For $A = \{x, y\}$, we write also $x \wedge y$ for their infimun, and $x \vee y$ for their supremum.   ◊

Lemma 3. *For any two formal concepts (A_1, B_1) and (A_2, B_2) of some formal context we obtain*

– *the infimum (**greatest common subconcept**) of (A_1, B_1) and (A_2, B_2) as*

$$(A_1, B_1) \wedge (A_2, B_2) := (A_1 \cap A_2, (B_1 \cup B_2)''),$$

– *the supremum (**least common superconcept**) of (A_1, B_1) and (A_2, B_2) as*

$$(A_1, B_1) \vee (A_2, B_2) := ((A_1 \cup A_2)'', B_1 \cap B_2).$$

It is not difficult to prove that what is suggested by this definition is in fact true: $(A_1, B_1) \wedge (A_2, B_2)$ is in fact a formal concept (of the same context), $(A_1, B_1) \wedge (A_2, B_2)$ is a subconcept of both (A_1, B_1) and (A_2, B_2), and any other common subconcept of (A_1, B_1) and (A_2, B_2) is also a subconcept of $(A_1, B_1) \wedge (A_2, B_2)$. Similarly, $(A_1, B_1) \vee (A_2, B_2)$ is a formal concept, it is a superconcept of (A_1, B_1) and of (A_2, B_2), and it is a subconcept of any common superconcept of these two formal concepts.

---

[4] This reconstruction is assured by the Basic Theorem given below.

With some practice, one can read off infima and suprema from the lattice diagram. Choose any two concepts from Fig. 4 and follow the descending paths from the corresponding nodes in the diagram. There is always a highest point where these paths meet, that is, a highest concept that is below both, namely, the infimum. Any other concept below both can be reached from the highest one on a descending path. Similarly, for any two formal concepts there is always a lowest node (the supremum of the two), that can be reached from both concepts via ascending paths. And any common superconcept of the two is on an ascending path from their supremum.

3.7   Complete Lattices

The operations for computing with formal concepts, infimum and supremum, are not as weird as one might suspect. In fact, we obtain with each concept lattice an algebraic structure called a "lattice", and such structures occur frequently in mathematics and computer science. "Lattice theory" is an active field of research in mathematics. A *lattice* is an algebraic structure with two operations (called "meet" and "join" or "infimum" and "supremum") that satisfy certain natural conditions:[5]

Definition 9. A partially ordered set $\mathbb{V} := (V, \leq)$ is called a **lattice**, if their exists, for every pair of elements $x, y \in V$, their infimum $x \wedge y$ as well as their supremum $x \vee y$.   ◇

We shall not discuss the algebraic theory of lattices in this lecture. Many universities offer courses in lattice theory, and there are excellent textbooks.[6]

Concept lattices have an additional nice property: they are **complete lattices**. This means that the operations of infimum and supremum do not only work for an input consisting of two elements, but for arbitrary many. In other words: each collection of formal concepts has a greatest common subconcept and a least common superconcept. This is even true for infinite sets of concepts. The operations "infimum" and "supremum" are not necessarily binary, they work for any input size.

Definition 10. A partially ordered set $\mathbb{V} := (V, \leq)$ is a **complete lattice**, if for every set $A \subseteq V$, there exists its infimum $\bigwedge V$ and its supremum $\bigvee A$.   ◇

Note that the definition requests the existence of infimum and supremum for every set A, hence also for the empty set $A := \emptyset$. Following the definition, we obtain that $\bigwedge \emptyset$ has to be the (unique) largest element of the lattice. It is denoted by $\mathbf{1_V}$. Dually, $\bigvee \emptyset$ has to be the smallest element of the lattice; it is denoted by $\mathbf{0_V}$.

---

[5] Unfortunately, the word "lattice" is used with different meanings in mathematics. It also refers to generalized grids.

[6] *An introduction to lattices and order* by B. Davey and H. Priestley is particularly popular among CS students.

The arbitrary arity of infimum and supremum is very useful, but will make essentially no difference for our considerations, because we shall mainly be concerned with finite formal contexts and finite concept lattices. Well, this is not completely true. In fact, although the concept lattice in Fig. 4 is finite, its ten objects are representatives for *all* possibilities to combine two unit squares. Of course, there are infinitely many such possibilities. It is true that we shall consider finite concept lattices, but our examples may be taken from an infinite reservoir.

3.8   The Basic Theorem of FCA

We give now a mathematically precise formulation of the algebraic properties of concept lattices. The theorem below is not difficult, but basic for many other results. Its formulation contains some technical terms that we have not mentioned so far.

In a complete lattice, an element is called **supremum-irreducible** if it cannot be written as a supremum of other elements, and **infimum-irreducible** if it can not be expressed as an infimum of other elements. It is very easy to locate the irreducible elements in a diagram of a finite lattice: the supremum-irreducible elements are precisely those from which there is exactly one edge going downward. An element is infimum-irreducible if and only if it is the start of exactly one upward edge. In Fig. 4, there are precisely nine supremum-irreducible concepts and precisely five infimum-irreducible concepts. Exactly four concepts have both properties, they are **doubly irreducible**.

A set of elements of a complete lattice is called **supremum-dense**, if every lattice element is a supremum of elements from this set. Dually, a set is called **infimum-dense**, if the infima that can be computed from this set exhaust all lattice elements.

The notion of isomorphism defined next essentially captures the idea of two lattices being the same up to a renaming of the elements.

Definition 11. Two lattices $\mathbb{V}$ and $\mathbb{W}$ are **isomorphic** ($\mathbb{V} \cong \mathbb{W}$), if there exists a bijective mapping $\varphi \colon V \to W$ with $x \leq y \iff \varphi(x) \leq \varphi(y)$. The mapping φ is then called **lattice isomorphism** between $\mathbb{V}$ and $\mathbb{W}$.   ◇

Now we have defined all the terminology necessary for stating the main theorem of Formal Concept Analysis.

Theorem 4 (The Basic Theorem of Formal Concept Analysis). *The concept lattice of any formal context* (G, M, I) *is a complete lattice. For an arbitrary set* $\{(A_i, B_i) \mid i \in J\} \subseteq \mathfrak{B}(G, M, I)$ *of formal concepts, the supremum is given by*

$$\bigvee_{i \in J} (A_i, B_i) = \left(\left(\bigcup_{i \in J} A_i \right)'', \bigcap_{i \in J} B_i \right)$$

and the infimum is given by

$$\bigwedge_{i \in J}(A_i, B_i) = \left(\bigcap_{i \in J} A_i, \left(\bigcup_{i \in J} B_i\right)''\right).$$

A complete lattice L is isomorphic to $\mathfrak{B}(G, M, I)$ precisely if there are mappings $\tilde{\gamma} : G \to L$ and $\tilde{\mu} : M \to L$ such that $\tilde{\gamma}(G)$ is supremum-dense and $\tilde{\mu}(M)$ is infimum-dense in L, and for all $g \in G$ and $m \in M$

$$g \, I \, m \iff \tilde{\gamma}(g) \le \tilde{\mu}(m).$$

In particular, $L \cong \mathfrak{B}(L, L, \le)$.

The theorem is less complicated than it may first seem. We give some explanations below. Readers in a hurry may skip these and continue with the next section.

The first part of the theorem gives the precise formulation for infimum and supremum of arbitrary sets of formal concepts. The second part of the theorem gives (among other information) an answer to the question if concept lattices have any special properties. The answer is "no": every complete lattice is (isomorphic to) a concept lattice. This means that for every complete lattice, we must be able to find a set G of objects, a set M of attributes and a suitable relation I, such that the given lattice is isomorphic to $\mathfrak{B}(G, M, I)$. The theorem does not only say how this can be done, it describes in fact *all* possibilities to achieve this.

In Fig. 4, every object is attached to a unique concept, the corresponding *object concept*. Similarly for each attribute there corresponds an *attribute concept*. These can be defined as follows:

Definition 12. Let (G, M, I) be some formal context. Then

– for each object $g \in G$ the corresponding **object concept** is

$$\gamma g := (\{g\}'', \{g\}'),$$

– and for each attribute $m \in M$ the **attribute concept** is given by

$$\mu m := (\{m\}', \{m\}'').$$

The set of all object concepts of (G, M, I) is denoted γG, the set of all attribute concepts is μM. ◇

Using Definition 5 and Proposition 1, it is easy to check that these expressions in fact define formal concepts of (G, M, I).

We have that $\gamma g \le (A, B) \iff g \in A$. A look at the first part of the Basic Theorem shows that each formal concept is the supremum of all the object concepts below it. Therefore, the set γG of all object concepts is supremum-dense. Dually, the attribute concepts form an infimum-dense set in $\mathfrak{B}(G, M, I)$. The Basic Theorem says that, conversely, any supremum-dense set in a complete lattice L can be taken as the set of objects and any infimum-dense set be taken as a set of attributes for a formal context with concept lattice isomorphic to L.

We conclude with a simple observation that often helps to find errors in concept lattice diagrams. The fact that the object concepts form a supremum-dense set implies that every supremum-irreducible concept must be an object concept (the converse is not true). Dually, every infimum-irreducible concept must be an attribute concept. This yields the following rule for concept lattice diagrams:

Proposition 5. *Given a formal context (G, M, I) and a finite order diagram, labeled by the objects from G and the attributes from M. For $g \in G$ let $\tilde{\gamma}(g)$ denote the element of the diagram that is labeled with g, and let $\tilde{\mu}(m)$ denote the element labeled with m. Then the given diagram is a correctly labeled diagram of $\mathfrak{B}(G, M, I)$ if and only if it satisfies the following conditions:*

1. *The diagram is a correct lattice diagram,*
2. *every supremum-irreducible element is labeled by some object,*
3. *every infimum-irreducible element is labeled by some attribute,*
4. *$g \, I \, m \iff \tilde{\gamma}(g) \leq \tilde{\mu}(m)$ for all $g \in G$ and $m \in M$.*

The definitions of lattices and complete lattices are self-dual: If $(V, \leq)$ is a (complete) lattice, then $(V, \leq)^d := (V, \geq)$ is also a (complete) lattice. If a theorem holds for a (complete) lattice, then the 'dual theorem' also holds, i.e., the theorem where all occurrences of $\leq, \vee, \wedge, \bigvee, \bigwedge, \mathbf{0_V}, \mathbf{1_V}$ etc. are replaced by $\geq, \wedge, \vee, \bigwedge, \bigvee, \mathbf{1_V}, \mathbf{0_V}$, resp.

For concept lattices, their dual can be obtained by "flipping" the formal context:

Lemma 6. *Let (G, M, I) be a formal context and $\mathfrak{B}(G, M, I)$ its concept lattice. Then $(\mathfrak{B}(G, M, I))^d \cong \mathfrak{B}(M, G, I^{-1})$, with $I^{-1} := \{(m, g) \mid (g, m) \in I\}$.*

3.9　Computing All Concepts of a Context

There are several algorithms that help drawing concept lattices. We shall discuss some of them below. But we find it instructive to start by some small examples that can be drawn by hand. For computing concept lattices, we will investigate a fast algorithm later. We start with a naive method before proceeding to a method which is suitable for manual computation.

In principle, it is not difficult to find all the concepts of a formal context. The following proposition summarizes the naive possibilities of generating all concepts.

Lemma 7. *Each concept of a context (G, M, I) has the form (X'', X') for some subset $X \subseteq G$ and the form (Y', Y'') for some subset $Y \subseteq M$. Conversely, all such pairs are concepts. Every extent is the intersection of attribute extents and every intent is the intersection of object intents.*

The first part of the lemma suggests a first algorithm for computing all concepts: go through all subsets X of G and record (X'', X') as concept (skipping duplicates). However, this is rather inefficient, and not practicable even for relatively small contexts. The second part of the proposition at least yields the possibility to calculate the concepts of a small context by hand.

The following method is more efficient, and is recommended for computations by hand. It is based on the following observations:

1. It suffices to determine all concept extents (or all concept intents) of (G, M, I), since we can always determine the other part of a formal concept with the help of the derivation operators.
2. The intersection of arbitrary many extents is an extent (and the intersection of arbitrary intents is an intent). This follows easily from the formulae given in the Basic Theorem. By the way: a convention that may seem absurd on the first glance allows to include in "arbitrary many" also the case "zero". The convention says that the intersection of zero intents equals M and the intersection of zero extents equals G.
3. One can determine all extents from knowing all **attribute extents** $\{m\}'$, $m \in M$ (and all intents from all **object intents** $\{g\}'$, $g \in G$) because every extent is an intersection of attribute extents (and every intent is the intersection of object intents). This follows from the fact that the attribute concepts are infimum-dense and the object concepts are supremum-dense.

These observations give rise to the following procedure.

Instruction for determining all formal concepts of a small formal context

1. Initialize a list of concept extents. To begin with, write for each attribute $m \in M$ the attribute extent $\{m\}'$ to this list (if not already present).
2. For any two sets in this list, compute their intersection. If the result is a set that is not yet in the list, then extend the list by this set. With the extended list, continue to build all pairwise intersections.
3. If for any two sets in the list their intersection is also in the list, then extend the list by the set G (provided it is not yet contained in the list). The list then contains all concept extents (and nothing else).
4. For every concept extent A in the list compute the corresponding intent A' to obtain a list of all formal concepts (A, A') of (G, M, I).

Example 1. We illustrate the method by means of an example from elementary geometry. The objects of our example are seven triangles. The attributes are five standard properties that triangles may or may not have:

Triangles				Attributes	
abbreviation	coordinates	diagram		symbol	property
T1	(0,0) (6,0) (3,1)			a	equilateral
				b	isoceles
T2	(0,0) (1,0) (1,1)			c	acute angled
T3	(0,0) (4,0) (1,2)			d	obtuse angled
				e	right angled
T4	(0,0) (2,0) (1,$\sqrt{3}$)				
T5	(0,0) (2,0) (5,1)				
T6	(0,0) (2,0) (1,3)				
T7	(0,0) (2,0) (0,1)				

We obtain the following formal context

	a	b	c	d	e
$T1$		×		×	
$T2$		×			×
$T3$			×		
$T4$	×	×	×		
$T5$				×	
$T6$		×	×		
$T7$					×

Following the above instruction, we proceed:

1. Write the attribute extents to a list.

No. extent	found as
$e_1 := \{T_4\}$	$\{a\}'$
$e_2 := \{T_1, T_2, T_4, T_6\}$	$\{b\}'$
$e_3 := \{T_3, T_4, T_6\}$	$\{c\}'$
$e_4 := \{T_1, T_5\}$	$\{d\}'$
$e_5 := \{T_2, T_7\}$	$\{e\}'$

2. Compute all pairwise intersections, and

3. add G.

No.	extent	found as
$e_1 := \{T_4\}$		$\{a\}'$
$e_2 := \{T_1, T_2, T_4, T_6\}$		$\{b\}'$
$e_3 := \{T_3, T_4, T_6\}$		$\{c\}'$
$e_4 := \{T_1, T_5\}$		$\{d\}'$
$e_5 := \{T_2, T_7\}$		$\{e\}'$
$e_6 := \varnothing$		$e_1 \cap e_4$
$e_7 := \{T_4, T_6\}$		$e_2 \cap e_3$
$e_8 := \{T_1\}$		$e_2 \cap e_4$
$e_9 := \{T_2\}$		$e_2 \cap e_5$
$e_{10} := \{T_1, T_2, T_3, T_4, T_5, T_6, T_7\}$		step 3

4. Compute the intents.

Concept No.	(extent , intent)
1	$(\{T_4\}\,,\{a, b, c\})$
2	$(\{T_1, T_2, T_4, T_6\}\,,\{b\})$
3	$(\{T_3, T_4, T_6\}\,,\{c\})$
4	$(\{T_1, T_5\}\,,\{d\})$
5	$(\{T_2, T_7\}\,,\{e\})$
6	$(\varnothing\,,\{a, b, c, d, e\})$
7	$(\{T_4, T_6\}\,,\{b, c\})$
8	$(\{T_1\}\,,\{b, d\})$
9	$(\{T_2\}\,,\{b, e\})$
10	$(\{T_1, T_2, T_3, T_4, T_5, T_6, T_7\}\,,\varnothing)$

We have now computed all ten formal concepts of the triangles–context. The last step can be skipped if we are not interested in an explicit list of all concepts, but just in computing a line diagram.

3.10 Drawing Concept Lattices

Based on one of the lists 3 or 4, we can start to draw a diagram. Before doing so, we give two simple definitions.

Definition 13. Let (A_1, B_1) and (A_2, B_2) be formal concepts of some formal context (G, M, I). We say that (A_1, B_1) is a **proper subconcept** of (A_2, B_2) (written as $(A_1, B_1) < (A_2, B_2)$), if $(A_1, B_1) \leq (A_2, B_2)$ and $(A_1, B_1) \neq (A_2, B_2)$. We call (A_1, B_1) a **lower neighbour** of (A_2, B_2) (written as $(A_1, B_1) \prec (A_2, B_2)$), if $(A_1, B_1) < (A_2, B_2)$, but no formal concept (A, B) of (G, M, I) exists with $(A_1, B_1) < (A, B) < (A_2, B_2)$. $\Diamond$

Instruction how to draw a line diagram of a small concept lattice

5. Take a sheet of paper and draw a small circle for every formal concept, in the following manner: a circle for a concept is always positioned higher than the all circles for its proper subconcepts.
6. Connect each circle with the circles of its lower neighbors.
7. Label with attribute names: attach the attribute m to the circle representing the concept $(\{m\}', \{m\}'')$.
8. Label with object names: attach each object g to the circle representing the concept $(\{g\}'', \{g\}')$.

We now follow these instructions.

5. Draw a circle for each of the formal concepts:

○ 10

○ 5 ○ 2 ○ 4   ○ 3

○ 9   ○ 8   ○ 7

○ 1

○ 6

6. Connect circles with their lower neighbours:

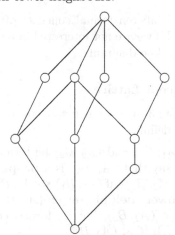

7. Add the attribute names:

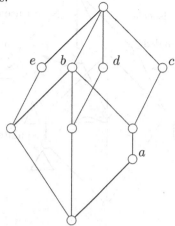

8. Determine the object concepts

object g	object intent $\{g\}'$	no. of concept
T_1	$\{b, d\}$	8
T_2	$\{b, e\}$	9
T_3	$\{c\}$	3
T_4	$\{a, b, c\}$	1
T_5	$\{d\}$	4
T_6	$\{b, c\}$	7
T_7	$\{e\}$	5

and add the object names to the diagram:

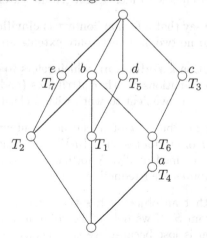

Done! Usually it takes some attempts before a nice, readable diagram is achieved. Finally we can make the effort to avoid abbreviations and to increase the readability. The result is shown in Fig. 5.

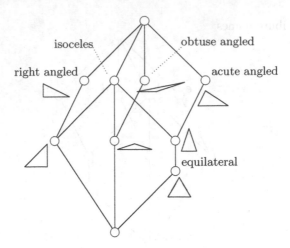

Fig. 5. A diagram of the concept lattice of the triangle context.

3.11   Clarifying and Reducing a Formal Context

There are context manipulations that simplify a formal context without changing the diagram, except for the labeling. It is usually advisable to do these manipulations first, before starting computations.

The simplest operation is **clarification**, which refers to identifying "equal rows" of a formal context, and "equal columns" as well. What is meant is that if a context contains objects $g_1, g_2, \ldots$ with $\{g_i\}' = \{g_j\}'$ for all i, j, that is, objects which have exactly the same attributes, then these can be replaced by a single object, the name of which is just the list of names of these objects. The same can be done for attributes with identical attribute extent.

Definition 14. We say that a formal context is **clarified** if no two of its object intents are equal and no two of its attribute extents are equal.   ◊

A stronger operation is **reduction**, which refers to omitting attributes that are equivalent to combinations of other attributes (and dually for objects). For defining reduction it is convenient to work with a clarified context.

Definition 15. An attribute m of a clarified context is called **reducible** if there is a set $S \subseteq M$ of attributes with $\{m\}' = S'$, otherwise it is **irreducible**. Reduced objects are defined dually. A formal context is called **reduced**, if all objects and all attributes are irreducible.   ◊

$\{m\}' = S'$ means that an object g has the attribute m if and only if it has all the attributes from S. If we delete the column m from our cross table, no essential information is lost because we can reconstruct this column from the data contained in other columns (those of S). Moreover, deleting that column does not change the number of concepts, nor the concept hierarchy, because $\{m\}' = S'$ implies that m is in the intent of a concept if and only if S is contained in that intent. The same is true for reducible objects and concept extents. Deleting a reducible object from a formal context does not change the structure of the concept lattice.

It is even possible to remove several reducible objects and attributes simultaneously from a formal context without any effect on the lattice structure, as long as the number of removed elements is finite.

Definition 16. Let (G, M, I) be a finite context, and let G_{irr} be the set of irreducible objects and M_{irr} be the set of irreducible attributes of (G, M, I). The context $(G_{irr}, M_{irr}, I \cap G_{irr} \times M_{irr})$ is the **reduced context** corresponding to (G, M, I).

For a finite lattice L let $J(L)$ denote the set of its supremum-irreducible elements and let $M(L)$ denote the set of its infimum-irreducible elements. Then $(J(L), M(L), \leq)$ is the **standard context** for the lattice L. ◇

Proposition 8. *A finite context and its reduced context have isomorphic concept lattices. For every finite lattice L there is (up to isomorphism) exactly one reduced context, the concept lattice of which is isomorphic to L, namely its standard context.*

3.12   Additive and Nested Line Diagrams

In this section, we discuss possibilities to generate line diagrams both automatically or by hand. A list of some dozens of concepts may already be quite difficult to survey, and it requires practice to draw good line diagrams of concept lattices with more than 20 elements.

The best and most versatile form of representation for a concept lattice is a well-drawn line diagram. It is, however, tedious to draw such a diagram by hand and one would wish an automatic generation by means of a computer. We know quite a few algorithms to do this, but none which provides a general satisfactory solution. It is by no means clear which qualities make up a *good* diagram. It should be transparent, easily readable and should facilitate the interpretation of the data represented. How this can be achieved in each individual case depends, however, on the aim of the interpretation and on the structure of the lattice. Simple optimization criteria (minimization of the number of edge crossings, drawing in layers, etc.) often bring about results that are unsatisfactory. Nevertheless, automatically generated diagrams are a great help: they can serve as the starting point for drawing by hand. Therefore, we will describe simple methods of generating and manipulating line diagrams by means of a computer.

3.12.1   Additive Line Diagrams

We will now explain a method where a computer generates a diagram and offers the possibility of improving it interactively. Programming details are irrelevant in this context. We will therefore only give a **positioning rule** which assigns points in the plane to the elements of a given ordered set $(P, \leq)$. If a and b are elements of P with $a < b$, the point assigned to a must be lower than the point assigned to b (i.e., it must have a smaller y-coordinate). This is guaranteed by our method. We will leave the computation of the edges and the checking for undesired coincidences of vertices and edges to the program. We do not even guarantee that our positioning is injective (which of course is necessary for a correct line diagram). This must also be checked if necessary.

Definition 17. A **set representation** of an ordered set $(P, \leq)$ is an order embedding of $(P, \leq)$ in the power-set of a set X, i.e., a map

$$\mathrm{rep} : P \to \mathfrak{P}(X)$$

with the property

$$x \leq y \iff \mathrm{rep}\, x \subseteq \mathrm{rep}\, y.$$

$$\Diamond$$

An example of a set representation for an arbitrary ordered set $(P, \leq)$ is the assignment

$$X := P, \quad a \mapsto \{x \mid x < a\}.$$

In the case of a concept lattice,

$$X := G, \quad (A, B) \mapsto A$$

is a set representation.

$$X := M, \quad (A, B) \mapsto M \setminus B$$

is another set representation, and both can be combined to

$$X := G \,\dot\cup\, M, \quad (A, B) \mapsto A \cup (M \setminus B).$$

It is sufficient to limit oneself to the irreducible objects and attributes.

For an **additive line diagram** of an ordered set $(P, \leq)$ we need a set representation $\mathrm{rep} : P \to \mathfrak{P}(X)$ as well as a **grid projection**

$$\mathrm{vec} : X \to \mathbb{R}^2,$$

assigning a real vector with a positive y-coordinate to each element of X. By

$$\mathrm{pos}\, p := n + \sum_{x \in \mathrm{rep}\, p} \mathrm{vec}\, x$$

we obtain a positioning of the elements of P in the plane. Here, n is a vector which can be chosen arbitrarily in order to shift the entire diagram. By only allowing positive y–coordinates for the grid projection we make sure that no element p is positioned below an element q with $q < p$.

Every finite line diagram can be interpreted as an additive diagram with respect to an appropriate set representation. For concept lattices we usually use the representation by means of the irreducible objects and/or attributes.

The resulting diagrams are characterized by a great number of parallel edges, which improves their readability. Experience shows that the set representation by means of the irreducible attributes is most likely to result in an easily interpretable diagram. Figure 5 for instance was obtaining by selecting the irreducible attributes for the set representation.

Since the second set representation given above is somehow unnatural, we introduce for this purpose the dual set representation.

Definition 18. A **dual set representation** of an ordered set $(P, \leq)$ is an order–inversing embedding of $(P, \leq)$ in the power-set of a set X, i.e., a map

$$\mathrm{rep}' : P \to \mathfrak{P}(X)$$

with the property

$$x \leq y \iff \mathrm{rep}'x \supseteq \mathrm{rep}'y.$$

$\diamond$

Now

$$X := M, \quad \mathrm{rep}' : (A, B) \mapsto B$$

is a dual set representation. We request now that the grid projection allows only negative y–coordinates. The following shows that the two ways are indeed equivalent: Let $\mathrm{vec}' : X \to \mathbb{R}^2$ be given by $\mathrm{vec}'(m) := (-x, -y)$ where $\mathrm{vec}(m) = (x, y)$ for all $m \in X$. Then all y–coordinates are indeed negative. We obtain then the following equality:

$$
\begin{aligned}
\mathrm{pos}(A, B) &= n + \sum_{m \in M \setminus B} \mathrm{vec}(m) \\
&= n + \sum_{m \in M} \mathrm{vec}(m) + \sum_{m \in B} -\mathrm{vec}(m) \\
&= n' + \sum_{m \in B} \mathrm{vec}'(m) \\
&= \mathrm{pos}'(A, B)
\end{aligned}
$$

where $n' := n + \sum_{m \in M} \mathrm{vec}(m)$.

It is particularly easy to manipulate these diagrams: If we change – the set representation being fixed – the grid projection for an element $x \in X$, this means that all images of the order filter $\{p \in P \mid x \in \mathrm{rep}\, p\}$ are shifted by the same distance and that all other points remain in the same position. In the case of the set representation by means of the irreducibles these order filters are precisely principal filters or complements of principal ideals, respectively. This means that we can manipulate the diagram by shifting principal filters or principal ideals, respectively, and leaving all other elements in position.

Even carefully constructed line diagrams loose their readability from a certain size up, as a rule from around 50 elements up. One gets considerably further with *nested line diagrams* which will be introduced next. However, these diagrams do not only serve to represent larger concept lattices. They offer the possibility to visualize how the concept lattice changes if we add further attributes.

3.12.2   Nested Line Diagrams

Nested line diagrams permit a satisfactory graphical representation of somewhat larger concept lattices. The basic idea of the nested line diagram consists of clustering parts of an ordinary diagram and replacing bundles of parallel lines between these parts by one line each. Thus, a nested line diagram consists of ovals, which contain clusters of the ordinary line diagram and which are connected by lines. In the simplest case, two ovals which are connected by a simple line are congruent. Here, the line indicates that corresponding circles within the ovals are direct neighbors, resp.

Furthermore, we allow that two ovals connected by a single line do not necessarily have to be congruent, but they may each contain a part of two congruent figures. In this case, the two congruent figures are drawn in the ovals as a "background structure", and the elements are drawn as solid circles if they are part of the respective substructures. The line connecting the two boxes then indicates that the respective pairs of elements of the background shall be connected with each other. An example is given in Fig. 6. It is a screenshot of a library information system which was set up for the library of the Center on Interdisciplinary Technology Research of Darmstadt University of Technology.

Nested line diagrams originate from partitions of the set of attributes. The basis is the following theorem:

Theorem.9. *Let (G, M, I) be a context and $M = M_1 \cup M_2$. The map*

$$(A, B) \mapsto (((B \cap M_1)', B \cap M_1), ((B \cap M_2)', B \cap M_2))$$

is a supremum-preserving order embedding of $\underline{\mathfrak{B}}(G, M, I)$ in the direct product of $\underline{\mathfrak{B}}(G, M_1, I \cap G \times M_1)$ and $\underline{\mathfrak{B}}(G, M_2, I \cap G \times M_2)$. The component maps

$$(A, B) \mapsto ((B \cap M_i)', B \cap M_i)$$

are surjective on $\underline{\mathfrak{B}}(G, M_i, I \cap G \times M_i)$.

In order to sketch a nested line diagram, we proceed as follows: First of all, we split up the attribute set: $M = M_1 \cup M_2$. This splitting up does not have to be disjoint. More important for interpretation purposes is the idea that the sets M_i bear meaning. Now, we draw line diagrams of the subcontexts $\mathbb{K}_i := (G, M_i, I \cap G \times M_i)$, $i \in \{1, 2\}$ and label them with the names of the objects and attributes, as usual. Then we sketch a nested diagram of the product of the concept lattices $\underline{\mathfrak{B}}(\mathbb{K}_i)$ as an auxiliary structure. For this purpose, we draw a large copy of the diagram of $\underline{\mathfrak{B}}(\mathbb{K}_1)$, representing the lattice elements not by small circles but by congruent ovals, which contain each a diagram of $\underline{\mathfrak{B}}(\mathbb{K}_2)$.

By Theorem 9 the concept lattice $\underline{\mathfrak{B}}(G, M, I)$ is embedded in this product as a $\bigvee$-semilattice. If a list of the elements of $\underline{\mathfrak{B}}(G, M, I)$ is available, we can enter them into the product according to their intents. If not, we enter the object concepts the intents of which can be read off directly from the context, and form all suprema.

This at the same time provides us with a further, quite practicable method of determining a concept lattice by hand: split up the attribute set as appropriate,

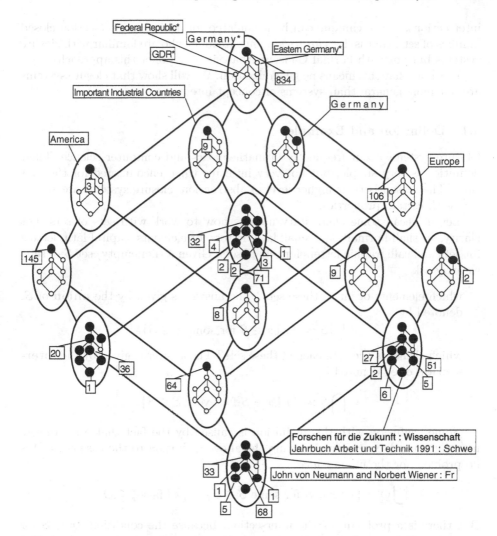

Fig. 6. Nested line diagram of a library information system

determine the (small) concept lattices of the subcontexts, draw their product in form of a nested line diagram, enter the object concepts and close it against suprema. This method is particularly advisable in order to arrive at a useful diagram quickly.

4   Closure Systems

The algorithm that will be one central theme of our course was developed for concept lattices, but can be rephrased without reference to Formal Concept Analysis. The reason is that the algorithm essentially relies on a single property of concept lattices, namely that the set of concept intents is closed under intersections. The technique can be formulated for arbitrary intersection closed families of sets, that is, for *closure systems*. Readers who are familiar with closure systems but not with Formal Concept Analysis may prefer this approach.

But note that this means no generalization. We will show that closure systems are not more general than systems of concept intents.

4.1 Definition and Examples

Closure systems occur frequently in mathematics and computer science. Their definition is very simple, but not very intuitive when encountered for the first time. The reason is their higher level of abstraction: closure systems are *sets of sets* with certain properties.

Let us recall some elementary notions how to work with sets of sets. For clarity, we shall normally use small latin letters for elements, capital latin letters for sets and calligraphic letters for sets of sets. Given a (nonempty) set $\mathcal{S}$ of sets, we may ask

- which elements occur in these sets? The answer is given by the **union** of $\mathcal{S}$, denoted by

$$\bigcup \mathcal{S} := \{x \mid x \in S \text{ for some } S \in \mathcal{S}\}.$$

- which elements occur in each of these sets? The answer is given by the **intersection** of $\mathcal{S}$, denoted by

$$\bigcap \mathcal{S} := \{x \mid x \in S \text{ for every } S \in \mathcal{S}\}.$$

Some confusion with this definition is caused by the fact that a set of sets may (of course) be empty. Applying the above definition to the case $\mathcal{S} := \emptyset$ is no problem for the union, since

$$\bigcup \emptyset = \{x \mid x \in S \text{ for some } S \in \mathcal{S}\} = \{x \mid \text{false}\} = \emptyset.$$

But there is a problem for the intersection, because the condition "$x \in S$ for every $S \in \mathcal{S}$" is satisfied by *all* x (because there is nothing to be satisfied). But there is no *set of all* x; such sets are forbidden in set theory, because they would lead to contradictions.

For the case $\mathcal{S} = \emptyset$ the intersection is defined only with respect to some base set M. If we work with the subsets of some specified set M (as we often do, for example with the set of all attributes of some formal context), then we define

$$\bigcap \emptyset := M.$$

A set M with, say, n elements, has 2^n subsets. The set of all subsets of a set M is denoted $\mathfrak{P}(M)$ and is called the **power set** of the set M. To indicate that $\mathcal{S}$ is a set of subsets of M, we may therefore simply write $\mathcal{S} \subseteq \mathfrak{P}(M)$.

A closure system on a set M is a set of subsets that contains M and is closed under intersections.

Definition 19. A **closure system** on a set M is a set $\mathcal{C} \subseteq \mathfrak{P}(M)$ satisfying

- $M \in \mathcal{C}$, and
- if $\mathcal{D} \subseteq \mathcal{C}$, then $\bigcap \mathcal{D} \in \mathcal{C}$.

$\Diamond$

Definition 20. A **closure operator** φ on M is a map $\mathfrak{P}(M) \to \mathfrak{P}(M)$ assigning a **closure** $\varphi X \subseteq M$ to each set $X \subseteq M$, which is

monotone: $X \subseteq Y \Rightarrow \varphi X \subseteq \varphi Y$,
extensive: $X \subseteq \varphi X$, and
idempotent: $\varphi \varphi X = \varphi X$.

(Conditions to be satisfied for all $X, Y \subseteq M$.) $\Diamond$

Closure operators are frequently met: their axioms describe the natural properties of a *generating process*. We start with some generating set X, apply the generating process and obtain the generated set, φX, the closure of X. Such generating processes occur in fact in many different variants in mathematics and computer science.

Closure systems and closure operators are closely related. In fact, there is a natural way to obtain from each closure operator a closure system and vice versa. It works as follows:

Lemma 10. *For any closure operator, the set of all closures is a closure system. Conversely, given any closure system $\mathcal{C}$ on M, there is for each subset X of M a unique smallest set $C \in \mathcal{C}$ containing X. Taking this as the closure of X defines a closure operator. The two transformations are inverse to each other.*

Thus closure systems and closure operators are essentially the same. We can add to this:

Theorem 11. *A closure system $\mathcal{C}$ on a set M can be considered as a complete lattice, ordered by set inclusion $\subseteq$. The infimum of any subfamily $\mathcal{D} \subseteq \mathcal{C}$ is equal to $\bigcap \mathcal{D}$, and the supremum is the closure of $\bigcup \mathcal{D}$. Conversely, we can find for any complete lattice L a closure system that is isomorphic to L.*

So closure systems and complete lattices are also very closely related. It comes as no surprise that concept lattices fit well into this relationship. It follows from the Basic Theorem (Theorem 4) that the set of all concept intents of a formal context is closed under intersections and thus is a closure system on M. Dually, the set of all concept extents always is a closure system on G. The corresponding closure operators are just the two operators $X \mapsto X''$ on M and G, respectively.

Conversely, given any closure system $\mathcal{C}$ on a set M, we can construct a formal context such that $\mathcal{C}$ is the set of concept intents. It can be concluded from the Basic Theorem that for example $(\mathcal{C}, M, \ni)$ is such a context. In particular, whenever a closure operator on some set M is considered, we may assume that it is the closure operator $A \mapsto A''$ on the attribute set of some formal context (G, M, I).

Thus, closure systems and closure operators, complete lattices, systems of concept intents, and systems of concept extents: all these are very closely related. It is not appropriate to say that they are "essentially the same", but it is true that all these structures have the same degree of expressiveness; none of them is a generalization of another. A substantial result proved for one of these structures can usually be transferred to the others, without much effort.

4.2   The Next Closure Algorithm

We present a simple algorithm that solves the following task: For a given closure operator on a finite set M, it computes all closed sets.

There are many ways to achieve this. Our algorithm is particularly simple. We shall discuss efficiency considerations below.

We start by endowing our base set M with an arbitrary linear order, so that

$$M = \{m_1 < m_2 < \cdots < m_n\},$$

where n is the number of elements of M. Then every subset $S \subseteq M$ can conveniently be described by its **characteristic vector**

$$\varepsilon_S : M \to \{0, 1\},$$

given by

$$\varepsilon_S(m) := \begin{cases} 1 & \text{if } m \in S \\ 0 & \text{if } m \notin S \end{cases}.$$

For example, if the base set is

$$M := \{a < b < c < d < e < f < g\},$$

then the characteristic vector of the subset $S := \{a, c, d, f\}$ is 1011010. In concrete examples we prefer to write a cross instead of a 1 and a blank or a dot instead of a 0, similarly as in the cross tables representing formal contexts. The characteristic vector of the subset $S := \{a, c, d, f\}$ will therefore be written as

×	.	×	×	.	×	.

.

Using this notation, it is easy to see if a given set is a subset of another given set, etc.

The set $\mathfrak{P}(M)$ of all subsets of the base set M is naturally ordered by the subset-order $\subseteq$. This is a complete lattice order, and $(\mathfrak{P}(M), \subseteq)$ is called the **power set lattice** of M. The subset-order is a *partial order*. We can also introduce a *linear* or *total* order of the subsets, for example the **lexicographic** or **lectic order** $\leq$, defined as follows: Let $A, B \subseteq M$ be two distinct subsets. We say that A is *lectically smaller* than B, if the smallest element in which A and B differ belongs to B. Formally,

$$A < B \quad :\Longleftrightarrow \quad \exists i.(i \in B \,\wedge\, i \notin A \,\wedge\, \forall j < i(j \in A \Longleftrightarrow j \in B)).$$

For example $\{a, c, e, f\} < \{a, c, d, f\}$, because the smallest element in which the two sets differ is d, and this element belongs to the larger set. This becomes even more apparent when we write the sets as vectors and interpret them as binary numbers:

$$1\ 0\ 1\ 0\ 1\ 1\ 0$$
$$\updownarrow$$
$$1\ 0\ 1\ 1\ 0\ 1\ 0 \ .$$

Note that the lectic order extends the subset-order, i.e.,

$$A \subseteq B \Rightarrow A \leq B.$$

The following notation is helpful:

$$A <_i B \quad :\Longleftrightarrow \quad i \in B \ \wedge \ i \notin A \ \wedge \ \forall j < i \ (j \in A \Longleftrightarrow j \in B)).$$

In words: $A <_i B$ iff i is the smallest element in which A and B differ, and $i \in B$.

Proposition 12. *1. $A < B$ if and only if $A <_i B$ for some $i \in M$.*
2. If $A <_i B$ and $A <_j C$ with $i < j$, then $C <_i B$.

We consider a closure operator

$$A \mapsto A''$$

on the base set M. To each subset $A \subseteq M$ it yields[7] its closure $A'' \subseteq M$. Our task is to find a list of all these closures. In principle, we might just follow the definition, compute for each subset $A \subseteq M$ its closure A'' and include that in the list. The problem is that different subsets may have identical closures. So if we want a list that contains each closure *exactly once*, we will have to check many times if a computed closure already exists in the list. Moreover, the number of subsets is exponential: a set with n elements has 2^n subsets. The naive algorithm "for each $A \subseteq M$, compute A'' and check if the result is already listed" therefore requires an exponential number of lookups in a list that may have exponential size.

A better idea is to generate the closures in some predefined order, thereby guaranteeing that every closure is generated only once. The reader may guess that we shall generate the closures in lectic order. We will show how to compute, given a closed set, the *lectically next* one. Then no lookups are necessary. Actually, it will not even be necessary to store the list. For many applications it will suffice to generate the list elements on demand. Therefore we do not have to store exponentially many closed sets. Instead, we shall store just *one*!

To find the next closure we define for $A \subseteq M$ and $m_i \in M$

$$A \oplus m_i := ((A \cap \{m_1, \ldots, m_{i-1}\}) \cup \{m_i\})''.$$

---

[7] For our algorithm it is not important *how* the closure is computed.

We illustrate this definition by an example: Let $A := \{a, c, d, f\}$ and $m_i := e$.

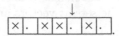

We first remove all elements that are greater or equal m_i from A:

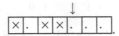

Then we insert m_i

and form the closure. Since we have not yet specified the closure operator $\cdot''$ (i. e., we have not given a formal context), the example stops here with

$$A \oplus e = \{a, c, d, e\}''.$$

Proposition 13. *1. If $i \notin A$ then $A < A \oplus i$.*
2. If B is closed and $A <_i B$ then $A \oplus i \subseteq B$, in particular $A \oplus i \leq B$.
3. If B is closed and $A <_i B$ then $A <_i A \oplus i$.

Theorem 14. *The smallest closed set larger than a given set $A \subset M$ with respect to the lectic order is*
$$A \oplus i,$$
i being the largest element of M with $A <_i A \oplus i$.

Now we are ready to give the algorithm for generating all extents of a given context (G, M, I): The lectically smallest extent is $\emptyset''$. For a given set $A \subset G$ we find the lectically next extent by checking all elements i of $G \setminus A$, starting from the largest one and continuing in a descending order until for the first time $A <_i A \oplus i$. $A \oplus i$ then is the "next" extent we have been looking for. These three steps are made explicit in Figs. 7, 8 and 9.

```
Algorithm First Closure
Input:   A closure operator X ↦ X'' on a finite set M.
Output:  The closure A of the empty set.
begin
   A := ∅'';
end.
```

Fig. 7. First Closure.

```
Algorithm NEXT CLOSURE
Input:    A closure operator X ↦ X'' on a finite set M,
              and a subset A ⊆ M.
Output:   A is replaced by the lectically next closed set.
begin
    i := largest element of M;
    i := succ(i);
    success := false;
    repeat
        i := pred(i);
        if i ∉ A then
        begin
            A := A ∪ {i};
            B := A'';
            if B \ A contains no element < i then
            begin
                A:= B;
                success := true;
            end;
        end else A := A \ {i};
    until erfolg or i = smallest element of M.
end.
```

Fig. 8. Next Closure.

```
Algorithm ALL CLOSURES
Input:    A closure operator X ↦ X'' on a finite set M.
Output:   All closed sets in lectic order.
begin
    First_Closure;
    repeat
        Output A;
        Next_Closure;
    until not success;
end.
```

Fig. 9. Generating all closed sets.

5   Implications

Have another look at the concept lattice shown in Fig. 4. The six attributes describe how two unit squares can be placed with respect to each other. Each of the ten objects is a pair of unit squares, representing a possible placement. These ten pairs are representatives for an infinite set of possible positions that such pairs of squares may have. It is not stated, but perhaps expected by the reader, that these ten examples cover all possible combinations of the given attributes.

Such a situation occurs often: attributes are given, but objects are not known, or too many to handle them completely. We then have to study the possible attribute combinations, the *attribute logic* of the respective situation.

Let M be some set. We shall call the elements of M *attributes*, so as if we consider a formal context (G, M, I). However we do not assume that such a context is given or explicitly known.

Definition 21. An **implication between attributes** in M is an expression of the form $A \to B$ where A and B are subsets of M. The set A is the **premise** of the implication and B is its **conclusion**.

A subset $T \subseteq M$ **respects** an implication $A \to B$ if $A \not\subseteq T$ or $B \subseteq T$. We then also say that T is a **model** of the implication $A \to B$, and denote this by $T \models A \to B$. T **respects a set** $\mathcal{L}$ of implications if T respects every single implication in $\mathcal{L}$. The implication $A \to B$ **holds** in a set $\{T_1, T_2, \ldots\}$ of subsets if each of these subsets respects $A \to B$. With

$$\mathrm{Imp}\{T_1, T_2, \ldots\}$$

we denote the set of all implications that hold in $\{T_1, T_2, \ldots\}$.   ◊

5.1   Implications of a Formal Context

Now let us consider the special case of implications of a formal context.

Definition 22. $A \to B$ **holds in a context** (G, M, I) if every object intent respects $A \to B$, that is, if each object that has all the attributes in A also has all the attributes in B. We then also say that $A \to B$ is an *implication of* (G, M, I).   ◊

Proposition 15. *An implication $A \to B$ holds in (G, M, I) if and only if $B \subseteq A''$, which is equivalent to $A' \subseteq B'$. It then automatically holds in the set of all concept intents as well.*

An implication $A \to B$ holds in (G, M, I) if and only if each of the implications

$$A \to m, \qquad m \in B,$$

holds ($A \to m$ is short for $A \to \{m\}$). We can read this off from a concept lattice diagram in the following manner: $A \to m$ holds if the infimum of the attribute concepts corresponding to the attributes in A is less or equal than the attribute concept for m, formally if

$$\bigwedge \{\mu a \mid a \in A\} \leq \mu m.$$

$A \to B$ holds in (G, M, I) if

$$\bigwedge \{\mu a \mid a \in A\} \leq \bigwedge \{\mu b \mid b \in B\}.$$

5.2   Semantic and Syntactic Implication Inference

As we will see, it is not necessary to store all implications of a formal context. We will discuss how implications can be derived from already known implications. First we discuss which kind of inference we want to model. This is given by the so-called *semantical inference*. Then we discuss a calculus (*syntactic inference*), and argue that the calculus is correct and complete with respect to our semantics.

5.3   When Does An Implication Follow from Other Implications (Semantically)?

Proposition 16. *If $\mathcal{L}$ is a set of implications in M, then*

$$\operatorname{Mod}\mathcal{L} := \{T \subseteq M \mid T \text{ respects } \mathcal{L}\}$$

is a closure system on M. If $\mathcal{L}$ is the set of all implications of a context, then $\operatorname{Mod}\mathcal{L}$ is the system of all concept intents.

The respective closure operator

$$X \mapsto \mathcal{L}(X)$$

can be described as follows: For a set $X \subseteq M$, let

$$X^{\mathcal{L}} := X \cup \bigcup \{B \mid A \to B \in \mathcal{L}, A \subseteq X\}.$$

Form the sets $X^{\mathcal{L}}$, $X^{\mathcal{L}\mathcal{L}}$, $X^{\mathcal{L}\mathcal{L}\mathcal{L}}$, ... until[8] a set $\mathcal{L}(X) := X^{\mathcal{L}\dots\mathcal{L}}$ is obtained with $\mathcal{L}(X)^{\mathcal{L}} = \mathcal{L}(X)$. Later on we shall discuss how to do this computation efficiently.

It is not difficult to construct, for any given set $\mathcal{L}$ of implications in M, a formal context such that $\operatorname{Mod}\mathcal{L}$ is the set of concept intents of this formal context. In fact, $(\operatorname{Mod}\mathcal{L}, M, \ni)$ will do.

Definition 23. An implication $A \to B$ **follows (semantically)** from a set $\mathcal{L}$ of implications in M if each subset of M respecting $\mathcal{L}$ also respects $A \to B$. A family of implications is called **closed** if every implication following from $\mathcal{L}$ is already contained in $\mathcal{L}$. A set $\mathcal{L}$ of implications of (G, M, I) is called **complete**, if every implication that holds in (G, M, I) follows from $\mathcal{L}$.   ◇

In other words: An implication $A \to B$ follows semantically from $\mathcal{L}$ if it holds in every model of $\mathcal{L}$.

5.4   When Does an Implication Follow from Other Implications (Syntactically)?

The semantic definition of implication inference has a syntactic counterpart. We can give sound and complete inference rules (known as **Armstrong rules** [1]) and an efficient algorithm for inference testing.

---

[8] If M is infinite, this may require infinitely many iterations.

Proposition 17. *A set $\mathcal{L}$ of implications in M is closed if and only if the following conditions are satisfied for all $W, X, Y, Z \subseteq M$:*

1. $X \to X \in \mathcal{L}$,
2. *If $X \to Y \in \mathcal{L}$, then $X \cup Z \to Y \in \mathcal{L}$,*
3. *If $X \to Y \in \mathcal{L}$ and $Y \cup Z \to W \in \mathcal{L}$, then $X \cup Z \to W \in \mathcal{L}$.*

Readers with a background in Computational Logic may prefer a different notation of these Armstrong rules:

$$\frac{}{X \to X}, \qquad \frac{X \to Y}{X \cup Z \to Y}, \qquad \frac{X \to Y, \quad Y \cup Z \to W}{X \cup Z \to W}.$$

The proposition says that a set of implications is the set of all implications of some context if and only if it is closed with respect to these rules. In other words, an implication follows from other implications if and only if it can be derived from these by successive applications of these rules. In particular, semantic and syntactic inference are the same.

However, these rules do not always suggest the best proof strategy. Instead, we may note the following:

Proposition 18. *An implication $X \to Y$ follows from a list $\mathcal{L}$ of implications if and only if $Y \subseteq \mathcal{L}(X)$.*

We give an algorithm that efficiently computes the closure $\mathcal{L}(X)$ for any given set X. Such algorithms are used in the theory of relational data bases for the study of functional dependencies.

We can give a rough complexity estimation of the algorithm in Fig. 10. Except for manipulations of addresses, the main effort is to apply the implications. Each implication is applied at most once, and each application requires a simple set operation. Therefore the time required by the closure algorithm is essentially *linear* in the size of the input $\mathcal{L}$.

Summarizing these considerations we learn that implication inference is easy: to check if an implication $X \to Y$ follows from a list $\mathcal{L}$ of implications, it suffices to check if $Y \subseteq \mathcal{L}(X)$ (by Proposition 18), and this can be done in time linear in the size of the input.

In other words: implications are easy to use, much easier than many other logical constructs. This may be a reason why implications are popular, and perhaps be part of an explanation why our simple theory of formal concepts is so useful.

5.5   The Stem Base

The number of implications that hold in a given situation can be very large. For example, if there is only one closed set, M, then *every* implication holds. If M has n elements, then these are some 2^{2n} implications. But this is ridiculous, because all these implications can be inferred from a single one, namely from $\emptyset \to M$.

```
Algorithm CLOSURE
Input:    A list L =: [L[1],...,L[n]] of implications in M
          and a set X ⊆ M.
Output:   The closure L(X) of X.
begin
  for all x ∈ M do
  begin
    avoid[x] := {1,...,n};
    for i := 1 to n do with A → B := L[i]
      if x ∈ A then avoid[x] :=avoid[x] \ {i};
  end;
  used_imps :=∅;
  old_closure :={-1};      (* some element not in M *);
  new_closure := X;
  while new_closure ≠ old_closure do
  begin
    old_closure := new_closure;
    T := M \new_closure;
    usable_imps := ∩_{x∈T} avoid[x];
    use_now_imps := usable_imps \ used_imps;
    used_imps := usable_imps;
    for all i ∈ use_now_imps with A → B := L[i] do
      new_closure := new_closure ∪ B;
  end;
  L(X) :=new_closure;
end.
```

Fig. 10. Algorithm CLOSURE.

We see from this trivial example that the set of *all* implications of a given formal context may be highly redundant. It is a natural question to ask for a small *implicational base*, from which everything else follows. More precisely we ask, for any given formal context (G, M, I), for a list $\mathcal{L}$ of implications that is

– sound (i.e., each implication in $\mathcal{L}$ holds in (G, M, I)),
– complete (i.e., each implication that holds in (G, M, I) follows from $\mathcal{L}$), and
– non-redundant (i.e., no implication in $\mathcal{L}$ follows from other implications in $\mathcal{L}$).

It is easy to see that (for finite M) such sets $\mathcal{L}$ exist. We may start with some sound and complete set of implications, for example, with the set of all implications that hold in (G, M, I). We then can successively remove redundant implications from this set, until we obtain a sound, complete, non redundant set.

But this is an unrealistic procedure. We therefore look for a better way to construct an implicational base. Duquenne and Guigues [5] have shown that there is a natural choice, the *stem base*.

The following recursive definition is rather irritating at the first glance. We define a pseudo-closed set to be a set which is not closed, but contains the closure of every pseudo-closed proper subset.

Definition 24. Let $X \mapsto X''$ be a closure operator on the finite set M. We call a subset $P \subseteq M$ **pseudo-closed**, if (and only if)

- $P \neq P''$, and
- if $Q \subset P$ is a pseudo-closed proper subset of P, then $Q'' \subseteq P$.

$\Diamond$

This *is* a valid recursive definition. It is not circular, because the pseudo-closed set Q must have fewer elements than P, because it is a proper subset.[9]

Reformulating this definition for formal contexts, we obtain the following definition.

Definition 25. Let (G, M, I) be a formal context with M finite. A subset $P \subseteq M$ is a **pseudo intent** of (G, M, I) iff

- P is not a concept intent, and
- if $Q \subset P$ is a pseudo-closed proper subset of P, then there is some object $g \in G$ such that $Q \subseteq g'$ but $P \not\subseteq g'$.

$\Diamond$

Theorem 19. *Let M be finite and let $X \mapsto X''$ be some closure operator on M. The set of implications*

$$\{P \rightarrow P'' \mid P \text{ pseudo-closed}\}$$

is sound, complete, and non redundant.

The implication set in the theorem deserves a name. It is sometimes called the *Duquenne–Guigues–base*. We simply call it the **stem base** of the closure operator (or the *stem base of a given formal context*, if the closure operator is given that way). In practice one uses a slightly different version of the stem base, namely

$$\{P \rightarrow P'' \setminus P \mid P \text{ pseudo-closed}\}.$$

The stem base is not the only implicational base, but it plays a special rôle. For example, no implicational base can consist of fewer implications, as the next proposition shows:

Proposition 20. *Every sound and complete set of implications contains, for every pseudo closed set P, an implication $A \rightarrow B$ with $A'' = P''$.*

---

[9] 'Recursive' is meant here with respect to set inclusion. Compare with the following recursive definition: A natural number is *prime* iff it is greater than 1 and not divisible by any smaller prime number.

5.6   Computing the Stem Base

As before, consider a closure operator $X \mapsto X''$ on a finite set M. We start with a harmless definition:

Definition 26. A set $Q \subseteq M$ is **•-closed** if it contains the closure of every •-closed set that is properly contained in Q.

Formally, Q is •-closed iff for each •-closed set $Q_0 \subset Q$ with $Q_0 \neq Q$ we have $Q_0'' \subseteq Q$. ◊

This is a simple (but convenient) renaming, as the next proposition shows.

Proposition 21. *A set is •-closed iff it is either closed or pseudo-closed.*

Observe that if Q contains the closure of every •-closed subset, then Q must be closed.

The first crucial step towards finding pseudo-closed sets is this:

Proposition 22. *The intersection of •-closed sets is •-closed.*

In other words: the •-closed sets form a closure system. We have described an algorithm to compute, for a given closure operator, all closed sets. We can apply this algorithm for computing all •-closed sets, provided that we can access the corresponding closure operator. This is easy. We prepare the result with a proposition that is an immediate consequence of Definition 26.

Proposition 23. Q *is •-closed iff* Q *satisfies the following condition:*

If $P \subset Q$, $P \neq Q$, is pseudo-closed, then $P'' \subseteq Q$.

Proposition 23 shows how to find the quasi closure of an arbitrary set $S \subseteq M$: As long as the condition in the proposition is violated, we (are forced to) extend the set S, until we finally reach a fixed point.

Let $\mathcal{L}$ be the stem base[10]. Define, for $X \subseteq M$,

$$X^{\mathcal{L}^\bullet} := X \cup \bigcup \{P'' \mid P \to P'' \in \mathcal{L}, P \subset X, P \neq X\},$$

iterate by forming

$$X^{\mathcal{L}^\bullet \mathcal{L}^\bullet}, X^{\mathcal{L}^\bullet \mathcal{L}^\bullet \mathcal{L}^\bullet}, \ldots$$

until a set

$$\mathcal{L}^\bullet(X) := X^{\mathcal{L}^\bullet \mathcal{L}^\bullet \ldots \mathcal{L}^\bullet}$$

is obtained that satisfies

$$\mathcal{L}^\bullet(X) = \mathcal{L}^\bullet(X)^{\mathcal{L}^\bullet}.$$

Proposition 24. $\mathcal{L}^\bullet(X)$ *is the smallest •-closed set containing X (Fig. 11).*

```
Algorithm L*–CLOSURE
Input:    A list L =: [L[1], ..., L[n]] of implications in M
            and a set X ⊆ M.
Output:   The closure L(X) of X.
begin
  for all x ∈ M do
  begin
    avoid[x] := {1, ..., n};
    for i := 1 to n do with A → B := L[i]
      if x ∈ A then avoid[x] := avoid[x] \ {i};
  end;
  used_imps := ∅;
  old_closure := {-1};     (* some element not in M *);
  new_closure := X;
  while new_closure ≠ old_closure do
  begin
    old_closure := new_closure;
    T := M \ new_closure;
    usable_imps := ∩_{x∈T} avoid[x] ∩ ∪_{x∈new_closure} avoid[x];
    use_now_imps := usable_imps \ used_imps;
    used_imps := usable_imps;
    for all i ∈ use_now_imps with A → B := L[i] do
      new_closure := new_closure ∪ B;
  end;
  L(x) := new_closure;
end.
```

Fig. 11. Computing the L^*–CLOSURE.

Note that in order to find the quasi closure, we use only pseudo-closed sets which are contained in the closure and therefore in particular are lectically smaller than the quasi closure. Thus, the same result is obtained if L is a subset of the stem base, containing those implications $P \to P''$ for which P is pseudo-closed and lectically smaller than $L^*(X)$.

Now it is easy to give an algorithm to compute all pseudo-closed sets for a given closure operator. We use the Next Closure algorithm applied to the closure system of •-closed sets. For short, we shall refer to this as the **next quasi closure** after a given set A, NEXT_L^*_CLOSURE(A). This produces all •-closed sets in lectic order. We record only those which are not closed. This yields a list of all pseudo-closed sets.

Since the •-closed sets are generated in lectic order, we have, at each step, the full information about the lectically smaller pseudo-closed sets. We have seen that this suffices to compute the "quasi closure" operator. The algorithm in Fig. 12 uses a dynamic list L. Whenever a pseudo-closed set P is found, the

---

[10] The reader might wonder why we use the stem base to construct the stem base. As we shall see soon, this works, due to the recursive definition of the stem base.

corresponding implication $P \rightarrow P''$ is included in the list. Since the pseudo-closed sets are found in lectic order, this makes sure that at any step we have sufficient information to compute the quasi closure.

```
Algorithm STEM BASE
Input:    A closure operator X ↦ X'' on a finite set M,
          for example given by a formal context (G, M, I).
Output:   The stem base L

begin
  L := ∅;
  A := ∅;
  while A ≠ M do
  begin
    if A ≠ A'' then L := L ∪ {A → A''};
    A := NEXT_L•_CLOSURE(A);
  end;
end.
```

Fig. 12. Computing the stem base for a given closure operator.

Example 2. We compute the stem base for the context of triangles given in Example 1. The steps are shown in Fig. 13. The first column contains all quasi closed sets, in lectic order. The pseudo-closed sets are precisely those which are not closed (see middle column). Each pseudo-closed set gives rise to an entry in the stem base (last column, short form).

•-closed set	closed ?	stem base implication
∅	yes	
$\{e\}$	yes	
$\{d\}$	yes	
$\{d, e\}$	no	$\{d, e\} \rightarrow \{a, b, c\}$
$\{c\}$	yes	
$\{c, e\}$	no	$\{c, e\} \rightarrow \{a, b, d\}$
$\{c, d\}$	no	$\{c, d\} \rightarrow \{a, b, e\}$
$\{b\}$	yes	
$\{b, e\}$	yes	
$\{b, d\}$	yes	
$\{b, c\}$	yes	
$\{a\}$	no	$\{a\} \rightarrow \{b, c\}$
$\{a, b, c\}$	yes	
$\{a, b, c, d, e\}$	yes	

Fig. 13. Steps in the stem base algorithm

Since the closure operator is given in terms of a formal context, we may speak of **quasi intent**s and **pseudo intent**s instead of •-closed sets or pseudo-closed sets. We see that the algorithm generates all quasi intents to find the stem base. In other words, to compute all pseudo intents we also compute all intents, possibly exponentially many. This looks like a rather unefficient method. Unfortunately, we do not know of a better strategy. It is an open problem to find a better algorithm for the stem base. In practice, the algorithm is not fast, but nevertheless very useful.

6   Conclusion

Much more can be said about FCA, here we only dealt with the foundations of the discipline. Over the past decades, the field has expanded in many directions. We give a few examples of central topics in FCA which weren't discussed here.

- **Conceptual Scaling.** One can argue that formal contexts are only able to represent very limited information (essentially yes/no). The idea to allow for proper entries in the tables rather than just crosses or blanks leads to the notion of *multi-valued contexts* which are closer to database tables typically encountered in practice. A very generic method to make the machinery of FCA applicable to data represented as multi-valued contexts is called *conceptual scaling*, where this data is transformed back into plain formal contexts.
- **Association Rules.** Real world data is often noisy and error-prone. In order to still extract meaningful implicational information from such data, one needs to formalize the notion of implications which do not hold always (but often). *Association rules* are such implications that come with two values, *support* (the fraction of all objects where the implication is applicable and valid) and *confidence* (the fraction of objects satisfying the implication among all objects where it is applicable). For association rules, one can again characterize semantic and syntactic consequences and establish implicational bases [7].
- **Triadic FCA.** The readers might ask themselves, why the incidence relation used to characterize data in FCA is a binary one. Couldn't it have a higher arity? Indeed, people have investigated *triadic FCA* [6], where the incidence relation is a ternary one between objects, attributes and conditions. Some of the notions of FCA can be nicely generalized to the triadic case (and even to incidence relations of higher arity, giving rise to *polyadic FCA* [8]) but others are specific to the binary case.
- **Attribute Exploration.** Sometimes the data of a domain is only partially recorded in a formal context, but there are experts who know the full domain of interest. In that case, algorithms exist which can complete the context and determine all the implications in an interactive process, where an expert is repeatedly asked questions about the domain [3].

Acknowledgments. We are grateful for the valuable feedback from the anonymous reviewers, which helped greatly to improve this work. Special thanks to Thomas Feller for his very careful proof-reading. This work has been funded by the European Research Council via the ERC Consolidator Grant No. 771779 (DeciGUT).

References

1. Armstrong, W.: Dependency structures of data base relationships. In: Proceedings of IFIP Congress, pp. 580–583 (1974)
2. Arnauld, A., Nicole, P.: La logique ou l'art de penser—contenant, outre les règles communes, plusieurs observations nouvelles, propres à former le jugement. Ch. Saveux, Paris (1668)
3. Ganter, B., Obiedkov, S.A.: Conceptual Exploration. Springer, Heidelberg (2016). https://doi.org/10.1007/978-3-662-49291-8
4. Ganter, B., Wille, R.: Formal Concept Analysis: Mathematical Foundations. Springer, Heidelberg (1997)
5. Guigues, J.L., Duquenne, V.: Familles minimales d'implications informatives resultant d'un tableau de données binaires. Mathématiques et Sciences Humaines **95**, 5–18 (1986)
6. Lehmann, F., Wille, R.: A triadic approach to formal concept analysis. In: Ellis, G., Levinson, R., Rich, W., Sowa, J.F. (eds.) ICCS-ConceptStruct 1995. LNCS, vol. 954, pp. 32–43. Springer, Heidelberg (1995). https://doi.org/10.1007/3-540-60161-9_27
7. Luxenburger, M.: Implications partielles dans un contexte. Mathématiques, Informatique et Sciences Humaines **113**(29), 35–55 (1991)
8. Voutsadakis, G.: Polyadic concept analysis. Order J. Theory Ordered Sets Appl. **19**(3), 295–304 (2002)

Logic-Based Learning of Answer Set Programs

Mark Law, Alessandra Russo[✉], and Krysia Broda

Department of Computing, Imperial College London, London, UK
{mark.law09,a.russo,k.broda}@imperial.ac.uk

Abstract. Learning interpretable models from data is stated as one of the main challenges of AI. The goal of logic-based learning is to compute interpretable (logic) programs that explain labelled examples in the context of given background knowledge. This tutorial introduces recent advances of logic-based learning, specifically learning non-monotonic logic programs under the answer set semantics. We introduce several learning frameworks and algorithms, which allow for learning highly expressive programs, containing rules representing non-determinism, choice, exceptions, constraints and preferences. Throughout the tutorial, we put a strong emphasis on the expressive power of the learning systems and frameworks, explaining why some systems are incapable of learning particular classes of programs.

Keywords: Non-monotonic Inductive Logic Programming ·
Generality of learning frameworks · Learning Answer Set Programs

1 Introduction

Over the last decade we have witnessed a growing interest in Machine Learning. In recent years Deep Learning has been demonstrated to achieve high-levels of accuracy in data analytics, signal and information processing tasks, bringing transformative impact in domains such as facial, image, speech recognition, and natural language processing. They have best performance on computational tasks that involve large quantities of data and for which the labelling process and feature extraction would be difficult to handle. However, they also suffer from two main drawbacks, which are crucial in the context of cognitive computing. They are not capable of supporting AI solutions that are good at more than one task. They are very effective when applied to single specific tasks (e.g. recognition of specific clues, objects in images, natural language translation). But applying the same technology from one task to another within the same class of problems would often require retraining, causing the system to possibly *forget* how to solve a previously learned task. Secondly, and most importantly, they are not *transparent*. Operating primarily as black boxes, deep learning approaches are not amenable to human inspection and human feedbacks, and the learned models

© Springer Nature Switzerland AG 2019
M. Krötzsch and D. Stepanova (Eds.): Reasoning Web 2019, LNCS 11810, pp. 196–231, 2019.
https://doi.org/10.1007/978-3-030-31423-1_6

are not explainable, leaving the humans agnostic of the cognitive and learning process performed by the system. This lack of transparency hinders human comprehension, auditing of the learned outcomes, and human active engagement into the learning and reasoning processes performed by the AI systems. This has become an increasingly important issue in view of the recent General Data Protection Regulation (GDPR) which requires actions taken as a result of a prediction from a learned model to be justified.

Within the last decade, there has been a growing interest in Machine Learning approaches whose learned models are explainable and human interpretable. The last ten years have witnessed a tremendous advancement in the field of logic-based machine learning, also referred to as Inductive Logic Programming (ILP) [28, 30], where the goal is the automated acquisition of knowledge (expressed as a logic program) from given (labelled) examples and existing background knowledge. The main advantage of these machine learning approaches is that the learned knowledge can be easily expressed into plain English and explained to a human user, so facilitating a closer interaction between humans and the machine. Although a well established field since the early '90s [28], logic-based machine learning has traditionally addressed the task of learning knowledge expressible in a very limited form [29] (definite clauses). Our logic-based machine learning systems [2, 7, 21] have extended this field to a wider class of formalisms for knowledge representation, captured by the answer set programming (ASP) semantics [14]. This ASP formalism is truly declarative, and due to its non-monotonicity it is particularly well suited to common-sense reasoning [9, 13, 27]. It allows constructs such as choice rules, hard and weak constraints, and support for default inference and default assumptions. Choice rules and weak constraints are particularly useful for modelling human preferences, as the choice rules can represent the choices available to the user, and the weak constraints can specify which choices a human prefers. The typical workflow in ASP is that a real world problem is encoded as an ASP program, whose *answer sets* – a special subset of the models of the program – correspond to the solutions of the original problem. Because of its expressiveness and efficient solving, ASP is also increasingly gaining attention in industry [10]; for example, in decision support systems [31], in e-tourism [38] and in product configuration [43].

In the recent years we have made fundamental contributions to the field of ILP by extending it to the learning of the full class of ASP programs [7, 21, 25, 33, 36, 40] and this tutorial provides an introduction to these results and to the general field of learning under the answer set semantics. In general, ASP programs can have one, many or even no answer sets. Early approaches to learning ASP programs can mostly be divided into two categories: *brave* learners aim to learn a program such that at least one answer set covers the examples; on the other hand, *cautious* learners aim to find a program which covers the examples in all answer sets. Most of the early ASP-based ILP systems were brave, and several of these are presented in Sect. 3 of this tutorial. In [21], we showed that some ASP programs cannot be learned using either the brave or the cautious settings, and in fact a combination of *both* brave and cautious semantics is needed.

This was the original motivation for the *Learning from Answer Sets* family of frameworks, which we have developed since then and have been shown to be able to learn any ASP program. Section 4 presents these Learning from Answer Sets frameworks and discusses the associated ILASP algorithms. The *generality* of the main ASP-based ILP frameworks was investigated, with the aim being to formally define the classes of problems that can be solved by each of these learning frameworks has also been investigated [25]. We re-present and discuss the main results of this investigation in Sect. 4.

The above is all presented in the context of learning tasks where all examples are assumed to be perfectly labeled, meaning that any inductive solution of a task must cover every example of that task. In practice, of course, examples are unlikely to be perfectly labeled. In real datasets, it is likely that there is *noise*, and a more realistic approach is to search for a hypothesis that covers the majority of examples, and balances the example coverage against the complexity of the hypothesis – dramatically increasing the hypothesis complexity in order to cover a few more examples is undesirable, as these examples may well be incorrectly labeled. We end the tutorial by discussing how ILP frameworks can be extended to learn from noisy examples.

The rest of this document is structured as follows. In the next section we recall the background material necessary for this tutorial. Section 3 covers the early approaches to learning under the answer set semantics. Section 4 introduces the more recent advances including the Learning from Answer Sets frameworks, the generality results for the frameworks and extensions for learning from noisy examples. Much of the material in this tutorial is based on [20].

2   Background

In this section we introduce the background material used in the tutorial.

2.1   Answer Set Programming

Given any atoms $h, h_1, \ldots, h_k, b_1, \ldots, b_n, c_1, \ldots, c_m$, a rule h :- $b_1, \ldots, b_n$, not c_1, $\ldots$, not c_m is called a *normal rule*, with h as the *head* and $b_1, \ldots, b_n$, not $c_1, \ldots$, not c_m (collectively) as the *body* ("not" represents negation as failure); a *constraint* is a rule :- $b_1, \ldots, b_n$, not $c_1, \ldots$, not c_m; and a *choice rule* is a rule of the form $l\{h_1, \ldots, h_k\}u$:- $b_1, \ldots, b_n$, not $c_1, \ldots$, not c_m where $l\{h_1, \ldots, h_k\}u$ is called an *aggregate*. In an aggregate l and u are integers and h_i, for $1 \leq i \leq k$, are atoms. For example, when learning a policy, we may need to learn that in a specific scenario, sc_1, at least one of a set of possible actions, $a_1, \ldots, a_n$, must be executed. This can be expressed in ASP with the following choice rule: $1\{\texttt{execute}(a_1), \ldots, \texttt{execute}(a_n)\}n$:- $\texttt{holds}(sc_1)$. This expresses that in a model whenever the scenario sc_1 is true, it must be the case that between 1 and n atoms $\texttt{execute}(a_1), \ldots, \texttt{execute}(a_n)$ are also true. In other words, whenever the scenario holds, at least one (but possibly more) of the actions must be executed.

A rule R is called *safe* if every variable in R occurs in at least one posi-tive literal in the body of R. For example, the rules $p(X):-q(Y), not\ r(Y)$ and $p:-q, not\ r(X)$ are not safe, as X does not occur in any positive literal in their respective body. Unless otherwise specified, an ASP program P is a finite set of safe normal rules, constraints and choice rules.

Given an ASP program P, the Herbrand Base of P, denoted as HB_P, is the set of ground (variable free) atoms that can be formed from the predicates and constants that appear in P. The subsets of HB_P are called the (Herbrand) inter-pretations of P. Informally, a model of an ASP program P, called *Answer Set* of P, is defined in terms of the notion of *reduct* of P, which is in turn constructed by applying four transformation steps (described below) to the grounding of P. As shown below. a reduct is a definite program. A model of a definite pro-gram is an interpretation I that makes every rule in the program true, and a model is *minimal* if it is the smallest such interpretation. Let's see how the reduct of an ASP program is constructed. We said that it uses the grounding of the program, so we can consider just the grounding of a given program P. A ground aggregate $l\{h_1, \ldots, h_k\}u$ is said to be satisfied by an interpretation I if and only if $l \leq |I \cap \{h_1, \ldots, h_k\}| \leq u$. As we restrict our programs to sets of normal rules, constraints and choice rules, we use the simplified definitions of the *reduct* for choice rules presented in [23]. Given a program P and an Her-brand interpretation $I \subseteq HB_P$, the reduct P^I is constructed from the grounding of P using the following four transformation steps: from the grounding of P we first remove rules whose bodies contain the negation of an atom in I; sec-ondly, we remove all negative literals from the remaining rules; thirdly, we set $\perp$ (note $\perp \notin HB_P$) to be the head to every constraint, and in every choice rule whose head is not satisfied by I we replace the head with $\perp$; and finally, we replace any remaining choice rule $l\{h_1, \ldots, h_m\}u:-b_1, \ldots, b_n$ with the set of rules $\{h_i:-b_1, \ldots, b_n \mid h_i \in I \cap \{h_1, \ldots, h_m\}\}$. Any $I \subseteq HB_P$ is an *answer set* of P if it is the minimal model of the *reduct* P^I. We denote an answer set of a program P with A and the set of answer sets of P with $AS(P)$. A program P is said to be satisfiable (resp. unsatisfiable) if $AS(P)$ is non-empty (resp. empty).

ASP also allows optimisation over the answer sets according to *weak con-straints*. These are rules of the form $:\sim b_1, \ldots, b_m, not\ b_{m+1}, \ldots, not\ b_n.[w@l, t_1, \ldots, t_k]$ where $b_1, \ldots, b_n$ are atoms called (collectively) the *body* of the rule, and $w, l, t_1 \ldots t_k$ are all terms with w called the weight and l the pri-ority level. We refer to $[w@l, t_1, \ldots, t_k]$ as the *tail* of the weak constraint. A ground instance of a weak constraint W is obtained by replacing all variables in W with ground terms. We assume that all weights and levels of all ground instances of weak constraints are integers. Unlike other ASP rules, *weak con-straints* do not affect what is (or is not) an answer set of a program. Instead, they create an ordering $\succ_P$ over $AS(P)$, specifying which answer sets are "pre-ferred" to others. Informally, at each *priority level* l, satisfying weak constraints with level l means discarding any answer set that does not minimise the sum of the weights of the ground weak constraints (with level l) whose bodies are satis-fied. Higher levels are minimised first. For example, the two weak constraints

:$\sim$ mode(L, walk), distance(L, D).[D@2, L] and :$\sim$ cost(L, C).[C@1, L] express a preference ordering over alternative journeys. The first constraint (at priority 2) expresses that the total walking distance (the sum of the distances of journey legs whose mode of transport is walk) should be minimised, and the second constraint expresses that the total cost of the journey should be minimised. As the first weak constraint has a higher priority level than the second, it is minimised first – so given a journey j_1 with a higher cost than another journey j_2, j_1 is still preferred to j_2 so long as the walking distance of j_1 is lower than that of j_2. The set $ord(P)$ captures the ordering of interpretations induced by P and generalises the $\succ_P$ relation, so it not only includes $\langle A_1, A_2, < \rangle$ if $A_1 \succ_P A_2$, but includes tuples for each binary comparison operator ($<, >, =, \leq, \geq$ and $\neq$).

2.2   Inductive Logic Programming

The most common setting for ILP is called *learning from entailment*, where a task consists of a background knowledge B (a pre-defined logic program, defining concepts which may be useful), and two sets of examples (usually atoms) called the positive and negative examples, E^+ and E^-, respectively. The goal is to find a *hypothesis* H such that $\forall e \in E^+$, $B \cup H \models e$ and $\forall e \in E^-$, $B \cup H \not\models e$. When learning definite logic programs, the notion of entailment ($\models$) is usually entailment in the unique minimal Herbrand model of $B \cup H$, but we will see that under the answer set semantics, it is interesting to explore the use of other notions of entailment.

Usually, the search for hypotheses is bounded by a *hypothesis space*, which is the set of all rules allowed to appear in H. In an ILP task, the expressivity of the hypothesis space is defined by a notion of *language bias* of the task. Mode declarations are a popular means of characterising the language bias [30]. They specify which literals may appear in the head and in the body of a hypothesis. Given a language bias the full hypothesis space, also called *search space* and denoted with S_M, is given by the finite set of all the rules that can be constructed according to the given bias. A language bias can be defined as a pair of mode declarations $\langle M_h, M_b \rangle$, where M_h (resp. M_b) are called the *head* (resp. *body*) *mode declarations*. Each mode declaration m_h (resp. m_b) is a literal whose abstracted arguments are either +t or −t or #t, for some constant t (referred to as a *type*). Informally, a literal is said to be *compatible* with a mode declaration m if every instance of +t and −t in m has been replaced with a variable, and every #t with a constant of type t. We say that a variable occurs as an *input* (resp. *output*) variable of type t if it replaces an argument +t (resp. −t). Given a mode bias M, S_M is the set of all rules which are compatible with the mode declarations M.

Definition 1. *Given a set of mode declarations $M = \langle M_h, M_b \rangle$, a normal rule R is in the search space S_M if and only if (i) the head of R is compatible with a mode declaration in M_h; (ii) each body literal of R is compatible with a mode declaration in M_b; (iii) every input variable in the body of R occurs earlier in R, either as an input variable in the head, or an output variable in the body; and (iv) no variable occurs with two different types.*

In the input to most ILP systems a mode declaration is written as atom `class(recall,m)`, where `class` is either `#modeh` or `#modeb`, specifying whether the mode declaration `m` is in M_h or M_b. The *recall* is an optional integer argument, which puts an upper bound on the number of times the mode declaration can be used in a single rule. In many ILP systems, the types of variables are "enforced" by adding an extra "type" atom to the body for each variable; for instance, for a variable `V` of type `bird`, the atom `bird(V)` is added.

The notion of mode bias given in Definition 1 is commonly used in the ILP literature, but it is not universal. ILASP uses a different notion of mode bias, which we will not present in this tutorial. We refer the reader to https://www.doc.ic.ac.uk/~ml1909/ILASP/language_biases_2018.pdf for further details. For simplicity, all the ILASP learning tasks presented in this paper will include an explicit hypothesis space defined in terms of sets of ASP (choice) rules and constraints rather than using a declarative mode bias.

2.3 Complexity Theory

We assume the reader is familiar with the fundamental concepts of complexity, such as Turing machines and reductions; for a detailed explanation, see [34]. $\mathcal{P}$ is the class of all problems which can be solved in polynomial time by a Deterministic Turing Machine (DTM); $\Sigma_0^{\mathcal{P}} = \Pi_0^{\mathcal{P}} = \Delta_0^{\mathcal{P}} = \mathcal{P}$; $\Delta_{k+1}^{\mathcal{P}} = \mathcal{P}^{\Sigma_k^{\mathcal{P}}}$ is the class of all problems which can be solved by a DTM in polynomial time with a $\Sigma_k^{\mathcal{P}}$ oracle. $\Sigma_{k+1}^{\mathcal{P}} = NP^{\Sigma_k^{\mathcal{P}}}$ is instead the class of all problems which can be solved by a non-deterministic Turing Machine in polynomial time with a $\Sigma_k^{\mathcal{P}}$ oracle. Finally, $\Pi_{k+1}^{\mathcal{P}} = \text{co-}NP^{\Sigma_k^{\mathcal{P}}}$ is the class of all problems whose complement can be solved by a non-deterministic Turing Machine in polynomial time with a $\Sigma_k^{\mathcal{P}}$ oracle. $\Sigma_1^{\mathcal{P}}$ and $\Pi_1^{\mathcal{P}}$ are *NP* and co-*NP* (respectively). Note, *NP* is the class of problems which can be solved by a non-deterministic Turing machine in polynomial time and co-*NP* is the class of problems whose complement is in *NP*. *DP* is the class of problems that can be mapped to a pair of problems D_1 and D_2 such that $D_1 \in NP$ and $D_2 \in \text{co-}NP$. It is well known that the following inclusions hold: $\mathcal{P} \subseteq NP \subseteq DP \subseteq \Delta_2^{\mathcal{P}} \subseteq \Sigma_2^{\mathcal{P}}$ and $\mathcal{P} \subseteq \text{co-}NP \subseteq DP \subseteq \Delta_2^{\mathcal{P}} \subseteq \Pi_2^{\mathcal{P}}$ [34].

3   Early Approaches to Logic-Based Learning Under the Answer Set Semantics

In ASP, there can be one, many or even no answer sets of a program. This leads to two different standard notions of entailment under the answer set semantics: *brave entailment* and *cautious entailment*. Consider, for instance, the following ASP program $P = \{1\{p,q\}1 :\text{-} r. \quad r. \quad :\text{- not } p,r.\}$. This program would accept exactly one answer set $A = \{r,p\}$. In this case r and p would be entailed bravely and cautiously. If the first choice rule was instead replaced with $1\{p,q\}2 :\text{-} r.$, the program would accept two answer sets $A_1 = \{r,p\}$ and $A_2 = \{r,p,q\}$. In this case r would be cautiously entailed but q would be only bravely entailed. If the additional constraint :- r was added to P, then the new program would have no answer set.

These two different notions of entailment naturally lead to two different frameworks for learning from entailment under the answer set semantics: *cautious induction* and *brave induction*. Early approaches to ILP under the answer set semantics tended to adopt cautious induction[1] [16,39,42], as this is closer to standard learning from entailment, where examples must be covered in every model. In [41], it was argued that in some cases cautious induction can be too strong as it would require that all positive examples must be true in all answer sets of a given background knowledge and learned hypothesis (this is illustrated in Example 2). In those cases a weaker form of induction – brave induction – is needed. It was in [41] that the notions of *brave* and *cautious* induction were first defined. Brave induction defines an inductive task where all of the examples should be covered in at least one answer set (i.e. entailed under brave entailment in ASP). Note that there should be at least one answer set that covers every example (rather than at least one answer set for each example). Therefore, brave induction cannot specify other brave learning tasks such as enforcing that two examples are both bravely entailed, but not necessarily in the same answer set (as brave induction requires all examples to be covered in the same answer set). In some cases, for instance, we might want to learn a hypothesis that would require to cover some positive example(s) in an answer set and other positive example(s) in other answer sets (of the same learned hypothesis when added to a given background knowledge). These examples would still be bravely entailed but brave induction would not be able to solve tasks requiring such type of coverage. Furthermore, brave induction can only reason about what should be true in at least one answer set of a learned hypothesis (together with the background knowledge). Therefore it cannot reason about what should be true in all answer sets of a program. For this reason, brave induction is incapable of learning constraints.

3.1   Cautious Induction

Cautious induction, first presented in [41], defines a learning task in which all examples should be covered in every answer set (i.e. entailed under cautious entailment in ASP) and $B \cup H$ should be satisfiable (have at least one answer

---

[1] As the notions had not been defined at the time, they did not call it cautious induction, but the definitions are the same.

set)[2]. Note that the satisfiability condition is crucial to avoid trivial solutions such as ":-.", which eliminate all answer sets.

Definition 2. A *cautious induction (ILP$_c$) task* T_c *is a tuple* $\langle B, S_M, \langle E^+, E^- \rangle \rangle$, *where* B *is an ASP program,* S_M *is a set of ASP rules and* E^+ *and* E^- *are sets of ground atoms. A hypothesis* $H \subseteq S_M$ *is an inductive solution of* T_c, *written* $H \in ILP_c(T)$, *if and only if* $AS(B \cup H) \neq \emptyset$ *and* $\forall A \in AS(B \cup H)$, $E^+ \subseteq A$ *and* $E^- \cap A = \emptyset$.

Example 1. Consider the ILP_c task $T = \langle B, S_M, \langle E^+, E^- \rangle \rangle$, where:

$$B = \left\{ \begin{array}{l} \texttt{bird(X):-penguin(X).} \\ \texttt{bird(X):-sparrow(X).} \\ \texttt{penguin(b1).} \\ \texttt{sparrow(b2).} \end{array} \right\} \qquad S_M = \left\{ \begin{array}{l} \texttt{h}_1 : \texttt{flies(X):-bird(X).} \\ \texttt{h}_2 : \texttt{flies(X):-bird(X),} \\ \qquad\qquad \texttt{not penguin(X).} \\ \texttt{h}_3 : \texttt{0\{flies(X)\}1:-bird(X).} \\ \texttt{h}_4 : \texttt{0\{flies(X)\}1:-bird(X),} \\ \qquad\qquad \texttt{not penguin(X).} \end{array} \right\}$$

$E^+ = \{\texttt{flies(b2)}\}$
$E^- = \{\texttt{flies(b1)}\}$

The background knowledge B has only one answer set $A = \{\texttt{penguin(b1)},$ $\texttt{sparrow(b2)}, \texttt{bird(b1)}, \texttt{bird(b2)}\}$.

- $\emptyset \notin ILP_c(T)$ as B has exactly one answer set, and it does not contain $\texttt{flies(b2)}$.
- $\{\texttt{h}_1\} \notin ILP_c(T)$ as $B \cup \{\texttt{h}_1\}$ has exactly one answer set, $A \cup \{\texttt{flies(b1)}\}$, which contains the negative example.
- $\{\texttt{h}_2\} \in ILP_c(T)$ as $B \cup \{\texttt{h}_2\}$ has exactly one answer set, $A \cup \{\texttt{flies(b2)}\}$ which contains the positive example $\texttt{flies(b2)}$ but not the negative example $\texttt{flies(b1)}$.
- $\{\texttt{h}_3\}$ and $\{\texttt{h}_4\}$ are not in $ILP_c(T)$, as they both have answer sets (when combined with B) that do not cover the examples. Specifically, $B \cup \{\texttt{h}_3\}$ has three answer sets: $A_1 = A$, $A_2 = A \cup \{\texttt{flies(b1)}\}$, and $A_3 = A \cup \{\texttt{flies(b2)}\}$. It is clearly not the case that all these three answer sets include the positive example and do not include the negative example. Similarly, $B \cup \{\texttt{h}_4\}$ has two answer sets, $A_1 = A$ and $A_2 = A \cup \{\texttt{flies(b2)}\}$ which also do not all include the positive example.

Limitations of Cautious Induction. Enforcing that examples are covered in every answer set is sometimes too *strong* a requirement, as shown in the following example.

---

[2] The original definitions of brave and cautious induction did not consider atoms which should not be present in an answer set (negative examples). Publicly available algorithms that realise brave induction, on the other hand, do allow for negative examples. We therefore upgrade the definitions in this tutorial to allow negative examples. Note that a negative example $\texttt{e}$ can be easily simulated by adding a rule $\texttt{a:- not e}$ to the background knowledge and giving $\texttt{a}$ as a positive example (where $\texttt{a}$ is a new atom that does not appear anywhere in the original task).

Example 2. Consider the background knowledge $B = \emptyset$ and the hypothesis space $S_M = \{\, h_1 : p :\text{-} \text{ not } q. \; ; \; h_2 : q :\text{-} \text{ not } p. \,\}$. There are only two atoms (p and q) in the Herbrand base of $B \cup S_M$. It is impossible to construct an ILP_c task T with background knowledge B and hypothesis space S_M, whatever example from the Herbrand base we consider, that would accept $\{h_1, h_2\}$ as solution and does not accept the empty set as solution. This can be seen as follows. Given that there are only two atoms (p and q) in the Herbrand base of $B \cup S_M$, there are only two atoms which would be meaningful as examples. Neither of them can be given as a positive example as for each atom there is an answer set of $B \cup \{h_1, h_2\}$ that does not contain it. Similarly, neither can be given as a negative example, as for each atom there is an answer set that contains it. This means that the only ILP_c task T such that $\{h_1, h_2\} \in ILP_c(T)$ is $\langle B, S_M, \langle \emptyset, \emptyset \rangle \rangle$. But, clearly for this task, as there are no examples in T, the empty set, $\emptyset$, would be also an inductive solution of T, and it would be the one that, in practice, caution ILP systems would return as they would always search for the shortest possible hypothesis. This means that no examples can be given such that a cautious induction system would return $\{h_1, h_2\}$, showing that caution induction is too restrictive.

3.2   Brave Induction

Brave induction (ILP_b) was also formalised in [41]. It defines an inductive task in which all of the examples should be covered in at least one answer set (i.e. entailed under brave entailment in ASP). Note that there should be at least one answer set that covers every example (rather than at least one answer set for each example).

Definition 3. *A brave induction (ILP_b) task T_b is a tuple $\langle B, S_M, \langle E^+, E^- \rangle \rangle$, where B is an ASP program, S_M is a set of ASP rules and E^+ and E^- are sets of ground atoms. A hypothesis $H \subseteq S_M$ is an inductive solution of T_b, written $H \in ILP_b(T)$, if and only if $\exists A \in AS(B \cup H)$ such that $E^+ \subseteq A$ and $E^- \cap A = \emptyset$.*

Example 3. Consider the ILP_b task $T = \langle B, S_M, \langle E^+, E^- \rangle \rangle$, where B, S_M, E^+ and E^- are defined as in Example 2:

$$
B = \left\{
\begin{array}{l}
\texttt{bird(X):-penguin(X).} \\
\texttt{bird(X):-sparrow(X).} \\
\texttt{penguin(b1).} \\
\texttt{sparrow(b2).}
\end{array}
\right\}
\quad
S_M = \left\{
\begin{array}{l}
h_1 : \texttt{flies(X):-bird(X).} \\
h_2 : \texttt{flies(X):-bird(X),} \\
\qquad\qquad \texttt{not penguin(X).} \\
h_3 : \texttt{0\{flies(X)\}1:-bird(X).} \\
h_4 : \texttt{0\{flies(X)\}1:-bird(X),} \\
\qquad\qquad \texttt{not penguin(X).}
\end{array}
\right\}
$$

$E^+ = \{\texttt{flies(b2)}\}$
$E^- = \{\texttt{flies(b1)}\}$

- $\emptyset \notin ILP_b(T)$ as B has exactly one answer set, and it does not contain $\texttt{flies(b2)}$.
- $\{h_1\} \notin ILP_b(T)$ as $B \cup \{h_1\}$ has exactly one answer set, and it contains $\texttt{flies(b1)}$.
- $\{h_2\}, \{h_3\}, \{h_4\} \in ILP_b(T)$ as each of $B \cup \{h_2\}$, $B \cup \{h_3\}$ and $B \cup \{h_4\}$ has the answer set $\{\texttt{penguin(b1)}, \texttt{sparrow(b2)}, \texttt{bird(b1)}, \texttt{bird(b2)}, \texttt{flies(b2)}\}$, which contains $\texttt{flies(b2)}$ but does not contain $\texttt{flies(b1)}$.

Limitations of Brave Induction. Brave induction can only reason about what should be true in at least one answer set of a program. It cannot reason about what should be true in all answer sets of a program. For this reason, brave induction is incapable of learning constraints, as illustrated in the following example. In particular, any solution of an ILP_b task T that includes a constraint is still a solution of T if the constraint is removed, indicating that brave induction omits searching for constraints when learning a solution.

Example 4. Consider the background knowledge $B = \{0\{p\}1.\}$ and a hypothesis space S_M, containing only the constraint :- p., $S_M = \{p.\}$. There is only one atom (p) in the Herbrand base of $B \cup S_M$. We show that it is impossible to construct a brave induction task T_b, with background knowledge B and hypothesis space SM, whatever example from the Herbrand base we consider, that accepts $\{:-p.\}$ as solution and does not accept the empty set as solution. This can be seen as follows. Given that there is only one atom (p) in the Herbrand base of $B \cup S_M$, there is then one atom which would be meaningful as an example. It must be given as a negative example (as $B \cup \{:-p.\}$ has only one answer set, and it does not contain p). But $B \cup \emptyset$ also covers this negative example, as it also has the answer set $\emptyset$, which does not contain p. Therefore for any ILP_b task that accepts the constraint $\{:-p.\}$ as a solution, $\emptyset$ is also a solution, meaning that in practice brave induction systems (searching for the shortest hypothesis) will never return the constraint as (part of) a solution.

XHAIL. One of the first logic-based machine learning systems under the answer set programming semantics is the *eXtended Hybrid Abductive Inductive Learning* (XHAIL) [36], which generalises the HAIL [35, 37] algorithm, defined for definite clauses, in order to solve ILP_b tasks for ASP programs with negation as failure. Similarly to HAIL, XHAIL combines abductive and deductive inference. Abductive inference is an *ampliative* form of inference, as it generates knowledge that is not included in the premises of the inference process. Specifically, abduction is the process of reasoning from examples (observations) to possible causes. An abductive inference task takes as input a background knowledge, a set of abducibles (i.e., ground atoms that are not defined in the background knowledge) and set of examples and returns as output possible explanations (i.e. cases in syllogistic terms), also called abductive solutions, that together with the background knowledge, entail (i.e. explain) the examples. It differs from inductive inference in the fact that explanations do not require a process of generalisation during the inference process, whereas induction aims at discovering new general rules from samples (positive and negative) of cases. In other words, abduction is the process of explanation – reasoning from effects to possible causes, whereas induction is the process of generalisation – reasoning from specific cases to general hypothesis.

The XHAIL learning system computes inductive solutions in three phases: an abductive phase; a deductive phase; and an inductive phase. The abductive step takes as abducibles ground atoms that conform with the **modeh** of the language bias of the given task and computes as abductive solution a set of abducible atoms, Δ, such that $B \cup \Delta \models_b (\bigwedge E^+) \wedge (\bigwedge\{ \text{not } e \mid e \in E^- \})$. An abductive solution becomes the heads of (ground instances of) rules in the final hypothesis. Next, in the deductive phase, XHAIL computes the set of all ground literals that could go in the body of the rules in the hypothesis. Each of these body atoms b is such that $B \cup \Delta \models_b$ b and b is a ground instance of an atom that conforms to at least one modeb declaration. The sets of ground rules with the heads from Δ and with bodies consisting of literals computed in the deductive phase is referred to as the (ground) Kernel Set $\mathcal{K}$.

Example 5 (from [36]). Consider the ILP_b task $T = \langle B, S_M, \langle E^+, E^- \rangle\rangle$, where B, M, E^+ and E^- are as follows:

$$B = \begin{cases} \texttt{bird(X):-penguin(X).} \\ \texttt{bird(a).} \\ \texttt{bird(b).} \\ \texttt{bird(c).} \\ \texttt{penguin(d).} \end{cases} \quad M = \begin{cases} \texttt{\#modeh(flies(+bird))} \\ \texttt{\#modeb(penguin(+bird))} \\ \texttt{\#modeb(not penguin(+bird))} \end{cases}$$

$$E^+ = \begin{cases} \texttt{flies(a),} \\ \texttt{flies(b),} \\ \texttt{flies(c)} \end{cases} \quad E^- = \{\, \texttt{flies(d)} \,\}$$

One abductive explanation of the examples is $\Delta = \{\texttt{flies(a)}, \texttt{flies(b)}, \texttt{flies(c)}\}$. This leads to the ground Kernel Set:

$$\mathcal{K} = \begin{cases} \texttt{flies(a):- not penguin(a).} \\ \texttt{flies(b):- not penguin(b).} \\ \texttt{flies(c):- not penguin(c).} \end{cases}$$

The unground Kernel Set that conforms to the mode bias and the declaration of input variables is given by:

$$\mathcal{K} = \{\, \texttt{flies(X):- not penguin(X).} \,\}$$

Note that although there are other potential body literals that are entailed by $B \cup \Delta$ and that are declared to be mode body predicates (i.e., $\texttt{penguin(d)}$) they are not added to the ground Kernel Set as they do not form part of a ground instance of a rule that conforms to the mode declarations. According to the declaration of input variables in the language bias body literals need to have the same variable as the one that appears in the head of the rule. The ground literal, $\texttt{penguin(d)}$, for instance, cannot be added to any of the three ground rules of the ground Kernel Set. This is because the unground version of the Kernel Set would give an unground rule with $\texttt{penguin(Y)}$ in the body, which violates the input variable constraint of the mode body declaration: an input variable in a mode body predicate must either appear in the head predicate of the rule, or appear as output variable in another body atom. In the mode declaration M given above, variables are only input variables, so any variable that appears in a body predicate must appear in the head predicate of the rule.

The final step of the XHAIL algorithm – the *inductive* step – computes a hypothesis H that conforms to the mode declarations, subsumes the unground Kernel Set and bravely entails the examples. This phase is also supported by an abductive inference process that takes as background knowledge a "transformed" unground Kernel Set, as abducibles ground instances of a predicate use, and as observation the same of examples used in the first phase. The abductive answer determines the literals in the body of the rules of the Kernel Set that need to be maintained in order for the examples to be covered.

Example 6. Consider again the unground Kernel Set computed in Example 5. This is transformed in the following ASP program:

$$\left\{ \begin{array}{l} \texttt{flies(X):-use(1,0),try(1,1,X).} \\ \texttt{try(1,1,X):-bird(X), not use(1,1).} \\ \texttt{try(1,1,X):-bird(X),use(1,1), not penguin(X).} \end{array} \right\}$$

The first and second arguments of each of the meta-level atoms use and try indicate respectively a unique identifier for the object-level rules (starting from 1) in the unground kernel Set, and a unique identifier for the literal in each of these rules (starting from 0 as identifier of the head atom). So, use(1, 0) means that the head atom flies(X) is being *used* (i.e. it is in the hypothesis). The try atoms are for testing whether the rule body is satisfied. If the head is being used, then flies(X) is true in two cases: (1), the literal not penguin(X) is not in the hypothesis (indicated by the first try rule); or (2), not penguin(X) is true (represented by the second try rule).

The transformed Kernel Set is augmented with the choice rule 0{use(1, 0), use(1, 1)}2. This phase computes an abductive solution. In the above example, XHAIL uses an ASP solver to compute the smallest abductive answer using the transformed Kernel Set, and this answer gives then hypothesis that subsumes the Kernel Set, conforms to the mode declarations and bravely entails the examples. In the above example, the abductive solution generated during the inductive phase would be $\Delta_i = \{\texttt{use(1,1)}\}$, which indicates that in the final hypothesis, constituted just by the first rule, the first body literal will have to be kept in order to cover the examples correctly.

One major difference between HAIL and XHAIL is that HAIL uses a cover loop approach, whereas XHAIL does not. This is due to the nonmonotonicity of negation as failure: in a cover loop approach, examples that were covered in previous iterations of the cover loop may not be covered in future iterations.

As in general there are many possible abductive solutions Δ, and not all Δ's lead to inductive solutions, XHAIL employs an iterative deepening approach, ordering the Δ's by size and terminating after processing the shortest Δ that leads to a solution. In general, this may not lead to the optimal solution being found, as there may be a large Δ that leads to a shorter hypothesis (e.g. with more individual rules, but fewer overall literals).

INSPIRE. Inspire [18], is an ILP system based on XHAIL, but with some modifications to aid scalability. The main modification is that some rules are "pruned" from the Kernel Set before the XHAIL's inductive phase. Both XHAIL and Inspire use a meta-level ASP program to perform the inductive phase, and the ground Kernel Set is generalised into a first order Kernel Set (using the mode declarations to determine which arguments of which predicates should become variables). Inspire prunes rules that have fewer than Pr instances in the ground Kernel Set (where Pr is a parameter of Inspire). The intuition is that if a rule is necessary to cover many examples then it is likely to have many ground instances in the Kernel Set. Clearly this is an approximation, so Inspire is not guaranteed to find the optimal hypothesis in the inductive phase. In fact, as XHAIL is not guaranteed to find the optimal inductive solution of the task (as it may pick the "wrong" abductive solution), this means that Inspire may be even further from computing optimal solutions.

ILED. ILED [17] is an incremental algorithm, based on XHAIL. It is targeted at learning Event Calculus [19] theories and, therefore, its examples are slightly different in that they are grouped into *time windows*. The examples are processed one at a time and at each timepoint the hypothesis is revised so that it covers all examples in all windows that have been processed so far.

ILED has been shown to be much more scalable than XHAIL when processing large numbers of examples divided into time windows [17]. On the other hand, like XHAIL, ILED is not guaranteed to find the optimal solution of a task. In fact, this incompleteness with respect to optimal solutions is more severe in ILED than in XHAIL, as it can also occur because of the incremental nature of the algorithm. Although at each step the revision may be optimal, the combination of every revision may result in a longer hypothesis than could have been found if all examples had been processed together.

ASPAL. The algorithms presented so far follow a bottom-up approach for searching for solutions within a given hypothesis space specified by a language bias. In the cases of XHAIL, ILED and INSPIRE the algorithms compute first a most specific (set of) clauses that cover the examples, which constitute the "bottom element" of the search space, and then try to generalise it searching for more general solutions within the search space. But alternative approaches to the search for inductive solutions have been proposed in the literature. These are referred to as meta-level approaches with top-down search. An example of such algorithms is the Top-directed Abductive Learning (TAL) system [6]. This system solves an ILP task by automatically translating it into a semantically equivalent abductive inference task, whose background knowledge is given by the background knowledge of the learning task augmented with *meta-level* representation of the hypothesis space, and the observation to explain is given, as in XHAIL, by the conjunction of the examples. The inference process performs a top-down abductive search [32] for abductive solutions that explain the observation. Such an abductive solution is then translated back into rules that are guaranteed to correspond to an inductive solution of the original brave inductive task. The transformation relies upon a one-to-one mapping that translates each (normal) rule of the hypothesis space into a meta-level representation and uses a "meta-program", called *top theory*, that captures possible ways of constructing such rules in terms of their meta-level representation. The main advantage of this approach is its ability to solve ILP tasks that require recursive rules as solutions or rules that are interdependent (e.g., predicates that appear in the body of a rule can also appear in the head of another rule belonging to the same solution). However, a drawback of this approach is its scalability. The abductive reasoning engine used by TAL is implemented in Prolog. As such, it suffers from redundant inference steps, which causes its computational time to be affected by the size of the hypothesis space and the number of examples of a given ILP task.

The ASPAL [7] algorithm is a brave induction system that aims at addressing the limitations of the TAL system by using an ASP implementation: as in TAL, an ILP task is translated into a meta-level program, but in ASPAL this is an ASP program. Given an ILP_b task $T_b = \langle B, S_M, \langle E^+, E^- \rangle \rangle$, where S_M is defined by a given set of mode declarations M, the first step is to compute a set of *skeleton rules* Sk_M. These are the set of rules R, such that there is an $R' \in S_M$, where each constant in R' is replaced by a placeholder variable in R.

Example 7. Consider the mode declarations M.

$$M = \left\{ \begin{array}{l} \texttt{\#modeh(penguin(+bird))} \\ \texttt{\#modeb(2, not can(+bird, \#ability))} \end{array} \right\}$$

The first argument of the mode body declaration is called the *recall* and it expresses the constraint that this mode declaration can be used at most twice per rule in the hypothesis space. There are three skeleton rules:

$$Sk_M = \left\{ \begin{array}{l} \texttt{penguin(X):-bird(X)} \\ \texttt{penguin(X):-bird(X), not can(X,C1)} \\ \texttt{penguin(X):-bird(X), not can(X,C1), not can(X,C2)} \end{array} \right\}$$

Note that the hypothesis space S_M consists of the rules in Sk_M but where $C1$ and $C2$ have been replaced with constants of type $\texttt{ability}$.

Each skeleton rule $R \in Sk_M$ is associated with a unique meta-level atom $\texttt{rule}(R_{id}, C_1, \ldots, C_n)$, denoted as R_{meta}, where $C_1, \ldots, C_n$ are the "constant placeholder" variables in R. For each rule $R' \in S_M$, we similarly write R'_{meta} to denote the ground atom representing R' (where each "constant placeholder" variable has been replaced with a constant of the correct type). Informally, consider for example the second rule $\texttt{penguin(X):-bird(X), not can(X,C1)}$ in Sk_M. Its associated meta-level atom would be $rule(2, C1)$. Now assuming, for the sake of the argument, that $ability(fly)$ is true in the background knowledge of the given learning task, the ASPAL computation may given rise for instance to a ground instance $rule(2, fly)$, which would then be used by ASPAL to generate the corresponding rule $R' \in S_M$ given by $\texttt{penguin(X):-bird(X), not can(X,fly)}$ in the final inductive solution.

Using this notion of skeleton rules, the ASPAL system automatically constructs an ASP meta-level representation of a learning task by adding to the background theory B of the learning task, the set of skeleton rules, each augmented with an additional body literal given by the associated meta-level atom. These meta-level atoms are considered to be abducible and their truth is determined by a choice rule which is used by the ASP solver to select the minimal number of such atoms (corresponding to the minimal number of rules $R' \in S_M$) so that the examples of the given learning task are covered. This is captured formally by the following definition.

Definition 4. *Let T be the ILP_b task $\langle B, S_M, \langle \{e_1^+, ..., e_n^+\}, \{e_1^-, ..., e_m^-\}\rangle\rangle$, where S_M is characterised by the set of mode declarations M. Let Sk_M be the set of skeleton rules derivable from M. The ASPAL meta-representation is the program consisting of the following components:*

- B
- $\mathtt{h\,:\text{-}\,b_1, \ldots b_{rl}, R_{meta}}$, *for each rule $R \in Sk_M$, where R is the rule* $\mathtt{h\,:\text{-}\,b_1, \ldots, b_{rl}}$.
- *A choice rule* $\mathtt{0\{ab_1, \ldots, ab_k\}k.}$, *where* $\{\mathtt{ab_1, \ldots, ab_k}\} = \{R_{meta} \mid R \in Sk_M\}$[3]
- *The rule* $\mathtt{goal\,:\text{-}\,e_1^+, \ldots, e_n^+, not\ e_1^-, \ldots, not\ e_m^-}$.
- *The constraint* $\mathtt{:\text{-}\,not\ goal}$.

We refer to the answer sets of this meta representation as meta-level answer sets, and the answer sets of $B \cup H$ as object-level answer sets. Each meta-level answer set A represents a single hypothesis H (defined by the $\mathtt{rule}$ atoms in A). Each meta-level answer set also contains exactly one object-level answer set of $B \cup H$ that contains all of the positive examples and none of the negative examples (enforced by the $\mathtt{goal}$ constraint).

Example 8. Consider the ILP_b task $T = \langle B, S_M, E^+, E^- \rangle$, where S_M is characterised by the mode declarations in Example 7.

$$B = \begin{cases} \mathtt{bird(a).} \\ \mathtt{bird(b).} \\ \mathtt{can(a, fly).} \\ \mathtt{can(b, swim).} \\ \mathtt{ability(fly).} \\ \mathtt{ability(swim).} \end{cases} \quad E^+ = \{\mathtt{penguin(b)}\}\ E^- = \{\mathtt{penguin(a)}\}$$

The ASPAL meta-level representation is shown in Fig. 1. Note that each skeleton rule R has been appended with the atom R_{meta}, where each of the arguments other than the identifier R_{id} is a variable representing a placeholder for a constant. The choice rule on the other hand contains atoms R'_{meta},

---

[3] This is a slight simplification. In the ASPAL algorithm, this is a choice rule using conditional literals, in order to delegate the grounding of the possible constants to the ASP solver. The ground version of ASPAL's choice rule is identical to the one presented in this definition.

```
bird(a).
bird(b).
can(a, fly).
can(b, swim).
ability(fly).
ability(swim).

penguin(X) :- bird(X), rule(1).
penguin(X) :- bird(X), not can(X, C1), rule(2, C1).
penguin(X) :- bird(X), not can(X, C1), not can(X, C2), rule(3, C1, C2).

0{rule(1), rule(2, fly), rule(2, swim), rule(3, fly, swim)}4.

goal :- penguin(b), not penguin(a).
:- not goal.
```

Fig. 1. The ASPAL meta-level representation for the learning task in Example 8.

generated by instantiating the constant placeholder variable with all possible constant values allowed by the constant type. In this way each ground R' is an instance of a skeleton rule R. The answer sets of this program can be mapped to the ILP_b inductive solutions of the task T_b. For example, the answer set {bird(a), bird(b), can(a, fly), can(b, swim), ability(fly), ability(swim), penguin(b), rule(2, fly), goal} shows that the hypothesis {penguin(X) :- bird(X), not can(X, fly).} is a solution of the ILP_b task T.

In the ASPAL algorithm, this meta representation is combined with an optimisation statement (similar to weak constraints in ASP), which orders the meta-level answer sets by the length of the hypothesis that they represent. This optimisation statement is equivalent to adding a weak constraint :∼ $R_{meta}.[|R|@1, R_{meta}]$ for each R in Sk_M, which means that the total penalty paid by a meta-level answer set at priority level 1 is the length of the hypothesis generated from the answer set. Note that when computing $|R|$ the "type" atoms in R (such as bird(X) in the rule above) are not counted.

ASPAL has been proven to be sound and complete with respect to the optimal inductive solutions of any brave induction task [5]. This means that, unlike XHAIL and XHAIL-based algorithms, ASPAL is guaranteed to return an optimal inductive solution of any brave induction task (resources permitting).

RASPAL. ASPAL scales poorly with respect to the size of $ground(B \cup S_M)$ [2]. One of the main factors in the size of this ground program is the number of body literals that are allowed to appear in a rule in the hypothesis space. RASPAL [2] addresses this limitation by iteratively refining a hypothesis until all of the examples in an ILP_b task are covered. At each step, the number of literals that are allowed to be added to the hypothesis is restricted, meaning that the grounding in RASPAL is often significantly smaller than the meta-level program in

ASPAL. In [1] it was shown that RASPAL significantly outperforms ASPAL on some learning tasks with large problem domains and large hypothesis spaces.

3.3   Induction of Stable Models

Induction of stable models [33] (ILP_{sm}), generalises ILP_b, in order to allow conditions to be set over multiple answer sets. The examples of an ILP_{sm} task are *partial interpretations*.

Definition 5. *A partial interpretation e is a pair of sets of atoms $\langle e^{inc}, e^{exc} \rangle$. We refer to e^{inc} and e^{exc} as the* inclusions *and* exclusions *respectively. An interpretation I is said to* extend e *if and only if $e^{inc} \subseteq I$ and $e^{exc} \cap I = \emptyset$.*

Example 9. Consider the partial interpretation $e = \langle \{p, q\}, \{r, s\} \rangle$.

- $\{p\}$ does not extend e, as it does not contain q.
- $\{p, q, r\}$ does not extend e, as it contains r.
- $\{p, q\}$ extends e, as it contains all of e's inclusions, and none of e's exclusions.
- $\{p, q, t\}$ extends e, as it contains all of e's inclusions, and none of e's exclusions.

Induction of stable models is formalised in Definition 6.

Definition 6. *An induction of stable models (ILP_{sm}) task T_{sm} is a tuple $\langle B, S_M, \langle E \rangle \rangle$, where B is an ASP program, S_M is the hypothesis space and E is a set of example partial interpretations. A hypothesis H is an inductive solution of T_{sm} if and only if $H \subseteq S_M$ and $\forall e \in E, \exists A \in AS(B \cup H)$ such that A extends e.*

Note that a brave induction task can be thought of as a special case of induction of stable models (with $|E| = 1$ and the inclusions and exclusions of the only partial interpretation example being the positive and negative examples of the brave task, respectively).

Example 10. Consider the ILP_{sm} task $T = \langle B, S_M, \langle E \rangle \rangle$, where:

$$B = \emptyset$$
$$S_M = \left\{ \begin{array}{l} \text{h}_1 : \text{p :- not q.} \\ \text{h}_2 : \text{q :- not p.} \end{array} \right\} \qquad E = \left\{ \begin{array}{l} \langle \{p\}, \{q\} \rangle, \\ \langle \{q\}, \{p\} \rangle \end{array} \right\}$$

$\{h_1, h_2\}$ is the only subset of the hypothesis space that is an inductive solution of T, as it is the only hypothesis that has answer sets that extend both of the examples.

Note that, although induction of stable models is a generalisation of brave induction, it is still incapable of learning constraints. This is because, similarly to brave induction, it can only give examples of what should be (in) an answer set, rather than examples of what should not be an answer set.

4   Learning from Answer Sets and ILASP

In the previous section, we presented the main frameworks for learning ASP programs, which fall into two categories: either the examples must be covered in at least one answer set of the learned program (brave induction [41] and induction of stable models [33]), or the examples must be covered in every answer set of the learned program (cautious induction [41]). Work on using brave induction (such as [36] and [7]) has often only considered learning stratified programs[4]. In general, however, ASP programs can have one, many or even no answer sets. Example 11 presents a program H describing the rules of Sudoku, and shows that no brave induction, induction of stable models or cautious induction task could possibly have H as an optimal solution.

Example 11. Consider a background knowledge B that contains definitions of the structure of a 4x4 Sudoku grid; i.e. definitions of cell, same_row, same_col and same_block (where same_row, same_col and same_block are true only for two *different* cells in the same row, column or block).

$$B = \left\{ \begin{array}{l} \texttt{cell}((1,1)). \quad \texttt{cell}((1,2)). \quad \ldots \quad \texttt{cell}((4,4)). \\ \texttt{same\_row}((\texttt{X1},\texttt{Y}),(\texttt{X2},\texttt{Y})) \texttt{:-} \texttt{cell}((\texttt{X1},\texttt{Y})),\texttt{cell}((\texttt{X2},\texttt{Y})),\texttt{X1} \neq \texttt{X2}. \\ \texttt{same\_col}((\texttt{X},\texttt{Y1}),(\texttt{X},\texttt{Y2})) \texttt{:-} \texttt{cell}((\texttt{X},\texttt{Y1})),\texttt{cell}((\texttt{X},\texttt{Y2})),\texttt{Y1} \neq \texttt{Y2}. \\ \texttt{block}((1,1),1). \quad \texttt{block}((1,2),1). \quad \texttt{block}((2,1),1). \quad \texttt{block}((2,2),1). \\ \texttt{block}((3,1),2). \quad \texttt{block}((3,2),2). \quad \texttt{block}((4,1),2). \quad \texttt{block}((4,2),2). \\ \texttt{block}((1,3),3). \quad \texttt{block}((1,4),3). \quad \texttt{block}((2,3),3). \quad \texttt{block}((2,4),3). \\ \texttt{block}((3,3),4). \quad \texttt{block}((3,4),4). \quad \texttt{block}((4,3),4). \quad \texttt{block}((4,4),4). \\ \texttt{same\_block}(\texttt{C1},\texttt{C2}) \texttt{:-} \texttt{block}(\texttt{C1},\texttt{B}),\texttt{block}(\texttt{C2},\texttt{B}),\texttt{C1} \neq \texttt{C2}. \end{array} \right\}$$

One hypothesis H that describes the correct rules of Sudoku is as follows:

$$H = \left\{ \begin{array}{l} \texttt{1\{value}(\texttt{C},1),\texttt{value}(\texttt{C},2),\texttt{value}(\texttt{C},3),\texttt{value}(\texttt{C},4)\}\texttt{1:-} \texttt{cell}(\texttt{C}). \\ \texttt{:-} \texttt{same\_row}(\texttt{C1},\texttt{C2}),\texttt{value}(\texttt{C1},\texttt{V}),\texttt{value}(\texttt{C2},\texttt{V}). \\ \texttt{:-} \texttt{same\_col}(\texttt{C1},\texttt{C2}),\texttt{value}(\texttt{C1},\texttt{V}),\texttt{value}(\texttt{C2},\texttt{V}). \\ \texttt{:-} \texttt{same\_block}(\texttt{C1},\texttt{C2}),\texttt{value}(\texttt{C1},\texttt{V}),\texttt{value}(\texttt{C2},\texttt{V}). \end{array} \right\}$$

Let S_M be a set of rules which contains the rules in H (for the purposes of this example, it does not matter which other rules it contains). There is no ILP_b, ILP_{sm} or ILP_c task such that H is a solution, and no subset of H is a solution. In practice, as ILP systems tend to search for a solution that is as short as possible (called an optimal solution), no system for ILP_b, ILP_{sm} or ILP_c will return H as the solution. We now show that no task exists, for any of the three frameworks, for which H is an optimal solution.

– Assume that there is an ILP_b task T_b with background knowledge B such that H is a solution of T_b. Then there must be at least one answer set of

---

[4] Both XHAIL [36] and ASPAL [7] support learning non-stratified programs, but the background knowledge and hypothesis space of each of the example tasks in [36] and [7] is stratified.

$B \cup H$ that contains all of the positive examples of T_b and none of the negative examples of T_b. But this answer set must also be an answer set of $B \cup \{1\{\mathtt{value(C, 1)}, \mathtt{value(C, 2)}, \mathtt{value(C, 3)}, \mathtt{value(C, 4)}\}1\mathtt{:-cell(C)}.\}$, as the constraints in H only rule out answer sets. Hence, $H' = \{1\{\mathtt{value(C, 1)},$ $\mathtt{value(C, 2)}, \mathtt{value(C, 3)}, \mathtt{value(C, 4)}\}1\mathtt{:-cell(C)}.\}$ must also be an inductive solution of T_b. As H' is shorter than H, this means that H cannot possibly be an optimal solution of T_b.

- The argument for ILP_{sm} is similar to ILP_b. Assume there is an ILP_{sm} task T_{sm} with background knowledge B such that H is a solution of T_{sm}. Then for each example e, there must be at least one answer set A_e of $B \cup H$, such that A_e extends e. In each case, A_e must also be an answer set of $B \cup \{1\{\mathtt{value(C, 1)}, \mathtt{value(C, 2)}, \mathtt{value(C, 3)}, \mathtt{value(C, 4)}\}1\mathtt{:-cell(C)}.\}$, as the constraints in H only rule out answer sets. Hence, the hypothesis $H' = \{1\{\mathtt{value(C, 1)}, \mathtt{value(C, 2)}, \mathtt{value(C, 3)}, \mathtt{value(C, 4)}\}1\mathtt{:-cell(C)}.\}$ must also be an inductive solution of T_{sm}. As H' is shorter than H, this means that H cannot possibly be an optimal solution of T_{sm}.

- If we use ILP_c to learn H, we have to give examples which are either true in every answer set of $B \cup H$, or false in every answer set. Therefore, we could not give any meaningful examples about the $\mathtt{value}$ predicate – for each atom $\mathtt{value(x, y)}$ (where $\mathtt{x}$ and $\mathtt{y}$ range from 1 to 4), there is at least one answer set of $B \cup H$ that contains $\mathtt{value(x, y)}$ and at least one that does not; this means that if $\mathtt{value(x, y)}$ is given as either a positive or negative example, H will not be a solution of the task. This means that for any ILP_c task $T_c = \langle B, S_M, E^+, E^- \rangle$ such that H is a solution, $E^+ \subseteq \{\mathtt{a} \mid \forall A \in AS(B), \mathtt{a} \in A\}$ and $E^- \subseteq \{\mathtt{a} \mid \forall A \in AS(B), \mathtt{a} \notin A\}$. Hence, for any such task, $\emptyset$ must be a solution of T_c, meaning that H cannot be an optimal solution.

The problem with using either brave or cautious induction to learn general ASP programs is that brave induction can only reason about what should be true in at least one answer set of the learned program, which can be far too weak a condition, and cautious induction can only express what should be true in all answer sets of a program, which can be far too strong a condition. Furthermore, examples in both frameworks are atoms. In ASP it is common [9] to represent a problem such that the answer sets are solutions (see Fig. 2(a)). In order to learn ASP programs, examples should therefore be of what should (or should not) be an answer set of the program (Fig. 2(b)). In the context of learning the rules of Sudoku using the representation in Example 11, this corresponds to giving examples of Sudoku grids rather than the values of individual cells.

In practice, there may be some atoms whose values are unknown before learning. It is therefore more practical to consider learning from partial interpretations rather than full interpretations. This setting, under the answer set semantics, is the basis of the *Learning from Answer Sets* framework.

A learning from answer sets task consists of an ASP background knowledge B, a hypothesis space and sets of positive and negative partial interpretation examples. The goal is to find a hypothesis H that has at least one answer set (when combined with B) that extends each positive example, and no answer set

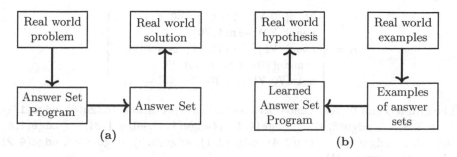

Fig. 2. (a) Shows the general paradigm of answer set programming [4]; (b) shows the general idea of Learning from Answer Sets.

that extends any negative examples. Note that each positive example could be extended by a different answer set of the learned program.

Definition 7. *A Learning from Answer Sets (ILP$_{LAS}$) task is a tuple $T = \langle B, S_M, \langle E^+, E^- \rangle \rangle$ where B is an ASP program, S_M a set of ASP rules and E^+ and E^- are finite sets of partial interpretations. A hypothesis $H \subseteq S_M$ is an inductive solution of T if and only if:*

1. $\forall e^+ \in E^+ \exists A \in AS(B \cup H)$ *such that A extends e^+*
2. $\forall e^- \in E^- \nexists A \in AS(B \cup H)$ *such that A extends e^-*

Example 12. Consider the problem of learning the definition of what it means for a graph to be Hamiltonian.[5] The background knowledge B defines what it means to be a graph, up to size 4.

$$B = \left\{ \begin{array}{l} \texttt{1\{size(1), size(2), size(3), size(4)\}1.} \\ \texttt{node(1..S) :- size(S).} \\ \texttt{0\{edge(N1, N2)\}1 :- node(N1), node(N2).} \end{array} \right\}$$

The answer sets of B exactly represent the graphs of size 1 to 4. For example, the answer set $\{\texttt{size(4)}, \texttt{node(1)}, \texttt{node(2)}, \texttt{node(3)}, \texttt{node(4)}, \texttt{edge(1, 2)}, \texttt{edge(2, 3)}, \texttt{edge(3, 4)}, \texttt{edge(4, 1)}\}$ represents the graph G:

The program H can be used to determine whether a graph is Hamiltonian or not. The answer sets of $B \cup H$ correspond exactly to the Hamiltonian graphs of size 1 to 4.

---

[5] A graph is Hamiltonian if it contains a cycle that visits each node exactly once.

$$H = \left\{ \begin{array}{l} \texttt{0\{in(V0,V1)\}1:-edge(V0,V1).} \\ \texttt{reach(V0):-in(1,V0).} \\ \texttt{reach(V1):-in(V0,V1),reach(V0).} \\ \texttt{:-node(V0), not reach(V0).} \\ \texttt{:-in(V0,V1),in(V0,V2),V1!=V2.} \end{array} \right\}$$

The graph G can be represented as a partial interpretation $\langle\{\texttt{size}(4), \texttt{edge}(1,2),$ $\texttt{edge}(2,3),$ $\texttt{edge}(3,4),$ $\texttt{edge}(4,1)\}, \{\texttt{edge}(1,1), \texttt{edge}(1,3),$ $\texttt{edge}(1,4),$ $\texttt{edge}(2,1), \texttt{edge}(2,2), \texttt{edge}(2,4), \texttt{edge}(3,1), \texttt{edge}(3,2), \texttt{edge}(3,3), \texttt{edge}(4,2),$ $\texttt{edge}(4,3),$ $\texttt{edge}(4,4)\}\rangle$.

Given sufficient positive and negative examples of Hamilton graphs, it is possible to learn the hypothesis H using the ILASP system for solving ILP_{LAS} tasks. Similarly to the Sudoku program in Example 11, it is impossible to learn H with any of the previous frameworks.

Since the original ILP_{LAS} framework was introduced in [21], it has been extended in several ways. The rest of this section presents each of these extensions.

4.1   Preference Learning in ASP

Preference Learning has received much attention over the last decade from within the machine learning community. A popular approach to preference learning is *learning to rank* [11,12], where the goal is to learn to rank any two objects given some examples of pairwise preferences (indicating that one object is preferred to another). While in previous work ILP systems such as TILDE [3] and Aleph [44] have been applied to preference learning [8,15], this has addressed learning ratings, such as good, poor and bad, rather than rankings over the examples. Ratings are not expressive enough if we want to find an optimal solution as we may rate many objects as good when some are better than others. ASP, on the other hand, allows the expression of preferences through *weak constraints*.

Weak constraints do not affect what is, or is not, an answer set of a program. Instead, they create a preference ordering over the answer sets of a program; i.e. they allow us to specify which answer sets are preferred to other answer sets. Example 13 shows how a set of preferences can be encoded as weak constraints.

Example 13. Consider the problem of using a user's preferences over alternative journeys, in order to select the optimal journey. Let A, B, C and D be the journeys represented by the following sets of attributes. Each journey is split into a number of *legs*, in which a single mode of transport is used.

$$\left\{\begin{array}{l} \texttt{leg\_mode(1, walk),} \\ \texttt{leg\_crime\_rating(1, 2),} \\ \texttt{leg\_distance(1, 500),} \\ \texttt{leg\_mode(2, bus),} \\ \texttt{leg\_crime\_rating(2, 4),} \\ \texttt{leg\_distance(2, 3000)} \end{array}\right\} \qquad \text{(A)}$$

$$\left\{\begin{array}{l} \texttt{leg\_mode(1, bus),} \\ \texttt{leg\_crime\_rating(1, 2),} \\ \texttt{leg\_distance(1, 4000),} \\ \texttt{leg\_mode(2, walk),} \\ \texttt{leg\_crime\_rating(2, 5),} \\ \texttt{leg\_distance(2, 1000)} \end{array}\right\} \qquad \text{(C)}$$

(B)

$$\left\{\begin{array}{l} \texttt{leg\_mode(1, bus),} \\ \texttt{leg\_crime\_rating(1, 5),} \\ \texttt{leg\_distance(1, 2000),} \\ \texttt{leg\_mode(2, walk),} \\ \texttt{leg\_crime\_rating(2, 1),} \\ \texttt{leg\_distance(2, 2000)} \end{array}\right\}$$

(D)

$$\left\{\begin{array}{l} \texttt{leg\_mode(1, bus),} \\ \texttt{leg\_crime\_rating(1, 2),} \\ \texttt{leg\_distance(1, 400),} \\ \texttt{leg\_mode(2, bus),} \\ \texttt{leg\_crime\_rating(2, 4),} \\ \texttt{leg\_distance(2, 3000)} \end{array}\right\}$$

The following weak constraints H give a preference ordering to the journeys A to D.

$$H = \left\{\begin{array}{l} \texttt{:\sim leg\_mode(L, walk), leg\_crime\_rating(L, C), C > 4.[1@3, L, C]} \\ \texttt{:\sim leg\_mode(L, bus).[1@2, L]} \\ \texttt{:\sim leg\_mode(L, walk), leg\_distance(L, D).[D@1, L, D]} \end{array}\right\}$$

The first weak constraint in H means that the user would like to avoid walking through an area with a crime rating higher than 4. A journey pays a penalty of 1 at priority level 3 for each leg of the journey that involves walking though such an area. As there is no weak constraint in H with a priority level higher than 3, this preference is the most important. The second weak constraint (at priority level 2) means that the user would like to take as few buses as possible. The third weak constraint (at priority level 1) means that the user would like to minimise the distance that they have to walk. Note that, as a penalty of the distance is paid for each leg where the user has to walk, the total penalty is equal to the total walking distance of the journey. Given these preferences, A is the best journey, followed by D, then C and then B.

The hypothesis in Example 13 could be learned by giving examples of which journeys are preferred to which other journeys. For the preferences to be learned as weak constraints, this would require examples of pairs of answer sets, such that the first is preferred to the second. In fact, each ordering example contains two partial interpretations, rather than two complete answer sets. Examples can also be given with any of the operators $<, \leq, =, \neq, >$ or $\geq$. The $<$ operator, for example, indicates that the first partial interpretation is preferred to the second; whereas the $=$ operator specifies that the two partial interpretations are equal.

Definition 8. *An* ordering example *is a tuple* $o = \langle e_1, e_2, op \rangle$ *where* e_1 *and* e_2 *are partial interpretations and op is a binary comparison operator* $(<, >, =, \leq, \geq$ *or* $\neq)$.

As ordering examples contain two partial interpretations, rather than two full interpretations, there are two possible semantics to give to the examples. The *brave* semantics indicates that there should be at least one pair of answer sets extending the pair of partial interpretations, which are ordered according to the operator. The *cautious* semantics, on the other hand, indicates that every pair of answer sets that extend the pair of partial interpretations should be ordered according to the operator.

Definition 9. *Let* $o = \langle e_1, e_2, op \rangle$ *be an ordering example. An ASP program* P bravely respects o *iff* $\exists A_1, A_2 \in AS(P)$ *such that all of the following conditions hold: (i)* A_1 *extends* e_1; *(ii)* A_2 *extends* e_2; *and (iii)* $\langle A_1, A_2, op \rangle \in ord(P)$. P cautiously respects o *iff* $\nexists A_1, A_2 \in AS(P)$ *such that all of the following conditions hold: (i)* A_1 *extends* e_1; *(ii)* A_2 *extends* e_2; *and (iii)* $\langle A_1, A_2, op \rangle \notin ord(P)$.

Definition 10 defines the notion of *Learning from Ordered Answer Sets* (ILP_{LOAS}).

Definition 10. *A* Learning from Ordered Answer Sets *task is a tuple* $T = \langle B, S_M, \langle E^+, E^-, O^b, O^c \rangle \rangle$ *where* B *is an ASP program,* S_M *is a set of ASP rules,* E^+ *and* E^- *are finite sets of partial interpretations and* O^b *and* O^c *are finite sets of ordering examples over* E^+ *called brave and cautious orderings. A hypothesis* $H \subseteq S_M$ *is an inductive solution of* T *if and only if:*

1. $H \in ILP_{LAS}(\langle B, S_M, \langle E^+, E^- \rangle \rangle)$
2. $\forall o \in O^b$ $B \cup H$ *bravely respects* o
3. $\forall o \in O^c$ $B \cup H$ *cautiously respects* o

Note that the orderings are only over positive examples. The justification behind this restriction is that there does not appear to be any scenario where a hypothesis would need to respect an ordering of a pair of partial interpretations that are not extended by any pair of answer sets of $B \cup H$.

Example 14. Recall the journey preferences in Example 13. Consider the background knowledge B, which defines a set of possible journeys.

$$B = \left\{ \begin{array}{l} \texttt{1\{leg(1),\dots,leg(5)\}5.} \\ \texttt{1\{leg\_mode(L,walk),leg\_mode(L,bus)\}1:-leg(L).} \\ \texttt{1\{leg\_crime\_rating(L,1),\dots,leg\_crime\_rating(L,4000)\}1:-leg(L).} \\ \texttt{1\{leg\_distance(L,0),\dots,leg\_distance(L,4000)\}1:-leg(L).} \end{array} \right\}$$

Journeys A to D of Example 13 can be represented by the four positive examples e_A to e_D.

$$\left\langle \left\{ \begin{array}{l} \texttt{leg\_mode}(1,\texttt{walk}), \\ \texttt{leg\_crime\_rating}(1,2), \\ \texttt{leg\_distance}(1,500), \\ \texttt{leg\_mode}(2,\texttt{bus}), \\ \texttt{leg\_crime\_rating}(2,4), \\ \texttt{leg\_distance}(2,3000) \end{array} \right\}, \left\{ \begin{array}{l} \texttt{leg}(3), \\ \texttt{leg}(4), \\ \texttt{leg}(5) \end{array} \right\} \right\rangle$$

$$(e_A)$$

$$\left\langle \left\{ \begin{array}{l} \texttt{leg\_mode}(1,\texttt{bus}), \\ \texttt{leg\_crime\_rating}(1,2), \\ \texttt{leg\_distance}(1,400), \\ \texttt{leg\_mode}(2,\texttt{bus}), \\ \texttt{leg\_crime\_rating}(2,4), \\ \texttt{leg\_distance}(2,3000) \end{array} \right\}, \left\{ \begin{array}{l} \texttt{leg}(3), \\ \texttt{leg}(4), \\ \texttt{leg}(5) \end{array} \right\} \right\rangle$$

$$(e_C)$$

$$\left\langle \left\{ \begin{array}{l} \texttt{leg\_mode}(1,\texttt{bus}), \\ \texttt{leg\_crime\_rating}(1,2), \\ \texttt{leg\_distance}(1,4000), \\ \texttt{leg\_mode}(2,\texttt{walk}), \\ \texttt{leg\_crime\_rating}(2,5), \\ \texttt{leg\_distance}(2,1000) \end{array} \right\}, \left\{ \begin{array}{l} \texttt{leg}(3), \\ \texttt{leg}(4), \\ \texttt{leg}(5) \end{array} \right\} \right\rangle$$

$$(e_B)$$

$$\left\langle \left\{ \begin{array}{l} \texttt{leg\_mode}(1,\texttt{bus}), \\ \texttt{leg\_crime\_rating}(1,5), \\ \texttt{leg\_distance}(1,2000), \\ \texttt{leg\_mode}(2,\texttt{walk}), \\ \texttt{leg\_crime\_rating}(2,1), \\ \texttt{leg\_distance}(2,2000) \end{array} \right\}, \left\{ \begin{array}{l} \texttt{leg}(3), \\ \texttt{leg}(4), \\ \texttt{leg}(5) \end{array} \right\} \right\rangle$$

$$(e_D)$$

As these positive examples completely represent each journey, there is exactly one answer set of B that extends each example. Therefore there is no distinction between brave and cautious orderings in this case. Recall from Example 13 that journey A was preferred to journey D, which was preferred to journey C, which was preferred to journey B. This means that to learn the preferences in Example 13, we could give the orderings $\langle e_A, e_D, < \rangle$, $\langle e_D, e_C, < \rangle$ and $\langle e_C, e_B, < \rangle$ as either brave or cautious orderings.

4.2   Context-Dependent Learning from Answer Sets

Common to previous ILP frameworks is the underlying assumption that hypotheses should cover the examples with respect to one fixed given background knowledge. But, in practice, some examples may be context-dependent – different examples may need to be covered using different background knowledges. The journey preferences in Example 13 can be extended, for example, with contextual information (e.g. the weather).

Example 15. Reconsider the background knowledge and examples from Example 14. It may be that certain attributes of a journey are *context-dependent*; for instance, weather conditions may be important. Any of the ordering examples o in Example 14 could be extended with a context such as $C = \{\texttt{raining.}\}$. This would mean that for a brave ordering o, there should be a pair of answer sets of $B \cup H \cup C$ that extends the partial interpretations in o and that respects the ordering (w.r.t. the weak constraints in $B \cup H \cup C$).

In fact, the definition of a context-dependent ordering example given in this tutorial is slightly more general than in Example 15, as each partial interpretation in a context-dependent ordering example can have its own context. We will see that in addition to representing genuine contextual information, in some cases, contexts can be used in order to partition the background knowledge into pieces that are relevant to particular examples. We now formalise the notion of *context-dependent* examples. Similarly to ILP_{LOAS} examples, these are of two types: partial interpretations and ordering examples.

Definition 11. *A context-dependent partial interpretation (CDPI) is a pair $e = \langle e_{pi}, e_{ctx}\rangle$, where e_{pi} is a partial interpretation and e_{ctx} is an ASP^{ch} program (i.e. an ASP program with no weak constraints), called a context. Given a program P, an interpretation I is said to be an accepting answer set of e w.r.t. P if and only if $I \in AS(P \cup e_{ctx})$ and I extends e_{pi}. P is said to accept e if there is at least one accepting answer set of e w.r.t. P.*

Definition 12. *A context-dependent ordering example (CDOE) o is a tuple $\langle\langle e_{pi}^1, e_{ctx}^1\rangle, \langle e_{pi}^2, e_{ctx}^2\rangle, op\rangle$, where the first two elements are CDPIs and op is a binary comparison operator ($<, >, =, \leq, \geq$ or $\neq$). Given a CDOE $o = \langle e_1, e_2, op\rangle$, $inverse(o) = \langle e_1, e_2, op^{-1}\rangle$, where $<^{-1}$ is $\geq$, $\leq^{-1}$ is $>$, $=^{-1}$ is $\neq$, $\neq^{-1}$ is $=$, $>^{-1}$ is $\leq$ and $\geq^{-1}$ is $>$. A pair of interpretations $\langle I_1, I_2\rangle$ is said to be an accepting pair of answer sets of o wrt a program P if all of the following conditions hold: (i) I_1 is an accepting answer set of $\langle e_{pi}^1, e_{ctx}^1\rangle$; (ii) I_2 is an accepting answer set of $\langle e_{pi}^2, e_{ctx}^2\rangle$; and (iii) $\langle I_1, I_2, op\rangle \in ord(P, AS(P \cup e_{ctx}^1) \cup AS(P \cup e_{ctx}^2))$. A program P is said to bravely respect o if there is at least one accepting pair of answer sets of o. P is said to cautiously respect o if there is no accepting pair of answer sets of $inverse(o)$.*

Definition 13. *A Context-dependent Learning from Ordered Answer Sets $(ILP_{LOAS}^{context})$ task is a tuple $T = \langle B, S_M, \langle E^+, E^-, O^b, O^c\rangle\rangle$ where B is an ASP program, S_M is a set of ASP rules, E^+ and E^- are finite sets of CDPIs, and O^b and O^c are finite sets of CDOEs over E^+ called, respectively, brave and cautious orderings. A hypothesis $H \subseteq S_M$ is an inductive solution of T if and only if:*

1. $\forall e \in E^+$, $B \cup H$ accepts e
2. $\forall e \in E^-$, $B \cup H$ does not accept e
3. $\forall o \in O^b$, $B \cup H$ bravely respects o
4. $\forall o \in O^c$, $B \cup H$ cautiously respects o

Example 16. Reconsider the journey preference learning task of Example 14. The contextual information in Example 15 can be added to the examples e_1 and

e_2, to show the preference "in the case that it is raining e_1 is preferred to e_2, but otherwise it is the other way around" with the context dependent ordering examples o_1 and o_2:

$$o_1 = \langle \langle e_1, \{\, \texttt{raining.}\,\} \rangle, \langle e_2, \{\, \texttt{raining.}\,\} \rangle, < \rangle$$
$$o_2 = \langle \langle e_2, \emptyset \rangle, \langle e_1, \emptyset \rangle, < \rangle$$

4.3   ILASP

Inductive Learning of Answer Set Programs (ILASP) is a collection of algorithms for solving ILP_{LAS}, ILP_{LOAS} and $ILP_{LOAS}^{context}$ tasks. Similarly to ASPAL, each ILASP algorithm makes use of meta-level ASP programs. As we will see in Sect. 4.4, deciding whether a hypothesis is a solution of one of the ILP_b tasks solved by ASPAL is NP-complete in the propositional case, whereas the same decision problem for the tasks solved by ILASP is DP-complete. For this reason, ILASP 1 and 2 do not encode the search for solutions in a single meta-level ASP program (solving such a program is NP-complete in the propositional case), but instead employ an iterative algorithm, where a meta-level ASP program is solved repeatedly with new constraints added in each iteration, until the optimal answer sets of the meta-level program correspond to the optimal inductive solutions of the task.

The details of ILASP's meta-level programs are beyond the scope of this tutorial[6]. For the purposes of this tutorial, all the reader needs to know is that given any ILP_{LAS}, $ILP_{LAS}^{context}$ or $ILP_{LOAS}^{context}$ task T, both ILASP1(T) and ILASP2(T) return an optimal solution of T (resources permitting, of course).

Relevant Examples and ILASP2i. The ILASP1 and ILASP2 algorithms both scale poorly with respect to the number of examples as the number of rules in the grounding of their meta-level ASP programs is proportional to the number of examples. The ILASP2i algorithm [24] solves a task iteratively, by building up a set of *relevant examples*. The idea is that in real tasks, many examples may be similar and may therefore be covered by exactly the same set of hypotheses. If this is the case, it is sufficient to consider only a small set of examples that are representative of the full set – these are the relevant examples. ILASP2i constructs this set iteratively, by assuming that its current relevant example set is completely representative of the full set, and using ILASP2 to solve the task with only those examples. If the assumption holds, then the hypothesis returned by ILASP2 will be an inductive solution of the full task. If not, then there must be at least one example which is not covered by the hypothesis returned by ILASP2 – this is added to the relevant example set before the next iteration.[7]

---

[6] Details of the encodings can be found in [20–22].

[7] In Algorithm 1.1 the set *Relevant* is a pair of sets of examples, the first set being relevant positive examples and the second set relevant negative examples. The notation on Line 5 means to add example *re* to the appropriate set, depending on whether it is a positive or a negative example.

The *findRelevantExample* method is used to check whether a given hypothesis H is an inductive solution of the full task; if it is, then it returns `nil` (as there are no relevant examples to find); otherwise, it returns an example which is not covered by H.

Algorithm 1.1. ILASP2i

1: **procedure** ILASP2I($\langle B, S_M, E^+, E^- \rangle$)
2:     $Relevant = \langle \emptyset, \emptyset \rangle$; $H = \emptyset$;
3:     $re = findRelevantExample(\langle B, S_M, E^+, E^- \rangle, H)$;
4:     **while** $re \neq$ `nil` **do**
5:         $Relevant << re$;
6:         $H = ILASP2(\langle B, S_M, Relevant \rangle)$;
7:         **if** $H ==$ `nil` **then**
8:             **return** UNSATISFIABLE;
9:         **else**
10:             $re = findRelevantExample(\langle B, S_M, E^+, E^- \rangle, H)$;
11:         **end if**
12:     **end while**
13:     **return** H;
14: **end procedure**

In some ways, ILASP2i can be thought of as a non-monotonic variation on the idea of a cover-loop with three major differences: (1) just because an example is covered in one iteration, it is not guaranteed to be covered in future iterations (unless it is added to the set of relevant examples); (2) the learning starts from scratch in each iteration (rather than iteratively building a hypothesis); and (3) the full set of relevant examples is considered in each iteration (rather than a single current seed example).

4.4   The Complexity and Generality of Learning Answer Set Programs

Throughout this tutorial, we have discussed the six main frameworks for learning under the answer set semantics. As we introduced the early learning frameworks, we discussed some of their limitations, such as the fact that systems based on brave induction are unable to learn constraints. These limitations were some of the original motivations of the later frameworks such as ILP_{LAS}.

Although we have already demonstrated that there are programs which can be learned by ILP_{LAS} based systems that cannot be learned by systems based on earlier frameworks, it is more interesting to consider exactly which classes of programs can be learned by each framework. The aim is to characterise the class of ASP programs that a framework is capable of learning, if given sufficient examples. Language biases tend, in general, to impose their own restrictions on the classes of program that can be learned. They are primarily used to aid the performance of the computation, rather than to capture intrinsic properties of a learning framework. In this chapter we will therefore consider learning tasks with unrestricted hypothesis spaces: hypotheses can be constructed from any set of normal rules, choice rules and hard and weak constraints. We assume each learning framework $\mathcal{F}$ to have a task consisting of a pair $\langle B, E_{\mathcal{F}} \rangle$, where B is the (ASP) background knowledge and $E_{\mathcal{F}}$ is a tuple consisting of the examples for this framework; for example $E_{LAS}^8 = \langle E^+, E^- \rangle$ where E^+ and E^- are sets of partial interpretations.

In [25], the generality of the six main frameworks was investigated and three new measures of generality were presented, based on which of the hypotheses a framework can *distinguish* from other hypotheses. Roughly speaking, a hypothesis H_1 can be distinguished from another hypothesis H_2 (with respect to a given background knowledge B) if there is at least one set of examples E such that $B \cup H_1$ satisfies every example in E and $B \cup H_2$ does not. The following definition formalises the *one-to-one-distinguishability* class of a learning framework.

Definition 14. *The* one-to-one-distinguishability class *of a learning framework* $\mathcal{F}$ *(denoted* $\mathcal{D}_1^1(\mathcal{F})$*) is the set of tuples* $\langle B, H_1, H_2 \rangle$ *of ASP programs for which there is at least one task* $T_{\mathcal{F}} = \langle B, E_{\mathcal{F}} \rangle$ *such that* $H_1 \in ILP_{\mathcal{F}}(T_{\mathcal{F}})$ *and* $H_2 \notin ILP_{\mathcal{F}}(T_{\mathcal{F}})$. *For each* $\langle B, H_1, H_2 \rangle \in \mathcal{D}_1^1(\mathcal{F})$, $T_{\mathcal{F}}$ *is said to* distinguish H_1 *from* H_2 *with respect to* B.

Note that the one-to-one-distinguishability relationship is not symmetric; i.e there are pairs of hypotheses H_1 and H_2 such that, given a background knowledge B, H_1 can be distinguished from H_2, but H_2 can not be distinguished from H_1. This is illustrated by Example 17.

Example 17. Consider a background knowledge B that defines the concepts of cell, same_block, same_row and same_column for a 4x4 Sudoku grid (see Example 11).

Let H_1 be the incomplete description of the Sudoku rules:

```
1 { value(C, 1), value(C, 2), value(C, 3), value(C, 4) } 1 :- cell(C).
:- value(C1, V), value(C2, V), same_row(C1, C2).
:- value(C1, V), value(C2, V), same_col(C1, C2).
```

Also let H_2 be the complete description of the Sudoku rules:

```
1 { value(C, 1), value(C, 2), value(C, 3), value(C, 4) } 1 :- cell(C).
:- value(C1, V), value(C2, V), same_row(C1, C2).
:- value(C1, V), value(C2, V), same_col(C1, C2).
:- value(C1, V), value(C2, V), same_block(C1, C2).
```

ILP_b can distinguish H_1 from H_2 with respect to B. This can be seen using the task $\langle B, \langle \{\text{value}((1,1),1), \text{value}((2,2),1)\}, \emptyset \rangle \rangle$. On the other hand, ILP_b cannot distinguish H_2 from H_1. Whatever examples are given in a learning task to learn H_2, it must be the case that $E^+ \subseteq A$ and $E^- \cap A = \emptyset$, where A is an answer set of $B \cup H_2$. But answer sets of $B \cup H_2$ are also answer sets of $B \cup H_1$. So A is also an answer set of $B \cup H_1$, which implies that H_1 satisfies the same examples and is a solution of the same learning task.

---

[8] Note that to avoid cumbersome notation, we denote this E_{LAS} rather than $E_{ILP_{LAS}}$.

Table 1 gives conditions which are both sufficient and necessary for a tuple $\langle B, H, H_1 \rangle$ to appear in the one-to-one-distinguishability class of each learning framework.[9] Proofs of the correctness of these conditions are given in [25]. The conditions show that the following orderings hold:

- $\mathcal{D}_1^1(ILP_b) = \mathcal{D}_1^1(ILP_{sm}) \subset \mathcal{D}_1^1(ILP_{LAS}) \subset \mathcal{D}_1^1(ILP_{LOAS}) \subset \mathcal{D}_1^1(ILP_{LOAS}^{context})$
- $\mathcal{D}_1^1(ILP_c) \subset \mathcal{D}_1^1(ILP_{LAS})$

Table 1. A summary of the sufficient and necessary conditions in each learning framework for a hypothesis H_1 to be distinguishable from another hypothesis H_2 with respect to a background knowledge B.

Framework $\mathcal{F}$	Sufficient/necessary condition for $\langle B, H_1, H_2 \rangle$ to be in $\mathcal{D}_1^1(\mathcal{F})$
ILP_b	$AS(B \cup H_1) \not\subseteq AS(B \cup H_2)$
ILP_{sm}	$AS(B \cup H_1) \not\subseteq AS(B \cup H_2)$
ILP_c	$AS(B \cup H_1) \neq \emptyset \wedge (AS(B \cup H_2) = \emptyset \vee (\mathcal{E}_c(B \cup H_1) \not\subseteq \mathcal{E}_c(B \cup H_2)))$
ILP_{LAS}	$AS(B \cup H_1) \neq AS(B \cup H_2)$
ILP_{LOAS}	$(AS(B \cup H_1) \neq AS(B \cup H_2)) \vee (ord(B \cup H_1) \neq ord(B \cup H_2))$
$ILP_{LOAS}^{context}$	$(B \cup H_1 \not\equiv^s B \cup H_2) \vee (\exists C \in \mathcal{ASP}^{ch}$ s.t. $ord(B \cup H_1 \cup C) \neq ord(B \cup H_2 \cup C))$

If we view one-to-one-distinguishability as a measure of the generality of a learning framework, then ILP_b, ILP_{sm} and ILP_c are each strictly less general than ILP_{LAS}, and ILP_{LOAS} and $ILP_{LOAS}^{context}$ are more general still.

The One-to-Many-Distinguishability Class of a Learning Framework. In practice, an ILP task has a search space of possible hypotheses, and it is important to know the cases in which one particular hypothesis can be distinguished from the rest. In what follows, we analyse the conditions under which a learning framework can distinguish a hypothesis from *a set* of other hypotheses. This corresponds to the notion of *one-to-many-distinguishability class* of a learning framework, which is a generalisation of the notion of the *one-to-one-distinguishability class*.

Definition 15. *The* one-to-many-distinguishability class *of a learning framework $\mathcal{F}$ (denoted $\mathcal{D}_m^1(\mathcal{F})$) is the set of all tuples $\langle B, H, \{H_1, \ldots, H_n\} \rangle$ such that there is a task $T_{\mathcal{F}}$ that distinguishes H from each H_i with respect to B.*

---

[9] In Table 1 the following two notations are used. For programs P and Q the relation $P \equiv^s Q$ means that for any program R $AS(P \cup R) = AS(Q \cup R)$ and for a program P $\mathcal{E}_c(BP)$ is the set of conjunctions of literals in every answer set of P.

Given two frameworks $\mathcal{F}_1$ and $\mathcal{F}_2$, we say that $\mathcal{F}_1$ is at least as (resp. more) $\mathcal{D}_m^1$-general as (resp. than) $\mathcal{F}_2$ if $\mathcal{D}_m^1(\mathcal{F}_2) \subseteq \mathcal{D}_m^1(\mathcal{F}_1)$ (resp. $\mathcal{D}_m^1(F_2) \subset \mathcal{D}_m^1(F_1)$).

The one-to-many-distinguishability class tells us the circumstances in which a framework is general enough to distinguish some target hypothesis from a set of unwanted hypotheses. Note that, although the tuples in a one-to-many-distinguishability class that have a singleton set as the third argument correspond to the tuples in a one-to-one-distinguishability class of that framework, it is not always the case that if $\mathcal{F}_1$ is more $\mathcal{D}_m^1$-general than $\mathcal{F}_2$ then $\mathcal{F}_1$ is also more $\mathcal{D}_1^1$-general than $\mathcal{F}_2$. For example, we will see that ILP_{sm} is more $\mathcal{D}_m^1$-general than ILP_b, but we have already seen that the ILP_b and ILP_{sm} are equally $\mathcal{D}_1^1$-general.

Example 18. $\mathcal{D}_m^1(ILP_b) \subset \mathcal{D}_m^1(ILP_{sm})$. We can see this as follows. Firstly, clearly $\mathcal{D}_m^1(ILP_b) \subseteq \mathcal{D}_m^1(ILP_{sm})$, as any ILP_b task can be trivially mapped into an ILP_{sm} task. Thus, it remains to show that $\mathcal{D}_m^1(ILP_b) \neq \mathcal{D}_m^1(ILP_{sm})$.

Consider the programs $B = \emptyset$, $H = \{1\{\text{heads}, \text{tails}\}1.\}$, $H_1 = \{\text{heads}.\}$ and $H_2 = \{\text{tails}.\}$. $\langle B, H, \{H_1, H_2\}\rangle \in \mathcal{D}_m^1(ILP_{sm})$ $(\langle B, \langle\{\langle\{\text{tails}\}, \emptyset\rangle,$ $\langle\{\text{heads}\}, \emptyset\rangle\}\rangle\rangle)$ distinguishes H from H_1 wrt the background knowledge B). We now show that there is no task $T_b = \langle B, \langle E^+, E^-\rangle\rangle$ such that $H \in ILP_b(T_b)$ and $\{H_1, H_2\} \cap ILP_b(T_b) = \emptyset$.

Assume for contradiction that there is such a task T_b. As $H \in ILP_b(T_b)$ and $AS(B \cup H) = \{\{\text{heads}\}, \{\text{tails}\}\}$, $E^+ \subset \{\text{heads}, \text{tails}\}$ and $E^- \subset \{\text{heads}, \text{tails}\}$ (neither can be equal to $\{\text{heads}, \text{tails}\}$ or H would not be a solution).

Case 1: $E^+ = \emptyset$

    Case a: $E^- = \emptyset$

        Then H_1 and H_2 would be inductive solutions. This is a contradiction as $\{H_1, H_2\} \cap ILP_b(T_b) = \emptyset$.

    Case b: $E^- = \{\text{heads}\}$

        Then H_2 would be an inductive solution of T_b. Contradiction.

    Case c: $E^- = \{\text{tails}\}$

        Then H_1 would be an inductive solution of T_b. Contradiction.

Case 2: $E^+ = \{\text{heads}\}$

    heads $\notin E^-$ as otherwise the task would have no solutions (and we know that H is a solution). In this case H_1 would be an inductive solution (regardless of what else is in E^-). Contradiction.

Case 3: $E^+ = \{\text{tails}\}$

    Similarly to above case, tails $\notin E^-$ as otherwise the task would have no solutions. In this case H_2 would be an inductive solution (regardless of what else is in E^-). Contradiction.

Hence, there is no such task $T_b = \langle B, \langle E^+, E^-\rangle\rangle$ such that $H \in ILP_b(T_b)$ and $\{H_1, H_2\} \cap ILP_b(T_b) = \emptyset$. So, $\mathcal{D}_m^1(ILP_b) \neq \mathcal{D}_m^1(ILP_{sm})$.

In [25], it is shown that the following orderings hold.

- $\mathcal{D}_m^1(ILP_b) \subset \mathcal{D}_m^1(ILP_{sm}) \subset \mathcal{D}_m^1(ILP_{LAS}) \subset \mathcal{D}_m^1(ILP_{LOAS}) \subset$ $\mathcal{D}_m^1(ILP_{LOAS}^{context})$
- $\mathcal{D}_m^1(ILP_c) \subset \mathcal{D}_m^1(ILP_{LAS})$

[25] presents a further measure of generality, many-to-many-distinguishability. The many-to-many-distinguishability class of a framework is used to analyse which sets of hypotheses can be distinguished from other sets of hypotheses. However, the many-to-many-distinguishability class is outside the scope of this tutorial.

Complexity. Given the differences in generality between the various learning frameworks, an obvious question to ask is whether there is any price to pay in terms of computational complexity when using the more general frameworks. In this section, we consider three common decision problems when using the learning frameworks:

- *Verification:* deciding whether a given hypothesis is an inductive solution of a given learning task.
- *Satisfiability:* deciding whether a given learning task has any inductive solutions.
- *Optimum Verification:* deciding whether a given hypothesis is an optimal inductive solution of a given learning task.

Table 2 gives the complexity results for propositional versions of each of the learning frameworks (where the background knowledge, contexts of examples and hypothesis space is restricted to propositional ASP). Proofs of the results in Table 2 can be found in [20]. Interestingly despite the great difference in the generality of the various frameworks, for each of the three decision problems, $ILP_{LOAS}^{context}$ has the same complexity as ILP_c. The complexity of both ILP_b and ILP_{sm} is lower than any of the other frameworks, which suggests that in applications where the increased generality of the other frameworks is not needed, ILP_{sm} may be more suitable. It should be noted that ILASP may still be used to solve such tasks – ILP_{LAS} tasks with no negative examples are equivalent to ILP_{sm} tasks.

4.5 Learning Answer Set Programs from Noisy Examples

The learning from answer sets frameworks have recently been upgraded to support learning from noisy examples [26]. In this section, we present a generalisation of the idea to give a general way of upgrading any non-noisy learning framework with a notion of penalised examples. There are already several algorithms, predating these formal definitions, which adopt the approach of penalising examples (e.g. XHAIL [36] and Inspire [18]).

Given any learning framework ILP_F covered in this tutorial (ILP_b, ILP_c, ILP_{LAS}, etc) a task of the penalised framework $n(ILP_F)$ is of the same form as tasks for ILP_F, other than the fact that each example is of the form $e@p$,

Table 2. A summary of the complexity of the various learning frameworks. *Verification* corresponds to deciding whether a given hypothesis is a solution of a given learning task. *Satisfiability* corresponds to deciding whether a learning task has any solutions at all. *Optimum verification* corresponds to deciding whether a given hypothesis is the optimal (shortest) solution of a given task.

Framework	Verification	Satisfiablity	Optimum verification
ILP_b	NP-complete	NP-complete	DP-complete
ILP_{sm}	NP-complete	NP-complete	DP-complete
ILP_c	DP-complete	Σ_2^P-complete	Π_2^P-complete
ILP_{LAS}	DP-complete	Σ_2^P-complete	Π_2^P-complete
ILP_{LOAS}	DP-complete	Σ_2^P-complete	Π_2^P-complete
$ILP_{LOAS}^{context}$	DP-complete	Σ_2^P-complete	Π_2^P-complete

where e is an example of the previous framework and p is either ∞ (meaning the example must be covered) or it is a positive integer representing the *penalty* for not covering that example. This penalty is also often called a *weight* for the example.

Given any task T and hypothesis H, the score of H w.r.t. T, written $\mathcal{S}(H,T)$, is equal to $|H| + \sum_{e@p \in U} p$ where U is the set of examples in T that are not covered by H. In the case of brave induction (where each answer set of $B \cup H$ might suggest that different examples are covered), $\mathcal{S}(H,T)$ is assigned the minimum possible score. In the case of cautious induction, for the penalty of a hypothesis to be finite, it must also be satisifiable when it is combined with the background knowledge. The inductive solutions of a task are the hypotheses with a finite score. The optimal inductive solutions are the set of inductive solutions which minimise the score.

Example 19. Consider an extension of the ILP_b task from Example 3, $T' = \langle B, S_M, \langle E^+, E^- \rangle \rangle$, where:

$$B = \left\{ \begin{array}{l} \texttt{bird(X):-penguin(X).} \\ \texttt{bird(X):-sparrow(X).} \\ \texttt{penguin(b1).} \\ \texttt{penguin(b2).} \\ \texttt{penguin(b3).} \\ \texttt{sparrow(b4).} \end{array} \right\}$$

$E^+ = \{\texttt{flies(b1)@2, flies(b4)@2}\}$

$E^- = \{\texttt{flies(b2)@2, flies(b3)@2}\}$

$$S_M = \left\{ \begin{array}{l} \texttt{h}_1 : \texttt{flies(X):-bird(X).} \\ \texttt{h}_2 : \texttt{flies(X):-bird(X),} \\ \qquad\qquad \texttt{not penguin(X).} \end{array} \right\}$$

- $\mathcal{S}(\emptyset, T') = |\emptyset| + 4 = 4$.
- $\mathcal{S}(\{\texttt{h}_1\}, T') = |\{\texttt{h}_1\}| + 4 = 5$ (recall that the *type* atom $\texttt{bird(X)}$ does not count towards the length of the rule).
- $\mathcal{S}(\{\texttt{h}_2\}, T') = |\{\texttt{h}_2\}| + 2 = 4$.
- $\mathcal{S}(\{\texttt{h}_2\}, T') = |\{\texttt{h}_1, \texttt{h}_2\}| + 4 = 7$.

This task has two optimal inductive solutions: $\emptyset$ and $\{\texttt{h}_2\}$. The choice of penalty for the examples is important. If each of the examples in this task had penalty 1, $\emptyset$ would have been optimal; whereas if the penalties had all been 3, $\{h_2\}$ would have been optimal.

Note that we have used an extremely small hypothesis space here to keep things simple. In reality, the hypothesis space would usually be much bigger!

The ASPAL encoding shown in the previous section can be extended to solve noisy tasks. This is achieved using weak constraints to represent the penalties of the examples. The XHAIL and ILASP systems have also been extended to handle noise in a similar way by using optimisation in ASP. ASPAL and ILASP are both guaranteed to find an optimal inductive solution of any task; however, as shown in Example 20 XHAIL may not.

Example 20. Consider the following noisy task, in the XHAIL input format:

```
p(X) :- q(X, 1), q(X, 2).        #modeh r(+s).
p(X) :- r(X).                    #modeh q(+s2, +t).
s(a).    s(b).    s2(b).         #example not p(a)=50.
t(1).    t(2).                   #example p(b)=100.
```

This corresponds to a hypothesis space that contains two facts $F_1 = r(X)$, $F_2 = q(X, Y)$ (in XHAIL, these facts are implicitly "typed", so the first fact, for example, can be thought of as the rule $r(X) :- s(X)$). The two examples have penalties 50 and 100 respectively. There are four possible hypotheses: $\emptyset$, F_1, F_2 and $F_1 \cup F_2$, with scores 100, 51, 1 and 52 respectively. XHAIL terminates and returns F_1, which is a suboptimal hypothesis.

The issue is with the first step. The system finds the smallest abductive solution, $\{r(b)\}$ and as there are no body declarations in the task, the kernel set contains only one rule: $r(b) :- s(b)$. XHAIL then attempts to generalise to a first order hypothesis that covers the examples. There are two hypotheses which are subsets of a generalisation of $r(b)$ (F_1 and $\emptyset$); as F_1 has a lower score than $\emptyset$, XHAIL terminates and returns F_1. The system does not find the abductive solution $\{q(b, 1), q(b, 2)\}$, which is larger than $\{r(b)\}$ and is therefore not chosen, even though it would eventually lead to a better solution than $\{r(b)\}$.

It should be noted that XHAIL does have an *iterative deepening* feature for exploring non-minimal abductive solutions, but in this case using this option XHAIL still returns F_1, even though F_2 is a more optimal hypothesis. Even when iterative deepening is enabled, XHAIL only considers non-minimal abductive solutions if the minimal abductive solutions do not lead to any non-empty inductive solutions.

Although ILASP1, ILASP2 and ILASP2i are all guaranteed to find optimal inductive solutions of any $n(ILP_{LOAS}^{context})$ task, they do not perform well when solving tasks with noise. ILASP3 is specifically targetted at learning tasks with noisy examples; however, a discussion of ILASP3 is beyond the scope of this tutorial. An in depth discussion of ILASP3 can be found in [20], and an evaluation of ILASP3 on several noisy datasets can be found in [26].

5   Conclusion

This tutorial has presented an introduction to logic based learning under the answer set semantics. We have introduced the six main frameworks for learning ASP programs, and presented generality results highlighting the flaws in early frameworks and showing that to learn some ASP programs the recent, more general, frameworks are required. The development of learning frameworks has been matched by the development of more sophisticated algorithms, of which we have given an overview in this tutorial. The most recent ILASP system supports learning ASP programs including normal rules, choice rules and hard and weak constraints, even from noisy examples. However, there are still challenges to be addressed, particularly with respect to scalability, which is the focus of our current research.

References

1. Athakravi, D.: Inductive logic programming using bounded hypothesis space. Ph.D. thesis, Imperial College London (2015)
2. Athakravi, D., Corapi, D., Broda, K., Russo, A.: Learning through hypothesis refinement using answer set programming. In: Zaverucha, G., Santos Costa, V., Paes, A. (eds.) ILP 2013. LNCS (LNAI), vol. 8812, pp. 31–46. Springer, Heidelberg (2014). https://doi.org/10.1007/978-3-662-44923-3_3
3. Blockeel, H., De Raedt, L.: Top-down induction of first-order logical decision trees. Artif. Intell. **101**(1), 285–297 (1998)
4. Brain, M., Cliffe, O., De Vos, M.: A pragmatic programmer's guide to answer set programming. In: Answer Set Programming, p. 49 (2009)
5. Corapi, D., Russo, A.: ASPAL. Proof of soundness and completeness. Technical report, Department of Computing (DTR11-5), Imperial College, London (2011)
6. Corapi, D., Russo, A., Lupu, E.: Inductive logic programming as abductive search. In: ICLP (Technical Communications), pp. 54–63 (2010)
7. Corapi, D., Russo, A., Lupu, E.: Inductive logic programming in answer set programming. In: Muggleton, S.H., Tamaddoni-Nezhad, A., Lisi, F.A. (eds.) ILP 2011. LNCS (LNAI), vol. 7207, pp. 91–97. Springer, Heidelberg (2012). https://doi.org/10.1007/978-3-642-31951-8_12
8. Dastani, M., Jacobs, N., Jonker, C.M., Treur, J.: Modeling user preferences and mediating agents in electronic commerce. In: Dignum, F., Sierra, C. (eds.) Agent Mediated Electronic Commerce. LNCS (LNAI), vol. 1991, pp. 163–193. Springer, Heidelberg (2001). https://doi.org/10.1007/3-540-44682-6_10
9. Eiter, T., Ianni, G., Krennwallner, T.: Answer set programming: a primer. In: Tessaris, S., et al. (eds.) Reasoning Web 2009. LNCS, vol. 5689, pp. 40–110. Springer, Heidelberg (2009). https://doi.org/10.1007/978-3-642-03754-2_2
10. Erdem, E., Gelfond, M., Leone, N.: Applications of answer set programming. AI Mag. **37**(3), 53–68 (2016)
11. Fürnkranz, J., Hüllermeier, E.: Pairwise preference learning and ranking. In: Lavrač, N., Gamberger, D., Blockeel, H., Todorovski, L. (eds.) ECML 2003. LNCS (LNAI), vol. 2837, pp. 145–156. Springer, Heidelberg (2003). https://doi.org/10.1007/978-3-540-39857-8_15

12. Geisler, B., Ha, V., Haddawy, P.: Modeling user preferences via theory refinement. In: Proceedings of the 6th International Conference on Intelligent User Interfaces, pp. 87–90. ACM (2001)
13. Gelfond, M., Kahl, Y.: Knowledge Representation, Reasoning, and the Design of Intelligent Agents: The Answer-Set Programming Approach. Cambridge University Press, Cambridge (2014)
14. Gelfond, M., Lifschitz, V.: The stable model semantics for logic programming. In: ICLP/SLP, vol. 88, pp. 1070–1080 (1988)
15. Horváth, T.: A model of user preference learning for content-based recommender systems. Comput. Inform. **28**(4), 453–481 (2012)
16. Inoue, K., Kudoh, Y.: Learning extended logic programs. In: IJCAI, no. 1, pp. 176–181 (1997)
17. Katzouris, N., Artikis, A., Paliouras, G.: Incremental learning of event definitions with inductive logic programming. Mach. Learn. **100**(2–3), 555–585 (2015)
18. Kazmi, M., Schüller, P., Saygın, Y.: Improving scalability of inductive logic programming via pruning and best-effort optimisation. Expert Syst. Appl. **87**, 291–303 (2017)
19. Kowalski, R., Sergot, M.: A logic-based calculus of events. New Gener. Comput. **4**(1), 67–95 (1986)
20. Law, M.: Inductive learning of answer set programs. Ph.D. thesis, Imperial College London (2018)
21. Law, M., Russo, A., Broda, K.: Inductive learning of answer set programs. In: Fermé, E., Leite, J. (eds.) JELIA 2014. LNCS (LNAI), vol. 8761, pp. 311–325. Springer, Cham (2014). https://doi.org/10.1007/978-3-319-11558-0_22
22. Law, M., Russo, A., Broda, K.: Learning weak constraints in answer set programming. Theory Pract. Log. Program. **15**(4–5), 511–525 (2015)
23. Law, M., Russo, A., Broda, K.: Simplified reduct for choice rules in ASP. Technical report, Department of Computing (DTR2015-2), Imperial College London (2015)
24. Law, M., Russo, A., Broda, K.: Iterative learning of answer set programs from context dependent examples. Theory Pract. Log. Program. **16**(5–6), 834–848 (2016)
25. Law, M., Russo, A., Broda, K.: The complexity and generality of learning answer set programs. Artif. Intell. **259**, 110–146 (2018)
26. Law, M., Russo, A., Broda, K.: Inductive learning of answer set programs from noisy examples. In: Advances in Cognitive Systems (2018)
27. Mueller, E.T.: Commonsense Reasoning: An Event Calculus Based Approach. Morgan Kaufmann, San Francisco (2014)
28. Muggleton, S.: Inductive logic programming. New Gener. Comput. **8**(4), 295–318 (1991)
29. Muggleton, S.: Inverse entailment and progol. New Gener. Comput. **13**(3–4), 245–286 (1995)
30. Muggleton, S., et al.: ILP turns 20. Mach. Learn. **86**(1), 3–23 (2012)
31. Nogueira, M., Balduccini, M., Gelfond, M., Watson, R., Barry, M.: An A-Prolog decision support system for the space shuttle. In: Ramakrishnan, I.V. (ed.) PADL 2001. LNCS, vol. 1990, pp. 169–183. Springer, Heidelberg (2001). https://doi.org/10.1007/3-540-45241-9_12
32. Nuffelen, B.: Abductive constraint logic programming: implementation and applications. Ph.D. thesis, K.U. Leuven (2004)
33. Otero, R.P.: Induction of stable models. In: Rouveirol, C., Sebag, M. (eds.) ILP 2001. LNCS (LNAI), vol. 2157, pp. 193–205. Springer, Heidelberg (2001). https://doi.org/10.1007/3-540-44797-0_16

34. Papadimitriou, C.H.: Computational Complexity. Wiley, New York (2003)
35. Ray, O.: Hybrid abductive inductive learning. Ph.D. thesis, Imperial College London (2005)
36. Ray, O.: Nonmonotonic abductive inductive learning. J. Appl. Log. **7**(3), 329–340 (2009)
37. Ray, O., Broda, K., Russo, A.: Hybrid abductive inductive learning: a generalisation of progol. In: Horváth, T., Yamamoto, A. (eds.) ILP 2003. LNCS (LNAI), vol. 2835, pp. 311–328. Springer, Heidelberg (2003). https://doi.org/10.1007/978-3-540-39917-9_21
38. Ricca, F., et al.: A logic-based system for e-tourism. Fundam. Inform. **105**(1–2), 35–55 (2010)
39. Sakama, C.: Inverse entailment in nonmonotonic logic programs. In: Cussens, J., Frisch, A. (eds.) ILP 2000. LNCS (LNAI), vol. 1866, pp. 209–224. Springer, Heidelberg (2000). https://doi.org/10.1007/3-540-44960-4_13
40. Sakama, C.: Nonmonotomic inductive logic programming. In: Eiter, T., Faber, W., Truszczyński, M. (eds.) LPNMR 2001. LNCS (LNAI), vol. 2173, pp. 62–80. Springer, Heidelberg (2001). https://doi.org/10.1007/3-540-45402-0_5
41. Sakama, C., Inoue, K.: Brave induction: a logical framework for learning from incomplete information. Mach. Learn. **76**(1), 3–35 (2009)
42. Seitzer, J., Buckley, J.P., Pan, Y.: INDED: a distributed knowledge-based learning system. IEEE Intell. Syst. Appl. **15**(5), 38–46 (2000)
43. Soininen, T., Niemelä, I.: Developing a declarative rule language for applications in product configuration. In: Gupta, G. (ed.) PADL 1999. LNCS, vol. 1551, pp. 305–319. Springer, Heidelberg (1998). https://doi.org/10.1007/3-540-49201-1_21
44. Srinivasan, A.: The Aleph Manual. Machine Learning at the Computing Laboratory, Oxford University (2001)

Constraint Learning: An Appetizer

Stefano Teso[✉]

KU Leuven, Leuven, Belgium
stefano.teso@cs.kuleuven.be

Abstract. Constraints are ubiquitous in artificial intelligence and operations research. They appear in logical problems like propositional satisfiability, in discrete problems like constraint satisfaction, and in full-fledged mathematical optimization tasks. Constraint learning enters the picture when the structure or the parameters of the constraint satisfaction/optimization problem to be solved are (partially) unknown and must be inferred from data. The required supervision may come from offline sources or gathered by interacting with human domain experts and decision makers. With these lecture notes, we offer a brief but self-contained introduction to the core concepts of constraint learning, while sampling from the diverse spectrum of constraint learning methods, covering classic strategies and more recent advances. We will also discuss links to other areas of AI and machine learning, including concept learning, learning from queries, structured-output prediction, (statistical) relational learning, preference elicitation, and inverse optimization.

Keywords: Machine learning · Constraint satisfaction ·
Constraint optimization · Interactive learning

1 Introduction

Constraint learning is the task of acquiring constraint satisfaction or optimization problems from examples of solutions and non-solutions or other types of supervision. This document overviews selected topics in constraint learning and has no ambition of completeness. More specifically, we will discuss learning of *hard* constraints, which implicitly define a satisfaction problem; learning of *soft* constraints, where competing constraints are assigned different preferences; and *interactive* learning of hard or soft constraints, where the learning algorithm obtains supervision by interacting with an oracle (e.g. a human expert, a non-expert, a measurement apparatus). For a more in-depth guide to constraint learning and related areas, we refer the interested reader to the works by O'Sullivan [35], Bessiere et al. [9], Lombardi et al. [30], and De Raedt et al. [15]. Further pointers to the literature are provided in Sect. 6.

Why Constraint Learning?

Constraints are extremely popular from a modeling perspective, and appear in all sort of satisfaction and optimization problems in both artificial intelligence and

© Springer Nature Switzerland AG 2019
M. Krötzsch and D. Stepanova (Eds.): Reasoning Web 2019, LNCS 11810, pp. 232–249, 2019.
https://doi.org/10.1007/978-3-030-31423-1_7

operations research. Plenty of frameworks for modelling and solving problems involving constraints exist, from propositional satisfiability (SAT) and linear programming (LP), to constraint satisfaction [42], to more sophisticated alternatives that combine combinatorial and numerical elements, like mixed-integer linear programming (MILP). Given a formal specification of a constraint satisfaction or optimization problem of interest, it is often sufficient to feed it to an appropriate solver to obtain an appropriate solution[1].

A major bottleneck of this setup is that obtaining a formal constraint theory is non-obvious: designing an appropriate, working constraint satisfaction or optimization problem requires both domain and modeling expertise. For this reason, in many cases a modeling expert is hired and has to interact with domain expert to acquire informal requirements and turn them into a valid constraint theory. This process is can be expensive and time consuming.

The idea is then to replace or assist this process by acquiring a constraint theory directly from examples. In the simplest setting, when learning hard constraints—which implicitly define constraint satisfaction models like propositional concepts [54], satisfiability modulo theory formulas [5], and general constraint theories over discrete variables [9]—one is given a data set of positive and negative configurations and searches for a theory that covers (i.e. classifies as feasible) all of the positive examples and none of the negative ones. At its core, this is a form of concept learning [9]. Several variants and extensions of this setup have been proposed, including learning of soft constraints [41], which can be violated and are assigned different degrees of importance, and full-fledged optimization problems [36]. In the following, we will briefly cover the most prominent of these settings.

Of course, in practice the learned constraint theory may be approximate or wrong. In this case, two things can occur: either more data is provided and the theory is refined accordingly, or a human expert can revise the model via inspection and debugging. Here the goal is not to replace human experts, but rather to aid them by generating a reasonable initial theory consistent with all available supervision. Notice that in some settings the model does not have to be perfect. For instance, in recommendation the goal is to acquire a soft constraint theory that knows enough about the preferences of the target user to be able to provide reasonable personalized recommendations: so long as the model manages to identify some interesting products, the goal is met [39]. In this case, approximate models are acceptable and no human intervention is needed.

Dimensions of Constraint Learning

Formalisms and approaches to constraint learning can be roughly grouped based on three criteria:

---

[1] One should of course keep in mind that many constrained satisfaction/optimization problems can be NP-hard, so obtaining a solution in an acceptable time may still be tricky; see below for some examples.

(a) The type of constraints being learned. One can learn hard constraints, which define pure satisfaction models, soft constraints, which implicitly define optimization models, or (in principle) both. We will consider approaches to learning hard and soft constraints in Sects. 3 and 4, respectively. There is essentially no literature on learning both hard and soft constraints, so we will skip this topic.

(b) The technique used to represent and search over the set of candidate constraints or constraint theories. We will consider both search-based and syntax-guided synthesis approaches for learning hard constraints in Sect. 3, as well as optimization-based approaches in Sects. 4 and 5.

(c) Whether learning occurs from a pre-existing data set (passive learning) or by interactively asking questions to an oracle (interactive learning). Section 5 is dedicated to this last setting.

In the next Section, we proceed by introducing a simple, general formalism for expressing hard and soft constraint theories.

2   Constraint Theories

There are a number of successful formalisms for modeling and solving constraint programming problems [42]. In this overview we will restrict ourselves to a very general but minimal notation, to avoid as much overhead as possible.

Let us start by establishing some basic notation. In the following, variables will be written in upper-case X, constants in lower-case x, and value assignments $X = x$. For simplicity, all variables will take values in the same domain $\mathcal{X} \subset \mathbb{Z}$ or $\mathcal{X} = \mathbb{R}$, unless otherwise specified. Vectors of variables will be written in upper-case bold $\boldsymbol{X} = (X_1, \ldots, X_n)$ and total value assignments $X_1 = x_1 \wedge \ldots \wedge X_n = x_n$ simply as $\boldsymbol{X} = \boldsymbol{x}$. Total assignments are also called configurations. The indicator function $\mathbb{1}\{\varphi\}$ evaluates to 1 if condition φ holds and to 0 otherwise. Throughout the paper, we will use the terms "model", "constraint theory", and "constraint satisfaction/optimization problem" interchangeably.

In our simple framework, a constraint theory (or constraint network [9]) is defined by a set of variables $\boldsymbol{X}$ and by a set of constraints $c_1, \ldots, c_m$. A constraint with n arguments $c_j(X_1, \ldots, X_n)$ distinguishes between feasible and infeasible configurations. For instance, the constraint $c_j(X_1, X_2) = X_1 \vee X_2$, where X_1 and X_2 are Boolean, specifies that the configurations (true, true), (true, false), and (false, true) are feasible, while (false, false) is not. A constraint with n arguments can be more formally defined as an n-ary relation [9], but we will leave such technicalities aside. Further, constraints over continuous variables are also possible, e.g., consider the linear constraint $2X_1 - 3X_2 \leq 15$ or the mixed non-linear constraints $\neg\text{Slim} \implies (H < 110) \vee (\pi \cdot R^2 \geq 1000)$, where $X_1, X_2, H, R \in \mathbb{R}$ and Slim is Boolean. The set of constraints $c_1, \ldots, c_m$ is usually implicitly conjoined, but keep in mind that this does not hold for soft constraint theories; see Sects. 4 for some counter-examples.

In this overview, we will consider different kinds of constraint theories over both discrete and continuous variables, including:

- *Propositional logic theories* (or formulas) consist of Boolean variables, $\mathcal{X} = \{\texttt{true}, \texttt{false}\}$, combined with the usual logical connectives: conjunction $\wedge$, disjunction $\vee$, and negation $\neg$. Propositional formulas are most often encountered in conjunctive or disjunctive normal form. An example may be:

$$(\text{Saturday} \vee \text{Sunday}) \wedge \text{Sunny} \wedge \neg\text{Bored} \wedge \neg\text{Sick}$$

  which is a possible definition for the concept of "fun weekend". The main inference task is satisfiability (SAT) [21], where the goal is to find a total value assignment $\boldsymbol{X} = \boldsymbol{x}$ that renders the theory true. Extensions include MAX-SAT and weighted MAX-SAT, which will be discussed later.
- *Constraint networks* [9] leverage pure propositional logic to general discrete variables and arbitrary constraints, for instance inequality relations $\leq$, $=$, $\neq$, ... and global constraints like all-different. Inference is once again a form of satisfaction.
- *Satisfiability modulo theories* (SMT for short) extend propositional logic with one or more decidable theories $\mathcal{T}$ [5]. For instance, satisfiability modulo *linear real arithmetic* (SMT($\mathcal{LRA}$)) combines logic with linear arithmetic over the reals, and therefore introduces continuous variables and arbitrary linear constraints over them. An example SMT($\mathcal{LRA}$) formula for describing a happy outdoor weekend is:

$$(\text{Saturday} \vee \text{Sunday}) \wedge (\text{Rain} + \text{SoilHumidity} \leq 2)$$

  Other decidable theories include linear *integer* arithmetic, bit-vectors, and uninterpreted functions, but we will stick to $\mathcal{LRA}$ for simplicity. In this case inference is also a form of satisfaction.
- *Linear programs* (LP) include both a linear objective function of real variables and a set of implicitly conjoined linear constraints [52]. An LP in canonical form is written as:

$$\max_{\boldsymbol{x}} \boldsymbol{f}^{\top}\boldsymbol{x}$$
$$\text{s.t. } \boldsymbol{a}_j^{\top}\boldsymbol{x} \leq b_j \qquad\qquad \forall j = 1, \ldots, m$$

  Here $\boldsymbol{f}$ is a constant vector that defines the (gradient of) the objective function while $\boldsymbol{a}_1, \ldots, \boldsymbol{a}_m$ and $b_1, \ldots, b_m$ specify m linear constraints. Mixed-integer linear programs (MILP) have the same form, but allow both continuous and integer variables.

  Notice that, for both LPs and MILPs, inference is a form of optimization rather than satisfaction, that is, the model specifies not only a feasible space (like in standard satisfaction) but also a score over alternative feasible configurations.

Of course there are many other kinds of constraint theories, e.g., in database systems and spreadsheet software. In this case, variables can be strings or other objects. However, we will not consider these further, and refer the interested reader to [17,26] instead.

3   Learning Hard Constraints

Warmup: Learning k-CNF Theories

Let us start from the simplest constraint learning problem: learning a k-CNF formula. Such formulas are the conjunction of clauses (disjunctions) with at most k literals each, where a literal is either a variable or its negation. For instance, happy weekends are captured by the 2-CNF formula (Saturday $\vee$ Sunday) $\wedge$ ¬Rainy $\wedge$ ¬ImminentDeadline.

Now, let there be a hidden k-CNF theory φ^*. Given a set of example configurations labeled based on whether they are feasible with respective to φ^* (positive) or not (negative), we want to recover φ^* from the data only. Valiant's algorithm is a classic strategy to achieve this goal. The idea is simple. First, build the set of all candidate clauses of length at most k over the variables X. Then, taking each positive example in turn, remove from the set of candidates all of the clauses that are inconsistent with the example. This makes sure that, upon scanning over all positive examples, the set of candidates only contains clauses consistent with the data set. Upon termination, the learned formula is retrieved by taking the conjunction of all surviving clauses.

Despite its simplicity, Valiant's algorithm is PAC (probabilistically approximately correct) algorithm, meaning that the probability over all random data sets that the algorithm works is arbitrarily high so long as there are enough examples [54]. However, in order to work, Valiant's algorithm requires two key assumptions[2]: (a) the data set must be noiseless, i.e., there must be no measurement errors on the variables and no corruption on the labels, and (b) the hidden concept φ^* must be a k-CNF formula, which is not always known in advance. This is the so-called *realizable* setting. If these assumptions are not satisfied, then the algorithm gives unreliable results.

Encoding and Searching the Space of Candidates

Let us focus on discrete variables only for the time being. During learning, it is convenient to sort the space of candidate theories according to the *generality relation* $\succeq_g$. A constraint c is said to be more general than a constraint c', written $c \succeq_g c'$, if and only if the feasible set of c contains the feasible set of c', that is:

$$c \succeq_g c' \quad \Longleftrightarrow \quad \forall x . (x \models c') \implies (x \models c)$$

If c is more general than c', then c' is more specific than c, and vice versa. The generality relation induces a lattice over both constraints and constraint theories, which is perhaps the most common way to structure the space of candidates.

Once the set of candidates is given, the question becomes how to efficiently find a candidate theory compatible with the observed examples. In this sense,

---

[2] There are other technical assumptions over the distribution of the examples, which will be ignored for simplicity.

Valiant's algorithm can be viewed as a generate-and-test algorithm: it enumerates all candidates and then discards all of the ones incompatible with the data. In doing so, it keeps track of the *most specific* candidate theory. Beyond Valiant's algorithm, there exist notable examples of generate-and-test learners, led by the impressive ModelSeeker [7] approach to acquiring global constraints. But let us briefly consider alternative search strategies too.

There are three classic approaches to searching the space of candidates, all based on the above lattice structure:

- *General-to-specific* (aka top-down) approaches start from the most general concept (namely, true) and gradually specialize it according to the examples by introducing extra constraints that exclude the observed negatives from the feasible space of the learned theory.
- *Specific-to-general* (aka bottom-up) approaches, unsurprisingly, do the converse: they start from a most specific theory or set of theories and incrementally generalize them. It is often the case that the most specific theory is simply the disjunction of all positive examples—which, unless the data is inconsistent, excludes all negative examples. Generalization then boils down to removing constraints or removing conditions from constraints so to enlarge the feasible space of the candidate theory, while keeping all negatives outside of it.
- *Version space* approaches keep track of the whole set of candidates at once. The version space (VS) is indeed defined as the set of all theories that are consistent with respect to the dataset [33], and it is the sub-lattice (induced by the generality relation $\succeq_g$) between a set of most-general (top) and most-specific (bottom) candidates.
  In a discrete setting, VS learners leverage incremental bi-directional search, whereby an initial estimate of the VS is incrementally refined iterating over all examples, and for each example checking whether the most general theory wrongly covers it (if it is negative) or whether the most specific candidate wrongly excludes it (if it is positive). In either case the VS is updated.
  We note in passing that version spaces are not restricted to discrete variables, and that they are used in recent algorithms for both active learning [20] and preference elicitation [12].

In more general terms, learning of hard constraints can be viewed simply as a search problem, and therefore any search algorithm can in principle be used, including stochastic local search, genetic algorithms, *etc.*

It is worth remarking that, in most cases of interest, the set of candidates is exponential in the number of variables. For instance, for k-CNF, given v variables one can build $2v$ literals, and thus $(2v)^k$ potential clauses out of them. Learning approaches use smart strategies to avoid enumerating such a humongous set. Perhaps the simplest solution is to leverage the generality relation, by automatically excluding all constraints that are more general than an already excluded one. Alternative approaches include restricting the space of hypotheses by introducing background knowledge, e.g., by restricting the value of k or the initial set of candidates. Providing constraint templates to be filled in, as in sketching [48], is also an option. A sensible alternative is to compactly represent the space of candidates using a satisfaction or optimization problem. This strategy is at the core of syntax-guided synthesis (SyGuS) [1,2], a general framework for designing programs from specifications and examples. Two notable examples of SyGuS are the celebrated constraint learning Conacq [8], which encodes the version space using a propositional formula, and Incal [27], an approach for learning SMT($\mathcal{LRA}$) theories from examples that extends SyGuS to continuous variables.

4   Learning Soft Constraints

Soft constraints are a powerful tool for dealing with conflicting requirements, uncertain inputs, and imperfect specifications. Intuitively, soft constraints introduce two new rules: first, it is not mandatory to satisfy soft constraints, and second, some soft constraints are more important than others [10,42]. Therefore, if two soft constraints are incompatible, the most preferable one should be satisfied.

Soft Constraints with Linear Preferences

In their most general form, preferences over soft constraints can be encoded as a binary relation. Although very flexible, in the worst case encoding a relation over m soft constraints requires m^2 parameters. This is very cumbersome both when manually designing the constraint theory and when learning it from examples.

For this reason, we will focus on a more agile alternative, where the absolute importance of a constraint is determined by an objective function associated to it. The overall quality of a configuration is then given by the sum of the importances of the soft constraints that it satisfies. More formally, a theory in this form includes:

- m soft constraints $s_1, \ldots, s_m$, each associated to an objective function w_j : $\mathcal{X} \to \mathbb{R}$, $j = 1, \ldots, m$, and
- k hard constraints $h_1, \ldots, h_k$ that must be satisfied.

Finding the most preferable (aka optimal or highest scoring) configuration x^* is accomplished by maximizing the total weight $f_w(x)$ of the soft constraints it satisfies[3], as follows:

$$x^* = \operatorname*{argmax}_{x} f_w(x) := \sum_{j=1}^{m} w_j(x) \mathbb{1}\{x \models s_j\} \qquad (1)$$

$$\text{s.t. } x \models h_j \qquad\qquad \forall j = 1, \ldots, k \qquad (2)$$

---

[3] Notice that the optimal configuration may not be unique, and that all optima have the same score.

The computational complexity of this *inference* problem depends on the type of constraints and objective functions appearing above, but in most cases of interest (like MAX-SAT below) it is NP-hard or beyond.

Although more general alternatives exist, such as semiring-based constraints [10], we will stick to our simple framework because: (1) it captures many prominent settings, from weighted maximum satisfiability (weighted MAX-SAT) up to optimization modulo theories [44], and (2) soft constraint theories in the above format can be easily learned with high-quality machine learning algorithms, as discussed next.

Learning Weighted MAX-SAT from Annotated Configurations

In weighted MAX-SAT, the soft constraints $s_1, \ldots, s_m$ are arbitrary logic formulas and the per-constraint objective functions are constants $w_j(x) \equiv w_j \in \mathbb{R}$. In the simplest case, no hard constraints are present. Inference boils down to finding a configuration x that maximizes the total weight of the satisfied formulas:

$$\max_{x} f_x(x) := \sum_{j=1}^{m} w_j \mathbb{1}\{x \models s_j\} \tag{3}$$

This problem is notoriously NP-complete, but in can be solved efficiently in many practical cases [21].

Let us now consider perhaps the simplest possible learning scenario. We assume that there is a latent, unknown weighted MAX-SAT problem with parameters $w^* \in \mathbb{R}^m$. We cannot observe this latent model, but we do know the dictionary of soft constraints $s_1, \ldots, s_m$. Further, we are given a data set of example configurations $x_1, \ldots, x_n$ annotated with their own scores according to the unknown theory, i.e., $y_i = \sum_j w_j^* \mathbb{1}\{x_i \models s_j\}$ for all $i = 1, \ldots, n$. The configurations x_i are assumed to be sampled at random according to some underlying distribution, and no implicit guarantee is given as for their quality.

The goal of learning is to induce a model w that behaves similarly to the latent one w^*. For the time being, we will consider a model good so long as the estimated parameter vector w scores the examples similarly to the hidden model. Since we have access to the value of the true objective function y_i for all example configurations, finding an appropriate parameter vector can be cast as a regression problem, namely:

$$\hat{w} = \operatorname{argmin}_{w \in \mathbb{R}^m} \sum_{k=1}^{n} (y_i - f_w(x_i))^2 \tag{4}$$

Now, notice that the function $f_w(x)$ is linear with respect to the basis defined by the indicator functions $\{\mathbb{1}\{x \models s_j\} : j = 1, \ldots, m\}$. This means that Eq. 4 can be cast as linear regression and solved using standard regression techniques.

A very nice property of this setup is that linear regression works well even if the supervision y_i is moderately corrupted, and it provides a bridge to robust regression techniques for different kinds of noise.

One should keep in mind, however, that supervision on the per-instance scores y_i may not be readily available. This happens for instance when eliciting preferences from decision makers, who may be unable to state a numerical score. For this reason, we consider two alternative forms of supervision, namely pairwise rankings and input-output pairs, and show how to learn soft constraint theories from them.

Learning from Rankings

Let us start from pairwise rankings. In this case, we are given n pairs of configurations $\{(\boldsymbol{x}_i, \boldsymbol{x}'_i) : i = 1, \ldots, n\}$, where each pair is implicitly ranked according to the preference relation $\boldsymbol{x}_i \succeq \boldsymbol{x}'_i \Leftrightarrow f_{w^*}(\boldsymbol{x}_i) \geq f_{w^*}(\boldsymbol{x}'_i)$. The goal of learning is then to find a parameter vector $\boldsymbol{w}$ that ranks all of the example pairs correctly, or more formally:

$$\text{find } \boldsymbol{w} \tag{5}$$
$$\text{s.t. } f_{\boldsymbol{w}}(\boldsymbol{x}_i) - f_{\boldsymbol{w}}(\boldsymbol{x}'_i) \geq 0 \qquad \forall i = 1, \ldots, n \tag{6}$$

The above constraint can be shown to be linear (in the indicators) by rewriting it as $\sum_{j=1}^{m} w_j (\mathbb{1}\{\boldsymbol{x}_i \models s_j\} - \mathbb{1}\{\boldsymbol{x}'_i \models s_j\}) \geq 0$. Notice that, even though the model is acquired from ranking data, we use it to compute high-scoring configurations, as per Eq. 3.

One issue with the above formulation is that simply conforming to the supervision does not guarantee that the learned vector $\boldsymbol{w}$ generalizes well to unseen pairs. This means that the learned function $f_{\boldsymbol{w}}$ may fail to rank the true optima above all other configurations. In order to address this, following principles from statistical learning theory [43,55], it is customary to look for a vector $\boldsymbol{w}$ that correctly ranks all pairs by the largest possible margin, that is:

$$\max_{\boldsymbol{w}, \mu \geq 0} \mu \tag{7}$$
$$\text{s.t. } \mu \leq f_{\boldsymbol{w}}(\boldsymbol{x}_i) - f_{\boldsymbol{w}}(\boldsymbol{x}'_i) \qquad \forall i = 1, \ldots, n \tag{8}$$

where $\mu \in \mathbb{R}$ measures the margin. It turns out that maximizing μ is geometrically equivalent to minimizing the (squared) Euclidean norm of $\boldsymbol{w}$ ([43], Chap. 1), and so the above can be rewritten as the following quadratic convex optimization problem:

$$\min_{\boldsymbol{w}} \frac{1}{2} \|\boldsymbol{w}\|_2^2 \tag{9}$$
$$\text{s.t. } f_{\boldsymbol{w}}(\boldsymbol{x}_i) - f_{\boldsymbol{w}}(\boldsymbol{x}'_i) \geq 0 \qquad \forall i = 1, \ldots, n \tag{10}$$

Finally, if the observations $\boldsymbol{x}_i$ are noisy or their rankings are inconsistent, as it the case when the examples define cycles like $\boldsymbol{x}_1 \succeq \boldsymbol{x}_2 \succeq \ldots \succeq \boldsymbol{x}_1$, then it may be impossible to find a non-zero vector w that simultaneously satisfies Eq. 10 for all examples. A common solution is to introduce slack variables $\xi_i \in \mathbb{R}$ that measure the "degree of violation" for every example $k = 1, \ldots, n$:

$$\min_{w, \xi} \frac{1}{2} \|w\|_2^2 + \frac{\lambda}{2} \sum_{k=1}^{s} \xi_i \tag{11}$$

$$\text{s.t. } f_w(x_i) - f_w(x_i') \geq \xi_i \qquad \forall i = 1, \ldots, n \tag{12}$$

This is the well-known formulation of ranking support vector machine (SVM), a now classical machine learning algorithm for learning to rank, originally conceived for ranking results in search engines [23]. The constant $\lambda \geq 0$ controls the trade-off between generalization (first term) and error on the training set (second term), and it is assumed to be given.

Although earlier works focused on solving the above optimization problem (OP) in the dual (cf. [40]), current state-of-the-art approaches rely on gradient-based optimization in the primal [46]. Practitioners need not worry about these details, since efficient (ranking) SVM solvers are included by default in most machine learning libraries, like scikit-learn [37] and Weka [19].

Learning from Input-Output Pairs

Input-output pairs are another popular form of supervision. Let U and V partition the set of variables (i.e., $X = U \cup V$, $U \cap V = \varnothing$). The intuition is that the variables U act as inputs and thus are always known and fixed, while V are outputs and can be optimized over.

The data set, in this case, consists of n input-output pairs $\{(u_i, v_i) : i = 1, \ldots, n\}$, where for any partial assignment u_i, the output v_i is chosen optimally w.r.t. the latent parameters w^*, that is:

$$v_i = \operatorname*{argmax}_{v} f_{w^*}(u_i \circ v_i) \tag{13}$$

Here $\circ$ indicates vector concatenation. This kind of supervision is common in machine learning tasks that require to learn a map from structured inputs to structured outputs, like text parsing (where a sentence u is mapped to a parse tree v) or image segmentation (an image u is mapped to a set of labeled segments v), but it is also used in constraint learning, where the distinction between input and output variables is less well-defined [51].

In order to learn from input-output pairs, we adapt the training procedure of structured-output support vector machines (SSVM) [53], see [24] for a gentler introduction. Given an example (u_i, v_i) and an arbitrary vector w, let v' be the output of Eq. 4 when using u_i as input and w as parameters. Also, let $\Delta(v_i, v')$ be a distortion function[4] that measures the difference between the correct output v_i and the predicted one v'.

---

[4] For technical reasons, the distortion is often assumed to lie in the range $[0, 1]$, see [32].

The intuition behind structured-output SVMs is that the vector w should be chosen so that, for any example, the predicted output has low distortion. This can be formalized as follows[5]:

$$\max_{w,\xi} \frac{1}{2}\|w\|_2^2 + \frac{\lambda}{2}\sum_{k=1}^s \xi_i \qquad (14)$$

$$\text{s.t. } f_w(u_i \circ v_i) - f_w(u_i \circ v'') \geq \Delta(v_i, v'') - \xi_i \quad \forall v'' \neq v_i \,\forall i = 1, \ldots, n \qquad (15)$$

The objective function is the same as for ranking data. The constraint, on the other hand, is much more complex: it requires the correct output v_i to be scored higher than any alternative output $v'' \neq v_i$ by a margin proportional to the distortion. This takes care of enforcing an appropriate margin between v_i and the predicted output v' too. The per-example slacks ξ_i allow for scoring mistakes in noisy data sets.

The similarity to learning to rank is striking, but solving this optimization problem is trickier, because the number of alternative outputs v' can be very (exponentially) large. This means that Eq. 15 has to be enforced over an enormous amount of configurations. In order to solve this OP, the most straightforward approach is to employ cutting planes, which we will not discuss. We refer the interested reader to [24] instead.

Learning more General Constraint Theories

A striking property of the OPs described in the previous sections is that they are relatively agnostic to the particular choice of constraints and per-constraint objective functions. Indeed, regression-based learning has been used to learn arbitrary weighted CSPs [41], and SSVM-based learning for learning Optimization Modulo Theories [51]. In this last case, the restriction that the per-constraint objective functions $w_j(x)$ are constant is lifted—although the learning algorithm remains essentially unchanged.

5   Interactive Learning

Applications where supervision is scarce and expensive are not well suited for offline learning, because there are often too few examples to learn a reasonably accurate model. In this case, a sensible thing to do is to acquire supervision directly from an oracle by asking informative questions. This allows the learning algorithm to optimize the performance/example ratio, and thus to acquire good models at a small cost.

Notice that the oracle may be a human subject, like when eliciting constraints (knowledge) from a domain expert [41] or preferences from an end-user [39], or

---

[5] There exist several variants of structured-output SVM, here we opt for the simpler one; see the references for more details.

a full-fledged scientific apparatus, as in automated scientific experiments [25]. In the first case, the goal is to extract a (soft or hard) constraint theory from a domain expert, who is otherwise unable to formalize and model her knowledge upfront. The second case captures applications like interactive recommender systems, where very little expertise (or patience!) can be expected of the human counterpart, and any request for supervision has to be designed so to be easy to understand and answer.

The main questions that arise when designing an interactive learning algorithm are of course what *kind* of questions should be asked to the oracle, and how to pick good questions. The answer to both questions is very application- and oracle-specific, as we will see in the following.

Interactive Learning of Hard Constraints

The most straightforward approach to learning hard constraints is to trivially select each candidate constraint h_j in turn, $j = 1, \ldots, m$, and ask the oracle whether it appears in the latent constraint theory. Unfortunately, this naive approach is not very useful, as it requires to consider all constraints, which can be exponentially many in the number of variables. This procedure is therefore unfeasible even for theories of modest size, especially if the oracle is a human being.

A more appropriate procedure is to use *membership constraints*. In this setting, the learner chooses an instance x and asks the user whether it satisfies the hidden theory or not. This setup stands at the core of query-based learning [4], a venerable and sound approach to learning. Alternative query types will be considered later on.

Now, what is the best way to choose the query configuration x? Before proceeding, let us assume a realizable scenario, i.e., that (a) the set of hypotheses includes the latent concept, and (b) that the oracle always answers correctly. This is the case, for instance, if the learning problem is well engineered and the oracle is a human domain expert, e.g., an employee who has a vested interest in answering the questions asked by the algorithm. Under these (rather strong) assumptions, one can resort to halving approaches, which roughly work as follows.

Learning is interactive. At all iterations, the learner keeps track of the set of candidate theories that are consistent with respect to all answers observed so far—i.e., the version space. The intuition is that, in each iteration, the algorithm chooses a configuration x so that, regardless of the answer to the membership query, the version space is reduced as much as possible. In the best possible scenario, the version space halves at each iteration, and therefore ideally the number of queries necessary to find the correct concept is approximately $\log_2 |\mathcal{H}|$, where $\mathcal{H}$ is the set of candidate hypotheses.

Unfortunately, this theoretically appealing approach has several flaws. First, keeping track of the version space can be quite complicated. The best approaches to date make use of rather convoluted schemas [8] or only store the most general and most specific candidate theories in the version space [9]. Second, it turns out that in many interesting cases, choosing the (approximately) optimal instance x is computationally intractable [31]. A simple approximation to this schema is to choose an instance x that reduces the version space by *some* amount. This is accomplished in [9] by choosing an instance whose feasibility the most general and the most specific theories in the current version space disagree on. Therefore, if it turns out that x is feasible, the most specific hypothesis is wrong and it must be generalized. On the other hand, if x is actually unfeasbile, the converse is true and the most general hypothesis must be specialized. Thus, this strategy guarantees that the version space reduces at each iteration, and that it eventually contains only concepts that match all examples.

Interactive Learning of Soft Constraints

Consider the following toy application: A user wishes to buy a custom PC. The PC is assembled from individual components: CPU, HDD, RAM, etc. Valid PC configurations must satisfy constraints, e.g. CPUs only work with compatible motherboards [50]. In this setting, one is tasked with constructing a PC config-uration x that is both palatable to the customer and compatible with any hard constraints, e.g., that the CPU and the motherboard should be compatible.

As above, we cast this kind of problems as learning a weighted CSP that captures the preferences of the customer, and that can be used to generate high-scoring

In the following, we will consider three queries of three common kinds:

– *Scoring queries.* In this case, the oracle is asked to provide the true numerical score of configuration x chosen by the learner.
Queries of this type make sense when interacting with very precise oracles, such as measurement devices in automated scientific experiments [25], but not as much when interacting with human oracles. Indeed, it is very difficult—even for domain experts—to provide precise or approximate numerical scores. This is why most scoring systems use discrete ratings, such as star ratings. Regardless, depending on the application, other query types may be easier to answer.
– *Ranking queries.* In this case, the query consists of two unlabeled configura-tions x and x', and the oracle is tasked with indicating the most preferable of the two, i.e., whether $f_{w^*}(x) \geq f_{w^*}(x')$ holds.
These queries have found ample application in interactive preference elicita-tion and recommendation tools [39].
– *Improvement queries.* In this case, the algorithm chooses a single configuration x and asks the oracle to provide an improved version $\bar{x}$. Notice that the improvement is allowed to be small or partial [47].
It is easy to see that the pair $(\bar{x}, x)$ implicitly defines a pairwise ranking of the form $\bar{x} \succeq x$, and so the supervision is the same as for ranking queries. It is important to notice that, however, the interaction itself is different. Indeed, improvement queries only require the human oracle to observe and analyze a single configuration (rather than two), and synergize very well with direct manipulation interfaces like those used in computer-assisted design.

Notice that scoring queries lead to collecting a large data set of scoring information, and thus learning can be cast in terms of linear regression. In the other two cases, the collected data set contains pairwise ranking information, and therefore learning boils down to learning to rank. Thus the techniques discussed in the previous section can be immediately applied to the interactive case: it is just a matter of solving a regression, ranking, or structured-output learning every time a new answer is achieved. Although not entirely principled, this approach tends to work well in practice.

The question is, again, how to pick the right query. Ideally, the most informative question should be chosen. Unfortunately, evaluating the true informativeness of a query based on the information available during learning is tricky. A very simple solution is to choose a configuration x (or a pair of configurations) that is maximally *uncertain* according to the model. More specifically, an instance is uncertain if the entropy of the response variable (e.g. of the score, for scoring queries) is large. Measures based on the margin are also common [45]. The major down-side of uncertainty sampling is that the uncertainty provided by the model may be misleading, e.g., for instance if the learned theory is "overconfident". It is therefore customary to combine uncertainty sampling with other strategies that lessen its reliance on the model's estimates [45].

6 Further Reading

Learning and Optimization. The interplay between machine learning and satisfaction/optimization is not limited to constraint learning. In most of machine learning (excluding a large chunk of its Bayesian side), learning is framed as the task of optimizing some regularized loss function with respect to the data set [55]—hence the centrality of optimization in ML. Nowadays, the most popular solution approach are gradient descent techniques [29], but one should not forget that there are a plethora of valid alternatives, cf. [11,49]. Some learning frameworks go one step further, and make use of declarative constraint programming to model and solve learning tasks like program synthesis and pattern mining [2,18]. Another link between ML and optimization occurs in probabilistic graphical models, structured-output prediction, and constraint learning, where computing a prediction for an unseen instance is itself an optimization problem. There is also abundant literature on speed-up learning, i.e., on leveraging machine learning techniques for improving the run-time of satisfaction and optimization solvers; see for instance [22] for a method to accelerate branch & bound in mixed-integer linear programming solvers. It is easy to spot a recursion here: machine learning could in principle be used to accelerate an optimization step which is itself part of a learning problem. While true, diminishing returns make it difficult to exploit this loop.

Learning Hard Constraints. The task of acquiring hard constraints is surprisingly close to concept learning [54] and binary classification: in all of these, the learner has to acquire a hidden concept from examples. The main difference is whether the model in question is a constraint theory, and whether it is used in a purely predictive manner or also for inspection, interpretation, debugging, *etc.* A major advantage of constraints is that they can be verified by domain experts either manually or with appropriate tools—and modified, if necessary. Verification of models learned from data is often a required to guarantee the proper functioning of the model [3], and indeed verification techniques are being actively researched for non constraint-based models like neural networks [13].

From a search perspective, learning hard constraints is also closely related to feature selection, pattern mining, and structure learning of probabilistic graphical models [28]—e.g. Bayesian networks and Markov networks. All these tasks revolve around efficiently encoding and searching a very large set of candidates, and often employ similar techniques, e.g., version spaces, variants of grafting [38], and syntax-guided synthesis [6], albeit often under different names. Inductive logic programming, where the task is to learn first-order theories [34], can be handled analogously, but it involves even larger search spaces.

Learning soft constraints. As for soft constraints, we focused on learning techniques rooted in statistical learning theory and empirical risk minimization [55] and max-margin methods like support vector machines [43]. Learning from input-output examples stands at the core of structured-output prediction [24] and learning to search [14]. One link that is seldom made is to inverse (combinatorial) optimization, where the goal is to adjust a pre-existing (combinatorial) optimization model such that it adheres to a set of known optimal solutions. Unsurprisingly, this field has slowly been drifting closer to the structured-output setting, and these two problems have recently been tackled using similar strategies, see for instance [47] and [16].

Acknowledgments. The author is grateful to Luc De Raedt and Andrea Passerini for many insightful discussions. These lecture notes are partially based on material co-developed by LDR, AP and the author. This work has received funding from the European Research Council (ERC) under the European Union's Horizon 2020 research and innovation programme (grant agreement No. [694980] SYNTH: Synthesising Inductive Data Models).

References

1. Alur, R., et al.: Syntax-guided synthesis. In: 2013 Formal Methods in Computer-Aided Design, pp. 1–8. IEEE (2013)
2. Alur, R., Singh, R., Fisman, D., Solar-Lezama, A.: Search-based program synthesis. Commun. ACM **61**(12), 84–93 (2018)
3. Andrews, R., Diederich, J., Tickle, A.B.: Survey and critique of techniques for extracting rules from trained artificial neural networks. Knowl.-Based Syst. **8**(6), 373–389 (1995)
4. Angluin, D.: Queries and concept learning. Mach. Learn. **2**(4), 319–342 (1988)
5. Barrett, C.W., Sebastiani, R., Seshia, S.A., Tinelli, C.: Satisfiability modulo theories. Handb. Satisf. **185**, 825–885 (2009)

6. Bartlett, M., Cussens, J.: Integer linear programming for the Bayesian network structure learning problem. Artif. Intell. **244**, 258–271 (2017)
7. Beldiceanu, N., Simonis, H.: A model seeker: extracting global constraint models from positive examples. In: Milano, M. (ed.) CP 2012. LNCS, pp. 141–157. Springer, Heidelberg (2012). https://doi.org/10.1007/978-3-642-33558-7_13
8. Bessiere, C., Coletta, R., Koriche, F., O'Sullivan, B.: A SAT-based version space algorithm for acquiring constraint satisfaction problems. In: Gama, J., Camacho, R., Brazdil, P.B., Jorge, A.M., Torgo, L. (eds.) ECML 2005. LNCS (LNAI), vol. 3720, pp. 23–34. Springer, Heidelberg (2005). https://doi.org/10.1007/11564096_8
9. Bessiere, C., et al.: New approaches to constraint acquisition. In: Bessiere, C., De Raedt, L., Kotthoff, L., Nijssen, S., O'Sullivan, B., Pedreschi, D. (eds.) Data Mining and Constraint Programming. LNCS (LNAI), vol. 10101, pp. 51–76. Springer, Cham (2016). https://doi.org/10.1007/978-3-319-50137-6_3
10. Bistarelli, S., Montanari, U., Rossi, F.: Semiring-based constraint logic programming: syntax and semantics. ACM Trans. Program. Lang. Syst. (TOPLAS) **23**(1), 1–29 (2001)
11. Bottou, L., Curtis, F.E., Nocedal, J.: Optimization methods for large-scale machine learning. Siam Rev. **60**(2), 223–311 (2018)
12. Boutilier, C., Regan, K., Viappiani, P.: Simultaneous elicitation of preference features and utility. In: Twenty-Fourth AAAI Conference on Artificial Intelligence (2010)
13. Bunel, R.R., Turkaslan, I., Torr, P., Kohli, P., Mudigonda, P.K.: A unified view of piecewise linear neural network verification. In: Advances in Neural Information Processing Systems, pp. 4790–4799 (2018)
14. Daumé III, H., Marcu, D.: Learning as search optimization: approximate large margin methods for structured prediction. In: Proceedings of the 22nd International Conference on Machine Learning, pp. 169–176. ACM (2005)
15. De Raedt, L., Passerini, A., Teso, S.: Learning constraints from examples. In: Thirty-Second AAAI Conference on Artificial Intelligence (2018)
16. Dong, C., Chen, Y., Zeng, B.: Generalized inverse optimization through online learning. In: Advances in Neural Information Processing Systems, pp. 86–95 (2018)
17. Gulwani, S., Hernandez-Orallo, J., Kitzelmann, E., Muggleton, S.H., Schmid, U., Zorn, B.: Inductive programming meets the real world. Commun. ACM **58**(11), 90–99 (2015)
18. Guns, T., Dries, A., Tack, G., Nijssen, S., De Raedt, L.: Miningzinc: a modeling language for constraint-based mining. In: Twenty-Third International Joint Conference on Artificial Intelligence (2013)
19. Hall, M., Frank, E., Holmes, G., Pfahringer, B., Reutemann, P., Witten, I.H.: The weka data mining software: an update. ACM SIGKDD Explor. Newsl. **11**(1), 10–18 (2009)
20. Hanneke, S., et al.: Theory of disagreement-based active learning. Found. Trends® Mach. Learn. **7**(2–3), 131–309 (2014)
21. Hansen, P., Jaumard, B.: Algorithms for the maximum satisfiability problem. Computing **44**(4), 279–303 (1990)
22. He, H., Daume III, H., Eisner, J.M.: Learning to search in branch and bound algorithms. In: Advances in Neural Information Processing Systems, pp. 3293–3301 (2014)
23. Joachims, T.: Optimizing search engines using clickthrough data. In: Proceedings of the Eighth ACM SIGKDD International Conference on Knowledge Discovery and Data Mining, pp. 133–142. ACM (2002)

24. Joachims, T., Hofmann, T., Yue, Y., Yu, C.N.: Predicting structured objects with support vector machines. Commun. ACM **52**(11), 97 (2009)
25. King, R.D., et al.: The automation of science. Science **324**(5923), 85–89 (2009)
26. Kolb, S., Paramonov, S., Guns, T., De Raedt, L.: Learning constraints in spreadsheets and tabular data. Mach. Learn. **106**, 1–28 (2017)
27. Kolb, S., Teso, S., Passerini, A., De Raedt, L.: Learning SMT (LRA) constraints using SMT solvers. In: IJCAI, pp. 2333–2340 (2018)
28. Koller, D., Friedman, N.: Probabilistic Graphical Models: Principles and Techniques. MIT Press, Cambridge (2009)
29. LeCun, Y., Bengio, Y., Hinton, G.: Deep learning. Nature **521**(7553), 436 (2015)
30. Lombardi, M., Milano, M.: Boosting combinatorial problem modeling with machine learning. In: Proceedings of the 27th International Joint Conference on Artificial Intelligence, pp. 5472–5478. AAAI Press (2018)
31. Louche, U., Ralaivola, L.: From cutting planes algorithms to compression schemes and active learning. In: 2015 International Joint Conference on Neural Networks (IJCNN), pp. 1–8. IEEE (2015)
32. McAllester, D.: Generalization bounds and consistency. In: Predicting Structured Data, pp. 247–261 (2007)
33. Mitchell, T.M.: Generalization as search. Artif. Intell. **18**(2), 203–226 (1982)
34. Muggleton, S., De Raedt, L.: Inductive logic programming: theory and methods. J. Logic Program. **19/20**, 629–679 (1994)
35. O'Sullivan, B.: Automated modelling and solving in constraint programming. In: Twenty-Fourth AAAI Conference on Artificial Intelligence (2010)
36. Pawlak, T.P., Krawiec, K.: Automatic synthesis of constraints from examples using mixed integer linear programming. Eur. J. Oper. Res. **261**(3), 1141–1157 (2017)
37. Pedregosa, F., et al.: Scikit-learn: machine learning in Python. J. Mach. Learn. Res. **12**, 2825–2830 (2011)
38. Perkins, S., Lacker, K., Theiler, J.: Grafting: fast, incremental feature selection by gradient descent in function space. J. Mach. Learn. Res. **3**(03), 1333–1356 (2003)
39. Pigozzi, G., Tsoukias, A., Viappiani, P.: Preferences in artificial intelligence. Ann. Math. Artif. Intell. **77**(3–4), 361–401 (2016)
40. Platt, J.: Sequential minimal optimization: a fast algorithm for training support vector machines (1998)
41. Rossi, F., Sperduti, A.: Acquiring both constraint and solution preferences in interactive constraint systems. Constraints **9**(4), 311–332 (2004)
42. Rossi, F., Van Beek, P., Walsh, T.: Handbook of Constraint Programming. Elsevier, Amsterdam (2006)
43. Scholkopf, B., Smola, A.J.: Learning with Kernels: Support Vector Machines, Regularization, Optimization, and Beyond. MIT Press, Cambridge (2001)
44. Sebastiani, R., Tomasi, S.: Optimization modulo theories with linear rational costs. ACM Trans. Comput. Log. (TOCL) **16**(2), 12 (2015)
45. Settles, B.: Active learning. Synth. Lect. Artif. Intell. Mach. Learn. **6**(1), 1–114 (2012)
46. Shalev-Shwartz, S., Singer, Y., Srebro, N., Cotter, A.: Pegasos: primal estimated sub-gradient solver for svm. Math. Program. **127**(1), 3–30 (2011)
47. Shivaswamy, P., Joachims, T.: Coactive learning. J. Artif. Intell. Res. (JAIR) **53**, 1–40 (2015)
48. Solar-Lezama, A., Tancau, L., Bodik, R., Seshia, S., Saraswat, V.: Combinatorial sketching for finite programs. ACM Sigplan Not. **41**(11), 404–415 (2006)
49. Sra, S., Nowozin, S., Wright, S.J.: Optimization for Machine Learning. MIT Press, Cambridge (2012)

50. Teso, S., Dragone, P., Passerini, A.: Coactive critiquing: elicitation of preferences and features. In: AAAI (2017)
51. Teso, S., Sebastiani, R., Passerini, A.: Structured learning modulo theories. Artif. Intell. **244**, 166–187 (2017)
52. Todd, M.J.: The many facets of linear programming. Math. Program. **91**(3), 417–436 (2002)
53. Tsochantaridis, I., Hofmann, T., Joachims, T., Altun, Y.: Support vector machine learning for interdependent and structured output spaces. In: Proceedings of the Twenty-first International Conference on Machine Learning, p. 104. ACM (2004)
54. Valiant, L.: A theory of the learnable. Commun. ACM **27**, 1134–1142 (1984)
55. Vapnik, V.: An overview of statistical learning theory. IEEE Trans. Neural Netw. **10**(5), 988–999 (1999)

A Modest Markov Automata Tutorial

Arnd Hartmanns[1](✉) and Holger Hermanns[2,3]

[1] University of Twente, Enschede, The Netherlands
a.hartmanns@utwente.nl
[2] Saarland University, Saarland Informatics Campus, Saarbrücken, Germany
hermanns@cs.uni-saarland.de
[3] Institute of Intelligent Software, Guangzhou, China

Abstract. Distributed computing systems provide many important services. To explain and understand why and how well they work, it is common practice to build, maintain, and analyse models of the systems' behaviours. Markov models are frequently used to study operational phenomena of such systems. They are often represented with discrete state spaces, and come in various flavours, overarched by Markov automata. As such, Markov automata provide the ingredients that enable the study of a wide range of quantitative properties related to risk, cost, performance, and strategy. This tutorial paper gives an introduction to the formalism of Markov automata, to practical modelling of Markov automata in the MODEST language, and to their analysis with the MODEST TOOLSET. As case studies, we optimise an attack on Bitcoin, and evaluate the performance of a small but complex resource-sharing computing system.

1 Introduction

Distributed computing systems provide many important services, such as electronic banking, information and knowledge sharing, and social networking. They are enablers for innovation; for instance, blockchain technology is based on massively distributed computing. Since our societies increasingly depend on the services offered in this manner, it is important to ensure their performance, dependability, and correctness. The purpose of *performance evaluation* is to investigate and optimise the amount of useful work being accomplished. *Dependability evaluation* is concerned with assessing service continuity by means of measures such as reliability and availability. The evaluation of correctness—usually called *formal verification*—focusses on proving that the service delivered satisfies a formal specification of its behaviour. Usually, all of these techniques are based on a *model* of the system, which is an abstract representation of the system's behaviour.

Markov Chains. In numerical performance and dependability evaluation, by far the most prominent models used to represent the temporal dynamics of a system

Authors are listed alphabetically. This work has received financial support by DFG grant 389792660 as part of TRR 248 (see perspicuous-computing.science), by ERC Advanced Grant 69561 (POWVER), and by NWO VENI grant 639.021.754.

© Springer Nature Switzerland AG 2019
M. Krötzsch and D. Stepanova (Eds.): Reasoning Web 2019, LNCS 11810, pp. 250–276, 2019.
https://doi.org/10.1007/978-3-030-31423-1_8

are *Markov chains* [38]. In this model family, the system is supposed to occupy a state at any moment in time, with the set S of states (the state space) being finite or countably infinite. Markov chains come in two flavours, dependent on whether the time domain $\mathbb{T}$ is considered to be discrete ($\mathbb{T} = \mathbb{N} = \{0, 1, \dots\}$) or continuous ($\mathbb{T} = \mathbb{R}_+ = [0, \infty)$). The dynamics of a *discrete-time Markov chain* (DTMC) is determined by a mapping from states to probability distributions over (successor) states. For instance, if state s is mapped to probability distribution μ, then the system once occupying state s is understood to jump to state s' with probability $\mu(s')$ in one time step. Notably, the probability is assumed to be independent of any further information (such as any past behaviour) apart from the state identity of s. This is known as the *Markov* (or *memoryless*) *property*. A *continuous-time Markov chain* (CTMC) adheres to this property, too, but it now needs to be interpreted in *stochastic time*, i.e. on a continuous time line where probability mass flows continuously between states. For CTMC, the Markov property implies that neither the past history nor the time already spent in the current state s influences the flow of probability into some state s'. Instead, it is governed by a time-independent *rate* λ, a positive real value (or zero if no flow exists). Thus the overall behaviour of a CTMC is determined by a mapping from state pairs to rates in $\mathbb{R}_+$. CTMC are arguably better fit to the nature of distributed computing [5], where it is difficult to assume a common discrete time base. The time spent in state s before jumping to another state s' is usually called the *residence time* (or *sojourn time*) in s. Residence times are geometrically distributed in DTMC, and exponentially distributed in CTMC.

Labelled Transition Systems. In formal verification, other models appear: State-transition diagrams, automata, and similar formalisms describe the dynamic behaviour of systems here. They often appear in the specific form of *labelled transition systems* (LTS). A transition system consists of a set of states S and a set of possible state changes. The latter is given as a binary relation on states, i.e. a subset of the cross product $S \times S$. Intuitively, a pair of states $\langle s, s' \rangle$ is in this relation if it is *possible* to jump from s to s' in a single step. In LTS, state changes are associated with occurrences of *actions*. A state change from s to s' then implies the occurrence of a specific action a, which labels the transition— thus we have an <u>L</u>TS. If multiple transitions are possible in a state, then the decision of which one to take is usually interpreted as being *nondeterministic*. Nondeterminism is especially useful to represent concurrency, a crucial aspect of distributed computing systems. If two systems run concurrently and independently, this is best represented as the nondeterministic interleaving of their individual steps. LTS can thus be endowed with *parallel composition* operators to model concurrency and interaction of component LTS [6,40,44]. With further operators, this is convenient for a *compositional modelling* style, where the behaviour of components is the result of compositions of smaller building blocks.

Model Checking. Within the spectrum of techniques used in formal verification, model checking is an automated model-based technique to assess whether the possible system behaviours satisfy a property describing the desirable behaviour [3]. Typically, properties are expressed in temporal logics such as LTL or CTL. Model checking usually involves constructing an in-memory representation of the (part of the) state space (relevant to assess the property). It thus gives definitive answers, but faces the state space explosion problem. In the past decades, model checking has been extended to treat aspects such as discrete probabilities and stochastic time. It has become apparent that a joint consideration of performance, dependability and correctness is both possible and worthwhile [2].

This Paper. The purpose of this tutorial paper is to provide a gentle introduction to working with a mathematical formalism integrating the modelling aspects discussed above. We focus especially on the specification and modelling of real systems. The formalism we introduce is called *Markov automata* (MA), and it can best be described as an orthogonal and compositional superposition of DTMC, CTMC, and LTS. MA have been coined in [22,23]. They are expressive enough to give a semantics to generalised stochastic Petri nets (GSPN) in their full generality [20]. The theoretical properties of MA are the subject of the Ph.D. thesis of Christian Eisentraut [19]; a process-algebraic perspective is covered in the Ph.D. thesis of Mark Timmer [50]. Various algorithmic analysis methods for Markov automata have been developed over the past decade [8,9,14–16,21,28,29,36,37,52].

Using the mathematical formalism of MA directly to build complex models is, however, cumbersome. We instead need a higher-level *modelling language*. Aside from parallel composition, such languages typically provide variables over finite domains that can be used in expressions to e.g. enable or disable transitions, allowing to compactly describe very large models. In this paper, we use MODEST [30] to construct MA models. Rooted in process algebra, MODEST provides various composition operators that allow large models to be assembled from smaller, easier-to-understand components. After a formal definition of MA, parallel composition, and various types of properties (that we may want to compute for a given MA model) in Sect. 2, we introduce the basics of MODEST in a step-by-step fashion in Sect. 3. We compare it to alternative languages with respect to its succinctness, expressivity, and readability. We then guide the reader through the modelling and analysis of two very different applications: we optimise an *attack on Bitcoin* in Sect. 4, and we evaluate the performance of a small, but intricate resource-sharing *queueing system* in Sect. 5 with the MODEST TOOLSET. Algorithmic aspects of the analysis of MA with the MODEST TOOLSET are the subject of a companion paper [13].

Previous Work. Our presentation of MA in Sect. 2 is adapted and extended from [13], as is the text in Sects. 3.2 and 3.3. The Bitcoin models in Sect. 4 are inspired by [24], and the `bitcoin-attack.modest` model is part of the Quantitative Verification Benchmark Set [34]. The reentrant queueing system, of which we present a new MODEST model in Sect. 5, was first described in [36].

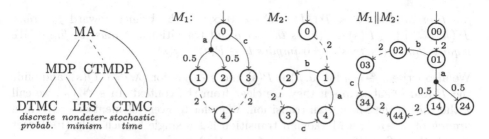

Fig. 1. The MA family tree   **Fig. 2.** Example Markov automata

2 Markov Automata

The mathematical formalism of Markov automata provides nondeterministic choices as in LTS, discrete probabilistic decisions as in DTMC, and stochastic time as in CTMC. The relationships between these and other formalisms are visualised in Fig. 1. The combination of DTMC and LTS leads to the model family of (discrete-time) Markov decision processes [46] (MDP, or probabilistic automata [49]) where transitions of the form $s \xrightarrow{a} \mu$ offer in state s a (nondeterministic) decision option (or choice option) labelled by action a that is followed by a probabilistic decision of where to jump according to probability distribution μ. The conceptually closest model in continuous time is that of continuous-time MDP [46] (CTMDP), where action-labelled transitions are of the form $s \text{-}^{a}\text{›} e$ with e mapping states to rates. Such a transition indicates that probability mass flows from state s to state s' with rate $e(s')$ provided action a is chosen in state s. Markov automata instead combine MDP and CTMC in an orthogonal manner by providing two types of transitions: $s \xrightarrow{a} \mu$ as in MDP, and $s \text{-}^{\lambda}\text{›} s'$ as in CTMC. We now define Markov automata formally and describe their semantics.

Preliminaries. We write $[a,b]$ for the real interval $\{x \in \mathbb{R} \mid a \leq x \leq b\}$, (a,b) for $\{x \in \mathbb{R} \mid a < x < b\}$, and analogously for half-open intervals. Given a set S, its powerset is 2^S. A (discrete) probability distribution over S is a function $\mu\colon S \to [0,1]$ such that its support $spt(\mu) \stackrel{\text{def}}{=} \{s \in S \mid \mu(s) > 0\}$ is countable and $\sum_{s \in spt(\mu)} \mu(s) = 1$. $Dist(S)$ is the set of all probability distributions over S, and $\mu_1 \otimes \mu_2$ is the product distribution of μ_1 and μ_2 defined by $(\mu_1 \otimes \mu_2)(\langle s_1, s_2 \rangle) = \mu_1(s_1) \cdot \mu_2(s_2)$. We refer to discrete random choices as *probabilistic* and to continuous ones as *stochastic*. We write $\{x_1 \mapsto y_1, \dots\}$ to denote the function that maps each x_i to y_i, and if necessary in some context, implicitly maps to 0 all x for which no explicit mapping is specified. Thus we can e.g. write $\{s \mapsto 1\}$ for the *Dirac distribution* that assigns probability 1 to s.

Definition 1. *A Markov automaton (MA) is a tuple $M = \langle S, s_0, A, P, Q, rr, br \rangle$ where S is a finite set of states with initial state $s_0 \in S$, A is a finite set of actions, $P\colon S \to 2^{A \times Dist(S)}$ is the probabilistic transition function, $Q\colon S \to 2^{\mathbb{Q} \times S}$ is the Markovian transition function, $rr\colon S \to [0,\infty)$ is the rate reward* *function, and* $br: S \times Tr(M) \times S \rightarrow [0, \infty)$ *is the* branch reward *function.* $Tr(M) \stackrel{\text{def}}{=} \bigcup_{s \in S} P(s) \cup Q(s)$ *is the set of all* transitions; *it must be finite. We require that* $br(\langle s, tr, s' \rangle) \neq 0$ *implies* $tr \in P(s) \cup Q(s)$.

We also write $s \xrightarrow{a}_P \mu$ for $\langle a, \mu \rangle \in P(s)$ and $s \xrightarrow{\lambda}_Q s'$ for $\langle \lambda, s' \rangle \in Q(s)$, and omit the P and Q subscripts if they are clear from the context. In $s \xrightarrow{\lambda}_Q s'$, we call λ the *rate* of the Markovian transition. We refer to every element of $spt(\mu)$ as a *branch* of $s \xrightarrow{a}_P \mu$; a Markovian transition has a single branch only (its target state). We define the *exit rate* of $s \in S$ as $E(s) = \sum_{\langle \lambda, s' \rangle \in Q(s)} \lambda$.

Example 1. Fig. 2 shows two MA M_1 and M_2 without rewards. We draw probabilistic transitions as solid, Markovian ones as dashed lines. If a transition leads to a single target state, we omit the intermediate probabilistic branching node. Thus, for $M_1 = \langle S, s_0, A, P, Q, rr, br \rangle$, we have five states in $S = \{0, 1, 2, 3, 4\}$, the initial state being $s_0 = 0$, two actions in $A = \{a, c\}$, two probabilistic transitions in $P = \{0 \mapsto \{\langle a, \{1 \mapsto 0.5, 2 \mapsto 0.5\}\rangle, \langle c, \{3 \mapsto 1\}\rangle\}\}$, and two Markovian transitions in $Q = \{1 \mapsto \{\langle 2, 4 \rangle\}, 3 \mapsto \{\langle 2, 4 \rangle\}\}$, both with rate 2.

Intuitively, the semantics of an MA is that, in state s, (1) the probability to take Markovian transition $s \xrightarrow{\lambda} s'$ and move to state s' within t model time units is $\lambda / E(s) \cdot (1 - e^{-E(s) \cdot t})$, i.e. the residence time in s follows the exponential distribution with rate $E(s)$ and the choice of transition is probabilistic, weighted by the rates; and (2) at any point in time, a probabilistic transition $s \xrightarrow{a} \mu$ can be taken with the successor state being chosen according to μ. An MA thus resolves some choices in a probabilistic (the choice of successor state of a probabilistic transition, the choice among Markovian transitions) or stochastic (the choice of residence time) way, while other choices are left open as *nondeterministic* (the timing of probabilistic transitions, and the choice among multiple available probabilistic transitions). Due to the presence of nondeterminism, an MA itself does not induce a probability measure over its possible behaviours. We refer the interested reader to e.g. [35] for a complete formal definition of this semantics.

An MA without Markovian transitions is an MDP; it is a DTMC if in addition P maps each state to a singleton set. An MA without probabilistic transitions is a CTMC. The co-existence of action-labelled probabilistic transitions of the form $s \xrightarrow{a} \mu$ and of Markovian transitions of the form $s \xrightarrow{\lambda} s'$ separates actions from timing. It enables parallel composition operators with action synchronisation for MA without the need to prescribe an ad-hoc operation for combining rates.

Definition 2. *Given two MA $M_i = \langle S_i, s_{0_i}, A_i, P_i, Q_i, rr_i, br_i \rangle$ $i \in \{1, 2\}$, a finite set A of actions, and a synchronisation relation*

$$sync \subseteq (A_1 \uplus \{\perp\}) \times (A_2 \uplus \{\perp\}) \times A,$$

their parallel composition is $M_1 \parallel M_2 \stackrel{\text{def}}{=} \langle S_1 \times S_2, \langle s_{0_1}, s_{0_2} \rangle, A, P, Q, rr, br \rangle$ where P is the smallest function that satisfies the inference rules

$$\frac{s_1 \xrightarrow{a_1}_{P_1} \mu \quad \langle a_1, \perp, a \rangle \in sync}{\langle s_1, s_2 \rangle \xrightarrow{a}_P \mu \otimes \{s_2 \mapsto 1\}} \qquad \frac{s_2 \xrightarrow{a_2}_{P_2} \mu \quad \langle \perp, a_2, a \rangle \in sync}{\langle s_1, s_2 \rangle \xrightarrow{a}_P \{s_1 \mapsto 1\} \otimes \mu}$$

$$\frac{s_1 \xrightarrow{a_1}_{P_1} \mu_1 \qquad s_2 \xrightarrow{a_2}_{P_2} \mu_2 \qquad \langle a_1, a_2, a \rangle \in sync}{\langle s_1, s_2 \rangle \xrightarrow{a}_P \mu_1 \otimes \mu_2},$$

Q is the smallest function that satisfies the inference rules

$$\frac{s_1 \xrightarrow{\lambda}_{Q_1} s_1'}{\langle s_1, s_2 \rangle \xrightarrow{\lambda}_Q \langle s_1', s_2 \rangle} \qquad \frac{s_2 \xrightarrow{\lambda}_{Q_2} s_2'}{\langle s_1, s_2 \rangle \xrightarrow{\lambda}_Q \langle s_1, s_2' \rangle},$$

and for all states $\langle s_1, s_2 \rangle$, we have $rr(\langle s_1, s_2 \rangle) = rr_1(s_1) + rr_2(s_2)$. Function br sums the values of br_1 and br_2 for the combinations of branches in synchronisation (third inference rule), and otherwise preserves the original branch rewards.

The first two inference rules for P allow the individual MA to proceed independently of each other if allowed by *sync*; the third rule covers the case where both automata synchronise on a pair of actions as determined by *sync*. The rules for Q simply state that Markovian transitions are always performed independently. An element of *sync* is called a *synchronisation vector*; we also write $\langle a_1, a_2 \rangle \mapsto a$ for vector $\langle a_1, a_2, a \rangle$. This form of parallel composition can be generalised to more than two automata in the straightforward way with longer synchronisation vectors. It is very flexible, allowing in particular the traditional CCS-style binary and CSP-style multi-way synchronisation patterns [40, 44] to be encoded. Originally established by CADP [26], it is today used for MA in the JANI format [12]. We refer to a general parallel composition of several MA as a *network* of MA.

Example 2. Fig. 2 includes the parallel composition of the example MA M_1 and M_2, where we write nm for state $\langle n, m \rangle$. The two automata synchronise on the shared actions a and c, i.e. we have $sync = \{ \langle a, a \rangle \mapsto a, \langle \bot, b \rangle \mapsto b, \langle c, c \rangle \mapsto c \}$.

We defined MA as *open* systems [10]: probabilistic transitions can interact with, wait for, and be blocked by other MA in parallel composition. For verification, we make the usual *closed system* and *maximal progress* assumptions: probabilistic transitions face no further interference and take place without delay. If multiple probabilistic transitions are available in a state, however, the choice between them remains nondeterministic. Since the probability that a Markovian transition is taken in zero time is 0, the maximal progress assumption allows us to remove all Markovian transitions from states that *also* have a probabilistic transition. In such *closed MA*, we can thus distinguish between Markovian states (where $P(s) = \varnothing$) and probabilistic states (where $Q(s) = \varnothing$). The behaviour of a closed, deadlock-free MA M is defined via its paths:

Definition 3. *Let M be a closed, deadlock-free MA M as above. A path π of M is an infinite sequence*

$$\pi = s_0 \, t_0 \, tr_0 \, s_1 \ldots \in (S \times [0, \infty) \times Tr(M))^\omega$$

such that, for all $i \in \{0, \ldots\}$, $Q(s_i) = \varnothing$ implies $t_i = 0$, $tr_i \in P(s_i) \cup Q(s_i)$, $tr_i = \langle a, \mu \rangle \in P(s_i)$ implies $\mu(s_{i+1}) > 0$, and $tr_i = \langle \lambda, s' \rangle \in Q(s_i)$ implies

$s' = s_{i+1}$. $\Pi(M)$ is the set of all paths of M. We write $\Pi_{fin}(M)$ for the set of all path prefixes π_{fin} ending in a state. The last state of π_{fin} is denoted $last(\pi_{fin})$. Let $\pi_{\leq j} \stackrel{\text{def}}{=} s_0 t_0 \ldots s_j$. The duration $dur(\pi_{fin})$ of a path prefix is the sum of its residence times t_i. A path's reward is

$$\text{rew}(\pi) \stackrel{\text{def}}{=} \sum\nolimits_{i=0}^{\infty} t_i \cdot rr(s_i) + br(s_i, tr_i, s_{i+1}).$$

It may be ∞, and is defined analogously for prefixes (where it is always finite).

A path comprises states s_i, times t_i spent in s_i, and transitions tr_i taken from s_i to s_{i+1}. It is a resolution of all nondeterministic, probabilistic, and stochastic choices. To define a probability measure, we resolve nondeterminism only:

Definition 4. Let M be a closed, deadlock-free MA as above. A scheduler is a function $\sigma\colon \Pi_{fin}(M) \to Tr(M)$ s.t. $\forall s \in S\colon \sigma(s) = tr$ implies $tr \in P(s) \cup Q(s)$. We write $\mathfrak{S}(M)$ for the set of all schedulers of M. A time-dependent scheduler is in $S \times [0, \infty) \to Tr(M)$; a memoryless scheduler is in $S \to Tr(M)$. Given a time bound $b \in [0, \infty)$, every time-dependent scheduler σ_t defines a corresponding scheduler σ by $\sigma(\pi_{fin}) = \sigma_t(\langle last(\pi_{fin}), b - dur(\pi_{fin})\rangle)$. Every memoryless scheduler σ_{ml} defines a corresponding scheduler σ by $\sigma(\pi_{fin}) = \sigma_{ml}(last(\pi_{fin}))$.

We define deterministic schedulers only since randomised schedulers are in practice only needed for multi-objective problems [47]. We note that CTMDP with early schedulers [48] can be encoded as closed MA. If we "apply" a scheduler to an MA, it removes all nondeterminism, and we are left with a fully stochastic process whose paths can be measured and assigned probabilities according to the rates and distributions in the (remaining) MA. Formally, these probability measures over sets of measurable paths are built via cylinder sets; we refer the interested reader to e.g. [35] for a fully formal definition. For all of the following types of properties, we are interested in the maximum (supremum) and minimum (infimum) values when ranging over all schedulers $\sigma \in \mathfrak{S}(M)$:

Reachability probabilities: Given goal states $G \subseteq S$, compute the probability of the set of paths that include a state in G. Memoryless schedulers suffice to achieve optimal results (i.e. the maximum and minimum probabilities).

Time-bounded reachability: Additionally restrict to paths where the duration of the prefix to the first state in G is below a bound $b \in [0, \infty)$. Time-dependent schedulers suffice.

Expected accumulated rewards: Compute the expected value of the random variable that assigns to π the value $\text{rew}(\pi_{fin})$ with π_{fin} being the shortest prefix of π with a state in G. This is well-defined if the maximum (minimum) probability to reach G is 1; otherwise, we define the minimum (maximum) expected accumulated reward to be ∞. Memoryless schedulers suffice.

Long-run average rewards: Compute the expected value of the random variable that assigns to path π the value $\lim_{i \to \infty} \text{rew}(\pi_{\leq i})/dur(\pi_{\leq i})$. Memoryless schedulers suffice.

Example 3. Consider MA $M_1 \parallel M_2$ of Fig. 2 and the probability to reach state $\langle 4, 4 \rangle$ within 1 time unit. In state $\langle 0, 1 \rangle$, we have to decide whether to choose action a or b. The optimal decision depends on the amount of time t that has passed in state $\langle 0, 0 \rangle$. In the plot on the right, we show the probability of reaching state $\langle 4, 4 \rangle$ within the time limit (y-axis) depending on the remaining time $1 - t$ (x-axis). The blue (initially upper) line represents the reachability probability for 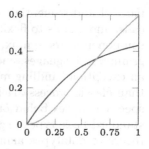 the memoryless scheduler that always chooses a and the red (initially lower) one is for the scheduler that always takes action b. A time-dependent scheduler can make better decisions than either of these two by determining the values of t for which a results in a higher probability than b and vice-versa. The optimal scheduler thus chooses a if and only if $1 - t \leq 0.63$ approximately.

We can extend MA with discrete *variables*: An MA with variables (MA^V) is an MA like in Definition 1 that additionally contains a finite set of variables. We call its states *locations*, its transitions *edges*, and their branches *destinations*. Every edge additionally has a *guard* and every destination has a set of *updates*. A guard is a Boolean expression over the variables that determines whether the edge is enabled, and a set of assignments modifies the values of the variables. Tools usually work with the semantics of an MA^V in terms of an MA: The MA^V M_V corresponds to the MA M with states $\langle \ell, v \rangle$, each consisting of a location ℓ of M_V and a valuation v that assigns a value to every variable. The transitions out of $\langle \ell, v \rangle$ are those edges out of ℓ in M_V whose guard is satisfied in v. The target state of a branch of a transition is $\langle \ell', v' \rangle$ with ℓ' the target location in M_V and v' obtained by executing the destination's assignments on v. Our parallel composition operator extends to MA with variables by using the conjunction of guards and the union of assignments for synchronising transitions. If we allow variables to be shared between MA^V, parallel composition does not distribute over semantics; we need to compose the MA^V before converting them to MA.

3 Modelling with Markov Automata

Tools for the automated analysis of MA need a syntax in which the model and the properties of interest are specified. As noted in Sect. 1, such a modelling language needs to provide a parallel composition operator (akin to the operator introduced in the previous section) such that large MA can be built from small specifications, and will typically support modelling with variables.

3.1 MODEST for Markov Automata

MODEST [4,30] is the <u>mo</u>delling and <u>des</u>cription language for <u>s</u>tochastic <u>t</u>imed systems. At its core, it is a process algebra: it provides various operations such as parallel and sequential composition, parameterised process definitions, process calls, and guards to flexibly construct complex models out of small and reusable components. Its syntax, however, borrows heavily from commonly used programming languages, and it provides high-level conveniences such as loops and an exception handling mechanism. As such, MODEST tends to be more verbose than classic process algebras, but also more readable and beginner-friendly. To specify complex behaviour in a succinct manner, MODEST provides variables of standard basic types (e.g. `bool`, `int`, or bounded `int`), arrays, and user-defined recursive datatypes akin to functional programming languages. Its syntax for expressions is aligned with C-like programming languages for ease of use.

Let us now introduce the MODEST language syntax step-by-step by using it to model our example MA shown in Fig. 2, starting with M_1. MODEST models are structured into *processes*, with each process consisting of *declarations* and a *behaviour*. The declarations introduce all named objects like actions, variables, exceptions, nested processes, etc., that are available for use in the behaviour and inside nested processes. A process' behaviour defines an MA with those variables[1]. To model M_1 as a MODEST process, we thus start by declaring the actions and a Boolean variable to later distinguish between states 1 and 2:

```
action a, c;
bool f = false; // to distinguish between states 1 and 2
```

The simplest behaviour in MODEST is to perform a (previously declared) action:

Semantically, this behaviour represents the MA with variables shown above on the right, where the one edge has guard expression *true*. Every location ℓ is uniquely identified by a behaviour such that the MA with ℓ as its initial location is the semantics of the behaviour. The checkmark ✓ is a special behaviour called *successful termination* that is not part of the syntax of MODEST, and whose semantics is a state with no outgoing edges. It receives special treatment by several other MODEST constructs. MODEST also contains a `stop` construct with the same semantics but without the special treatment.

Initially, automaton M_1 offers a choice between two probabilistic transitions. The `alt` construct combines multiple behaviours into a nondeterministic choice between them, thus the initial choice in M_1 can be represented as follows:

```
alt {
:: a
:: c
}
```

The semantic effect of the `alt` construct is simply to merge the initial states of the semantics of its child behaviours, the start of each of which is indicated by ::. Note that both edges lead to the same location here; this is because the semantics of both behaviours a and c end in the identical location ✓.

---

[1] Actually, the semantics of MODEST [30] is defined in terms of stochastic hybrid automata (SHA), of which MA are a special case; we restrict to that case in this paper.

Now, in M_1, the transition labelled a actually has two branches. The branching of probabilistic transitions can be represented in MODEST with the palt construct. Since it does not create a new transition, but only defines branches, it has to be prefixed by the transition's action:

```
alt {
 :: a palt {
    :1: {= f = true =}
    :1: {==}
    }
 :: c
}
```

Probabilities are specified as *weights* between colons :, i.e. the actual probability in the semantics is calculated as the given weight divided by the sum of all weights in the palt construct. The assignments for every branch are specified in {= =} blocks, and they are executed *atomically*, so e.g. the assignment block {= x = y, y = x =} performs an in-place swap of variables x and y. To create an edge labelled a with a single destination and assignments u, we can omit the palt and just write a {= u =}. Observe that, in the semantics of our example above, all destinations still lead to the same location. However, the semantics of this MA^V contains two states in location ✓: one where f is *true*, which is the target of the branch for the uppermost destination, and one where it is *false*. We will from now on omit *true* guards and empty assignment sets in MA^V.

Continuing to model M_1 in MODEST, we now add the Markovian transitions to state 4. We need two new constructs: for sequential composition, and for rates. First, the semantics of the sequential composition construct $P; Q$, for two behaviours P and Q, is to first behave like P, and upon successful termination of P (i.e. upon reaching location ✓), behave like Q. We thus get the following:

```
alt {
 :: a palt {
    :1: {= f = true =}; stop
    :1: {==}; tau
    }
 :: c; tau
}
```

tau is the predefined *silent action*, which does not take part in synchronisation (i.e. in a binary parallel composition, it is governed by synchronisation vectors $\langle \tau, \bot \rangle \mapsto \tau$ and $\langle \bot, \tau \rangle \mapsto \tau$, but cannot occur in any other vectors). To turn the τ-labelled probabilistic edge into a Markovian one, we simply specify rates:

```
alt {
 :: a palt {
    :1: {= f = true =}; stop
    :1: {==}; rate(2) tau
    }
 :: c; rate(2) tau
}
```

MODEST enforces the separation of probabilistic and Markovian transitions by requiring edges for which a rate is specified to have action tau. If this restriction is not met, the model is recognised as a CTMDP.

In the model above, the behaviour rate(2) tau occurs twice. We can eliminate this duplication by moving it out of the alt construct. At this point, let us also introduce the when construct to specify guards: instead of using stop to make the model deadlock in the upper destination, we use f to cause the deadlock in the semantics of the MA^V. The result is:

```
alt {
  :: a palt {
    :1: {= f = true =}
    :1: {==}
  }
  :: c
};
when(!f) rate(2) tau
```

The semantics of the MA^V on the right above is *almost* isomorphic to M_1; the difference is that states 1 and 3 are merged since they have the same behaviour.

In Fig. 3, we show the full MODEST model of the parallel composition of MA M_1 and M_2 of Fig. 2. It includes the model that we built for M_1 above as the body of the named process M1. Such processes can have parameters (specified between the parentheses in the declaration, not shown here) and local variables. A process call like M1() behaves exactly like the behaviour of M1, with all formal parameters being assigned the values of the actual arguments, and new variable instances created for all parameters and local variables to separate them from any other calls to M1. The semantics of the parallel composition construct par is the n-ary parallel composition of its child behaviours, with synchronisation vectors that implement CSP-style synchronisation for all actions declared with the action keyword (in this model, that is the vectors given in Example 2), and as described above for τ. The model also declares two properties for verification, P_Min and P_Max, which ask for the probability to reach state $\langle 4, 4\rangle$—made observable via the global variable succ, which is of bounded integer type[2] with range $\{0, 1, 2\}$—within time bound B akin to Example 3. B is an open parameter for which values can be specified at verification time.

At this point, we have covered most basic constructs of MODEST. There are many features not used in this small model; we will introduce more constructs in Sects. 4 and 5. The interested reader also finds additional MODEST MA models in the Quantitative Verification Benchmark Set (QVBS, [34]) at qcomp.org.

---

[2] MA model checking requires finite state spaces; thus all variables must be bounded. Indicating the bounds in the types is good practice to avoid accidentally creating infinite-state models and may improve performance, but it is not a requirement for the mcsta model checker (see Sect. 3.2) as long as only finitely many distinct values are ever assigned to the variables occurring in the model.

```
const real B;
int(0..2) succ = 0;

action a, b, c;

property P_Min = Pmin(<>[T<=B] (succ == 2));
property P_Max = Pmax(<>[T<=B] (succ == 2));

process M1()
{
  bool f = false;
  alt {
  :: a palt {
    :1: {==}
    :1: {= f = true =}
    }
  :: c
  };
  when(!f) rate(2) {= succ++ =}
}
process M2()
{
  rate(2) tau;
  alt {
  :: a {= succ++ =}
  :: b; rate(2) tau; c {= succ++ =}
  }
}
par {
:: M1()
:: M2()
}
```

Fig. 3. MODEST model for $M_1 \parallel M_2$

```
global succ:{0..2} = 0
DONE = done.DONE[]
M1  = a.psum(0.5 -> M1a[] ++ 0.5 -> DONE[])
      ++ c.M1a[]
M1a = <2>.setGlobal(succ, succ + 1)
      .DONE[]
M2  = <2>.(a.M2a[] ++ b.<2>.c.M2a[])
M2a = setGlobal(succ, succ + 1).DONE[]
init M1[] || M2[]
comm (a, a, a), (c, c, c)
reachCondition (succ = 2)
```

Fig. 4. MAPA process algebra

```
ma
const double B
module M1
  s1: [0..4];
  [a] s1=0 -> 0.5:(s1'=1) + 0.5:(s1'=2);
  [c] s1=0 -> 1:(s1'=3);
  <>  s1=1 | s1=3 -> 2:(s1'=4);
endmodule
module M2
  s2: [0..4];
  <>  s2=0 -> 2:(s2'=1);
  [a] s2=1 -> 1:(s2'=4);
  [b] s2=1 -> 1:(s2'=2);
  <>  s2=2 -> 2:(s2'=3);
  [c] s2=3 -> 1:(s2'=4);
endmodule
"P_Min": Pmin=? [F<=B (s1=4 & s2=4)];
"P_Max": Pmax=? [F<=B (s1=4 & s2=4)];
```

Fig. 5. PRISM dialect supporting MA

```
#INITIALS
s00
#GOALS
s44
#TRANSITIONS
s00 !
* s01 2
s01 a
* s02 1
s01 b
* s14 0.5
* s24 0.5
s14 !
* s44 2
s02 !
* s03 2
s03 c
* s34 1
s34 !
* s44 2
```

Fig. 6. IMCA state space format

3.2 The MODEST TOOLSET

The creation and analysis of MA with MODEST is supported by the MODEST TOOLSET [32], a comprehensive suite of tools for quantitative modelling and verification. Aside from MODEST, it also supports the JANI model interchange format [12] as an input language. MA are supported in the toolset's mosta, moconv[3], mcsta, and modes tools. mosta visualises the symbolic semantics of models (i.e. networks of MA^V before and after parallel composition as shown throughout Sect. 3.1) and is useful for model debugging. moconv transforms models between MODEST and JANI, and performs syntactic rewriting and optimisations. mcsta is

---

[3] moconv can also export CTMDP to JANI, but due to their lack of a natural parallel composition operator, the analysis of CTMDP is not supported in the other tools.

a fast explicit-state model checker that implements state-of-the-art MA-specific algorithms [13] and uses secondary storage to alleviate state space explosion [33]. modes [11] is a statistical model checker with automated rare event simulation capabilities. It implements lightweight scheduler sampling [43] for nondeterministic models, including MA [17]. The MODEST TOOLSET is written in C#, works on Linux, Mac OS, and Windows, and is freely available at modestchecker.net. All its tools share a common infrastructure for parsing and syntactic transformations. mcsta and modes build on the same state space exploration engine that compiles models to bytecode at runtime for memory efficiency and performance.

3.3 Alternative Modelling Languages

MODEST is not the only modelling language for MA. We now briefly contrast it to the currently available alternative modelling languages with support for MA.

State Space Files for IMCA. The first MA-specific algorithms were implemented in the IMCA tool [27]. Its only input language is a text-based explicit state space format as illustrated for our example of $M_1 \parallel M_2$ in Fig. 6. This is clearly not a useful modelling language, but a format to be automatically generated by tools.

Guarded Commands with STORM. The STORM model checker [18] provides many input languages, with MA being supported through a state space format similar to IMCA's, via JANI, as the semantics of generalised stochastic Petri nets [20] in GREATSPN format [1], and through an extension of the PRISM guarded command language. We show our example in the latter in Fig. 5. This is a very simple and small language that is easy to learn, however it completely lacks higher-level constructs to structure and compose models aside from the implicit parallel composition of its *modules*.

Process Algebra with SCOOP. MAPA [51] is a dedicated process algebra for MA. It is supported by SCOOP [51], which can linearise, reduce, and finally export MAPA models to IMCA for verification. We show the example of M_1 and M_2 in MAPA in Fig. 4. As a classic concise process algebra, MAPA tends to be very succinct, but also difficult to read. MAPA models can be much more flexibly composed than PRISM models, yet there is less syntactic structure than in MODEST—although the languages conceptually share many operators. MAPA notably has a predefined *queue* datatype, and users can specify custom non-recursive datatypes.

JANI [12] is a model interchange format designed to ease tool development and interoperation. It is JSON-based and thus human-debuggable, but not intended as human-writable. It represents networks of automata with variables symbolically. Since both the MODEST TOOLSET and STORM support JANI, it is possible to e.g. build MA models in the MODEST language, export them to JANI with moconv, and then verify them with STORM. Likewise in the other direction, we can e.g. create a Petri net with GREATSPN, convert to JANI with STORM, and analyse it with mcsta or modes. In this way, the most appropriate modelling language can be combined with the best analysis method and tool for every specific scenario. The JSON-based syntax however is too verbose to display the example in JANI format in this paper.

4   Optimising Attacks on Bitcoin

Bitcoin [45] is currently the most popular cryptocurrency. It is built on blockchain technology using the proof-of-work approach. Every block in the blockchain contains a *nonce* (a randomly chosen number), a set of (monetary) *transactions*, and a hash of the predecessor block in the chain. In this way, no past block can be changed without invalidating (the hashes in) all its successors. A block is valid if the hash of the block's contents falls below a *target value*. To create a valid block, a node in the Bitcoin network repeatedly selects a new nonce until it finds one that makes the block valid. Creating new blocks is called *mining*, and overall constitutes the *proof-of-work* approach since the repeated hashing is computationally (and thus environmentally) expensive. As the computational power used for mining (the *hash rate*) changes, the Bitcoin network periodically adjusts the target value such that the average time to find a new block (the *confirmation time*) is 10 min. In practice, the actual confirmation time varies; it was about 12 min in 2017 [24]. Every node in the network stores its own copy of the entire blockchain. Once a new node finds a new valid block, it broadcasts the block to the network. Due to network delays, multiple new blocks may propagate at the same time. Nodes add the first block they receive to their local chain. Thus multiple *forks* of the blockchain may exist on different nodes. Each node always considers the longest chain known to it as valid, and miners extend the longest chain. A transaction is *n-confirmed* with *confirmation depth* $n = 0$ if it is not part of any valid block and otherwise with $n > 1$ if there are $n - 1$ blocks in the chain beyond the block b that the transaction is part of. The amount of work to invalidate a fork that starts with b increases with n. Many services only accept Bitcoin payments once they are at least 6-confirmed [7].

In this section, we use MODEST and the MODEST TOOLSET to study two variants of a secret-fork attack on Bitcoin, inspired by the Andresen attack proposal and a study performed with UPPAAL SMC in [24]. The attackers secretly create a fork, keep mining on it until it reaches a certain length greater than that of the publicly known blockchain, and then publish it all at once. This would invalidate the public fork, with the private one becoming the valid blockchain. The original aim of the attack was to undermine the trust in Bitcoin; if it succeeds on the first attempted fork, it can equally be used for double spending by invalidating a specific transaction. For the attack to be feasible, the malicious attacker must control a significant fraction m of the hash rate.

4.1   Modelling and Evaluating the Double-Spending Attack

If the goal of the attacker is double spending, then it creates a transaction that spends some Bitcoin funds and announces it to the network for inclusion in the next block. At the same time, it starts mining on its own secret fork. Let *cd* be the confirmation depth after which a transaction is accepted by the receiver of the funds. If the attacker manages for its secret fork to become longer than the public fork, and longer than *cd*, then it can publish this fork immediately after the public one reaches length *cd*. At that point, the receiver of the funds has just accepted the transaction (and presumably fulfilled its part of the contract). The secret fork however invalidates the public one since it is longer, and thus invalidates the transaction. The attacker is now free to spend the same funds again. Due to the proof-of-work system, such an attack is possible, but—as long as the attacker controls less than 50% of the hash rate—has a low success probability and an immense computational cost.

Modelling the Attack in MODEST. We build an abstract model of mining in MODEST, reduced to the aspects relevant to the attack. The observation that a new block is mined every 12 min on average fits well with MA: we model block creation via Markovian transitions with a total rate of $\frac{1}{12}$. We abstract from network delays, i.e. blocks propagate instantaneously. We consider a single attacker, assuming that the rest of the world's miners behave in the normal "honest" manner and publish all mined blocks immediately.

Honest Mining Model. To start, we define a process HonestPool representing the pool of honest miners, which control $(1 - m) \cdot 100\%$ of the global computational resources used for mining, with m realised as model parameter M:

```
const real M;   // fraction of hash rate controlled by malicious mining pool
action sln;     // indicates that the honest pool mined a new block
process HonestPool()
{
    rate(1/12 * (1 - M)) tau;  // wait 12 / (1 − M) minutes on average
    sln;                       // signal that a new block was found
    HonestPool()               // repeat
}
```

Action sln models the propagation of a new block through the network, which can also be observed by the attacker. Due to the separation of timing and interaction in MA, we need two separate edges for mining delay and communication.

Attacker Model. We keep track of the length of the attacker's fork, and of the difference in length to the public fork. To make the MA finite, we identify all fork lengths greater than cd with the value $cd + 1$ (since we only need to know *whether* the fork is longer than cd, but not *how much* longer), and we assume that the attacker gives up on its fork once it is db blocks shorter than the public one. The attacker process is then as follows:

```
const int CD;              // confirmation depth required by victim
const int DB = CD;         // attacker gives up when this far behind
action cnt;                // indicates that the attacker continues
int(0..CD+1) m_len;        // length of the secret fork
int(-DB..CD+1) m_diff = 0; // length of secret fork minus honest fork
bool gup;                  // indicates whether the attacker gave up
process DoubleSpendingAttacker()
{
    do {
    :: rate(1/12 * M) {= m_len = min(CD+1, m_len + 1), m_diff++ =}
    :: sln {= m_diff-- =}; // public fork extended
```

```
      if(m_diff <= -DB) { tau {= gup = true =}; stop } // give up
      else { cnt }                                     // continue
   }
}
```

For illustration, we use the do construct to implement a loop here instead of the recursive process call used in HonestPool. A do loop is in essence a looping alt: There is an initial nondeterministic choice between the child behaviours; once the chosen behaviour successfully terminates, control loops back to the nondeterministic choice. do loops can be exited via the predefined break action. We also use the if shorthand: if(e) { P } else { Q } is syntactic sugar for alt{ :: when(e) P :: when(!e) Q }. Thus the behaviour of the attacker process is as follows: it waits until it either mines a new block itself (first child behaviour of the do loop), or until it observes a new block in the public fork. In both cases, it appropriately updates m_len and m_diff. In the second case, it then either gives up if it has fallen too far behind, or otherwise continues the attack.

Composition and Nondeterminism. The overall behaviour of our model is the parallel composition of the two processes, with synchronisation on sln:

```
par {
:: HonestPool()
:: DoubleSpendingAttacker()
}
```

Observe that the behaviour of neither of the two processes contains an actual nondeterministic choice: HonestPool is entirely sequential, and the choices in the attacker process are between a Markovian and a probabilistic edge (in do), i.e. the probability for both to be available at the same time is 0, and between two edges with disjoint guards (in if). Since the only probabilistic edge in HonestPool synchronises with the attacker, and is immediately followed by a Markovian edge, the parallel composition cannot introduce nondeterminism due to interleaving probabilistic transitions, either. Thus the entire model takes the form of an MA, but is in fact equivalent to a CTMC. MA that are equivalent to CTMC are a class of models that occurs frequently in practice. Several of the MA models in the QVBS belong to this class.

Evaluating the Attack. We are interested in the probability that the attacker eventually wins, and that it eventually gives up without winning. We expect it to eventually either win or give up, thus—due to the absence of nondeterminism—the probabilities should sum to 1. We declare the two properties in MODEST:

```
function bool win() = m_len > CD && m_diff > 0; // winning condition
property P_Win = Pmin(<> win());                // attacker wins
property P_GiveUp = Pmin(!win() U gup);         // attacker gives up
```

To avoid repeating the expression that characterises the winning condition, we encapsulate it in the user-defined function win(). Functions in MODEST can also take parameters, and they can be (mutually) recursive. The body of a function is an expression; since expressions in MODEST are free of side effects, functions provide for pure functional programming inside MODEST models. Combined with user-defined recursive datatypes (not shown in this paper), they make MODEST Turing-complete. Property P_Win is straightforward: we ask for the (minimum) probability to eventually (<>) enter a state that satisfies the winning condition. Since there is no nondeterminism, there is no difference between Pmin and Pmax for this model. Property P_GiveUp uses the until (U) operator to ask for the probability of those paths on which no state satisfies win() until a state where gup is *true* is reached. If we invoke mcsta on this model by executing

```
./modest mcsta bitcoin-ds.modest -E "M = 0.2, CD = 6"
```

we obtain probability ≈ 0.0087 for P_Win and ≈ 0.9913 for P_GiveUp: the attack is unlikely to succeed if the attacker controls only 20% of the hash rate. However, at $m = 0.4$, we get P_Win ≈ 0.343, and at $m = 0.5$, it is ≈ 0.719. It is not 1 here because the attacker gives up when falling behind too much. If we modify the model such that the attacker never gives up, it becomes an infinite-state MA since m_diff is no longer bounded from below. We cannot model-check this model, but due to the absence of nondeterminism, we can easily perform statistical model checking with modes by running

```
./modest modes bitcoin-ds-inf.modest -E "M=0.2, CD=6" --max-run-length 0
```

The output confirms our expectation that the probability is now 1, although we only know this with the statistical confidence provided by modes.

4.2   Optimising the Attack on Trust in Bitcoin

If the goal of the attack is to undermine the trust in the Bitcoin system by invalidating a large amount of work performed by the honest miners, the attacker gains some freedom in choices: Instead of having to give up when it gets too far behind, it can simply restart its attack from the then-current public fork. We thus keep the *cd* parameter, which now indicates the minimum desired length of the secret fork for it to be published. The winning condition becomes the length of the secret fork being greater than *or equal to cd*. Instead of only giving up (which now means resetting the secret fork) when *db* blocks behind, the attacker can additionally choose to continue the attack or reset its fork every time that the honest mining pool publishes a new block.

Modelling the Attack. Our new attacker process, which replaces the Double-SpendingAttacker process presented previously, is thus as follows:

```
action rst; // indicates that the attacker restarts from the public fork
process TrustAttacker()
{
    do {
    :: rate((1/12) * M) {= m_len = min(CD, m_len + 1), m_diff++ =}
    :: sln {= m_diff-- =}; // public fork extended
        alt { // strategy choice: restart or continue malicious fork
        :: rst {= m_len = 0, m_diff = 0 =} // can always restart
        :: when(m_diff > -DB) cnt // can continue if not too far behind
        }
    }
}
```

This model is nondeterministic due to the choice between `rst` and `cnt` in the attacker process. We use actions `rst` and `cnt` to indicate the choice made; they have no synchronisation partner, but will help understand the optimal scheduler.

Evaluation. The probability for the attacker to eventually win as expressed by an adjusted version of `P_Win` is now 1 since it can retry indefinitely. It is thus more interesting to investigate the expected time until it wins:

```
property T_WinMin = Xmin(T, m_len >= CD && m_diff > 0);
```

We ask for the *minimum* time here, i.e. for the attacker to make its choices such that the time to success is minimised, which arguably is its best strategy. `mcsta` reports that the value is ≈ 3735.94 minutes for $m = 0.2$, i.e. a little over two and a half days. Let us thus compute the probability to succeed in just two days:

```
property P_WinMax2 = Pmax(<>[T<=2880](m_len >= CD && m_diff > 0));
```

We now ask for the *maximum* probability, since this again corresponds to an optimal attack. The result that `mcsta` gives is ≈ 0.535. As originally discovered in [24], we thus have a more than 50% chance to undermine the trust in Bitcoin if we control only 20% of the hash rate and invest only two days of mining. According to blockchain.com/pools, on July 8, 2019, the BTC.com pool in fact controlled 21.6% of the global hash rate; it could thus perform the attack.

Optimising the Attack Strategy. While the above numbers tell us the time and probability for the attack to succeed, they do not give any information about the attack strategy: What are the points, in terms of the length of the secret and public forks, where we should restart in order to obtain these optimal times and probabilities? Probabilistic model checking as implemented by `mcsta`, however, implicitly computes the optimal choice for every state of the MA underlying the model it checks, and it can be instructed to write this scheduler to a file:

```
./modest mcsta bitcoin-attack.modest -E "M=0.2,CD=6" --scheduler sched.txt
```

The result is a text file `sched.txt` with entries of the form

```
+ State: (HonestPool.location = loc_1, TrustAttacker.location = loc_10,
          m_len = 1, m_diff = -2)
   Choice: rst
```

for every state; here, in a state where the secret fork's length is 1, and it is two blocks shorter than the public one, the attacker restarts. We processed the file by projecting to `m_len` and `m_diff` and then eliminating all subsequent duplicate entries to find that the optimal strategy is to restart the attack if

- the honest pool announces a block, but the secret fork is still empty,
- the secret fork has one block and the public fork adds a third block, or
- the secret fork has ≥ 2 blocks and gets 3 blocks shorter than the public one,

and to continue the attack in all other cases.

Summary. Throughout this section, we first built an MA model that was equivalent to a CTMC, and then a truly nondeterministic MA. However, even that model does not use all features of the MA formalism: it lacks discrete probabilistic branching. As such, it falls into the interactive Markov chain (IMC, [39]) subset of MA. In the next section, we will introduce a model that is a true MA.

5   Evaluating a Reentrant Queueing System

In the previous section, we considered quantitative aspects of attacks on a stochastic timed system. We now turn our attention to a prominent use of continuous-time Markov models: performance and dependability evaluation. A classic application is resource-sharing *queueing systems*, using various CTMC-based formalisms like (Jackson) queueing networks [41], with analytical or simulation-based techniques for the analysis. Yet these approaches are restricted, both in modelling and in analysis, to fully stochastic systems. MA as a model, and our analysis tools in the MODEST TOOLSET, sit right at the edge between performance evaluation and model checking [2]. In particular, they add the concept of nondeterminism, which is at the core of classic qualitative model checking, to modelling formalisms and analysis algorithms that directly apply to performance evaluation scenarios. We now study a queueing system with stochastic timing, discrete probabilistic choices, and nondeterministic decisions—its model is thus an MA that does not fall into any of the existing subsets.

We consider the system with two queues depicted in Fig. 7, originally presented in [36]. Both queues have the same capacity c. Jobs arrive with rate λ

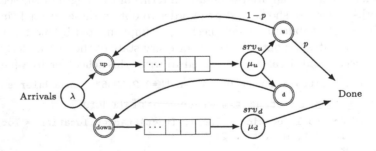

Fig. 7. A queuing system with postprocessing needs [36]

and enter one of the queues according to the standard join-the-shortest-queue strategy. This strategy is implicitly nondeterministic if both queues are equally filled. For each queue, jobs are processed by a dedicated server, serving jobs with rates μ_u and μ_d, respectively. Jobs leaving the lower server leave the system, while jobs once processed by the upper server are subject to an additional check. Dependent on the (nondeterministic) outcome thereof, they are either sent into the lower queue again (action d), or (action u) they may either leave the queue (with probability p) or reenter the upper queue (with probability $1 - p$).

A MODEST Model. As usual, we start our MODEST model by declaring all relevant constants, including the model parameters without specified values:

```
const int C;            // queue capacity
const int LAMBDA = 5;   // job arrival rate
const real MU_UP = 10;  // service rate of up server
const int MU_DOWN = 4;  // service rate of down server
const real P = 0.3;     // probability to be done after up server
```

In this model, we will use two *transient variables* to track when jobs are done, and when a job is dropped because both queues are full on arrival, or the queue in which it is due to re-enter after being processed by the up server is full:

```
transient int(0..1) done = 0;  // 1 when a job is done, otherwise 0
transient int(0..1) loss = 0;  // 1 when a job is dropped, otherwise 0
```

Unlike regular variables in MA^V, transient variables do not become part of the states. They can be used in assignments, but the assigned values are lost once the successor state is entered. However, the assigned value is visible to properties when the branch is taken, and we will make use of this later to define rewards.

We structure our model along the components shown in Fig. 7, defining a MODEST process for each of them. The arrivals process and the down server have the simplest behaviours:

```
action put, get;
process Arrivals()                 process ServerDown()
{                                  {
    rate(LAMBDA) tau;                  get;
    put;                               rate(MU_DOWN) tau {= done = 1 =};
    Arrivals()                         ServerDown()
}                                  }
```

Both processes synchronise with the input queues: `Arrivals` uses action `put` to enqueue a job that just arrived, and `ServerDown` uses action `get` to obtain a job to work on when idle, as soon as one is available. We will use synchronisation vectors to ensure that the synchronisation on `put` happens between `Arrivals` and exactly one of the two queues. Both queues use the same process definition:

```
int(0..C)[] q = [0, 0]; // array: number of jobs in the two queues
function bool isShortest(int id) = q[id] <= q[(id + 1) % 2];
process Queue(int id)
{
    do {
```

```
:: when(isShortest(id)) // enqueue
   put {= q[id] = min(q[id]+1, C), loss = q[id] == C ? 1 : 0 =}
:: when(q[id] > 0) get {= q[id]-- =} // dequeue
}
}
```

To distinguish the two queues, we use a process parameter id, and a two-element array storing the lengths of the queues that the processes index with their id. Function isShortest indicates whether the queue with the given id is no longer than the other one. A queue only accepts new jobs when isShortest(id) is *true*; if the queue is full in that case, the job is dropped, and loss is (temporarily) set to 1. The get action removes a job from a non-empty queue.

Finally, the up server has the most complicated structure, since it manages the reentry of jobs that it has finished serving into the two queues:

```
action u, d, rup, rdn;

process ServerUp()
{
   get;                     // get job from queue
   rate(MU_UP) tau;         // serve job
   alt {                    // nondeterministic choice between u and d:
   :: u palt {              // action u: probabilistic choice to either
      :1-P: {==}; rup       // reenter the up queue with probability 1-P,
      :  P: {= done = 1 =}  // or leave the system with probability P
      }
   :: d; rdn                // action d: reenter the down queue
   };
   ServerUp()
}
```

The nondeterministic choice between u and d is a choice between (d) making the job surely leave the system within a certain expected time, at the cost of processing by the slower down server, and (u) taking the chance for the job to leave the system immediately, at the risk of it reentering the up queue. The optimal choice will likely depend on the current lengths of both queues.

Now that we have specified all the necessary processes, we can put them into a parallel composition. We have rather different synchronisation requirements: put shall use a binary synchronisation between Arrivals and one of the two queues, with a nondeterministic choice if both have the same lengths; get in a queue shall synchronise only with the one server for that queue; and rup and rdn shall look like a put to the respective queues. We could declare put as a binary action, and cleverly use the relabel construct to rename the other actions in a way that makes MODEST create the correct synchronisation vectors internally. However, we can also just specify the desired vectors explicitly in a par:

```
par { put, put, get, get, put, put, u, d } {
   : put, put,  - ,  - ,  - ,  - , -, -  : Arrivals()
   : put,  - , get,  - , put,  - , -, -  : Queue(0)
   :  - ,  - , get,  - , rup, rdn, u, d  : ServerUp()
   :  - , put,  - , get,  - , put, -, -  : Queue(1)
   :  - ,  - ,  - , get,  - ,  - , -, -  : ServerDown()
}
```

If we read the "columns" in the above specification from bottom to top, we read the synchronisation vectors, with the topmost entry being the action that labels the synchronising edge in the composed MA^V, and - corresponding to $\perp$.

Performance Evaluation. We first add properties to investigate the probability and time until the queues are full, which is an undesirable condition that affects the dependability of the system by making it likely for jobs to be lost:

```
property ProbFullIsOne = Pmin(<> (q[0] == C && q[1] == C)) == 1;
property TminFull = Xmin(T, q[0] == C && q[1] == C);
property TmaxFull = Xmax(T, q[0] == C && q[1] == C);
property PminFull10 = Pmin(<>[T<=10] (q[0] == C && q[1] == C));
property PmaxFull10 = Pmax(<>[T<=10] (q[0] == C && q[1] == C));
```

We thus assert that the minimum probability for both queues to eventually be full is 1, which is a sanity check for the model; then we ask for the minimum and maximum of the expected time for both queues to be full, and of the probability for this to happen within 10 time units. By repeating the bottom two properties for different values of the time bound, we can obtain an approximation of the underlying cumulative distribution function over time. If we run mcsta with

```
./modest mcsta reentrant-q.modest -E"C=5" -O results.txt Minimal
```

we get an easy-to-parse file `results.txt` with the results:

```
"ProbFullIsOne": True
"TminFull": 7.165959461963808
"TmaxFull": 54.167593727326874
"PminFull10": 0.1338675853224175
"PmaxFull10": 0.7958342318163893
```

We see that the nondeterministic choices have a significant influence on the behaviour of the system; between the worst and best choices, the time to and probability for the undesirable event differs by a factor of 6 to 7. Since the standard probabilistic model checking algorithms implemented in mcsta are iterative numeric algorithms using double-precision floating-point numbers, every result is only an approximation of the true value despite the high number of decimal digits included in the output. The precision of mcsta is configurable.

Assume that we are designing a system of which our reentrant queueing system is an abstract model, and we have one parameter for which we must decide on a concrete value: the queue capacity c. We expect a higher capacity to improve throughput, utilisation, and reduce the number of lost jobs; however, it is also more costly to implement. We would thus like to find a good tradeoff between c and these quantities. We first specify properties that query for them:

```
property Throughput = Smax(S(done));
property Loss = Smax(S(loss));
property IdleOne = Smin(T(q[0] == 0 || q[1] == 0 ? 1 : 0));
property IdleBoth = Smin(T(q[0] == 0 && q[1] == 0 ? 1 : 0));
```

These queries are for long-run average rewards. The rewards are described by *accumulation expressions*: S(done) attaches to every branch (i.e. to every discrete step) the value of done after the branch's assignments have been executed (but before transient variables lose their values) as a branch reward. Expression T(q[0] == 0 || q[1] == 0 ? 1 : 0) sets the rate reward (accumulated over time) in every state to 1 if both queues are empty, and to 0 otherwise. We chose maximisation/minimisation as appropriate to correspond to the best possible strategy. We can ask mcsta to compute these quantities for many different values of c by specifying multiple experiments via the -E parameter:

```
./modest mcsta reentrant-q.modest -E "C=1" -E "C=2" -E "C=3" -E "C=4" \
    -E "C=5" -E "C=6" -E "C=7" -E "C=8" -E "C=9" -E "C=10" -E "C=11" \
    -E "C=12" -E "C=13" -E "C=14" -E "C=15" -E "C=16" -O perf.txt Minimal
```

We visualise the results in Fig. 8. The two lines converging to zero plot IdleOne (red, upper line) and IdleBoth (orange, lower). The other two lines plot Throughput (blue, upper) and Loss (purple, lower). We see that the fraction of time that the servers spent idle drops quickly with increasing c, whereas throughput and loss do not improve so much. Looking at this plot, we might choose c around 5 to 8.

Summary. In this section, we built a model for a queueing system that utilises all the features of the MA formalism. mcsta offers algorithms to calculate a variety of quantities (cf. Sect. 2), and we fully utilised them to evaluate the system from several perspectives.

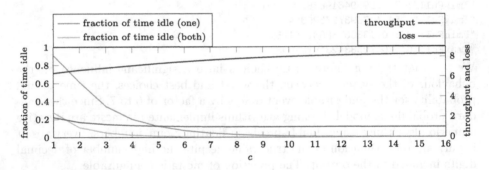

Fig. 8. Long-run average performance values for the reentrant queueing system

6   Conclusion

This tutorial paper has discussed how MODEST can be used as a convenient modelling language for Markov automata, together with some hints on what analysis is possible for such models. Markov automata can be considered as a central model family for studying the performance, dependability, and correctness of randomised and distributed systems.

We introduced all the basic and several advanced constructs of the MODEST language for MA. Among the features that we did not cover are exception handling (using the **throw** and **try-catch** constructs), the specification of values for transient variables in locations (using the `with` construct), dynamic array constructors, user-defined recursive datatypes (which allow the specification of, for example, unbounded list types), recursive functions, and `binary` and `broadcast` actions (which automatically generate appropriate synchronisation vectors, just like "normal" actions do for multi-way synchronisation). Going beyond MA, MODEST also supports the formalisms of probabilistic timed automata [42] (which add a `clock` type and time progress conditions via the `constrain` construct), stochastic timed automata [4] (which allow sampling values from continuous probability distributions in assignments; they are a generalisation of MA), and stochastic hybrid automata [25] (which add continuous variables of type `var` whose behaviour over time is specified via differential equations and inclusions using the `der` operator for derivatives). Further MODEST models are included in the MODEST TOOLSET download, available at modestchecker.net, and in the Quantitative Verification Benchmark Set at qcomp.org.

Data Availability. The models, example command lines, and results presented in this paper are archived and available at DOI 10.4121/uuid:5a73169e-b494-411b-b3a8-051e62efba9e [31].

Acknowledgments. The authors thank Michaela Klauck (Saarland University) for preparing an initial version of the MODEST model appearing in Sect. 5 and for helpful comments on a draft of this paper.

References

1. Amparore, E.G., Balbo, G., Beccuti, M., Donatelli, S., Franceschinis, G.: 30 years of GreatSPN. In: Fiondella, L., Puliafito, A. (eds.) Principles of Performance and Reliability Modeling and Evaluation. SSRE, pp. 227–254. Springer, Cham (2016). https://doi.org/10.1007/978-3-319-30599-8_9
2. Baier, C., Haverkort, B.R., Hermanns, H., Katoen, J.P.: Performance evaluation and model checking join forces. Commun. ACM **53**(9), 76–85 (2010)
3. Baier, C., Katoen, J.: Principles of Model Checking. MIT Press, Cambridge (2008)
4. Bohnenkamp, H.C., D'Argenio, P.R., Hermanns, H., Katoen, J.P.: MoDeST: a compositional modeling formalism for hard and softly timed systems. IEEE Trans. Softw. Eng. **32**(10), 812–830 (2006)
5. Bolch, G., Greiner, S., de Meer, H., Trivedi, K.S.: Queuing Networks and Markov Chains - Modeling and Performance Evaluation with Computer Science Applications, 2nd edn. Wiley, Hoboken (2006)
6. Bolognesi, T., Brinksma, E.: Introduction to the ISO specification language LOTOS. Comput. Netw. **14**, 25–59 (1987)
7. Bonneau, J., Miller, A., Clark, J., Narayanan, A., Kroll, J.A., Felten, E.W.: SoK: research perspectives and challenges for Bitcoin and cryptocurrencies. In: SP, pp. 104–121. IEEE Computer Society (2015)
8. Braitling, B., Fioriti, L.M.F., Hatefi, H., Wimmer, R., Becker, B., Hermanns, H.: MeGARA: menu-based game abstraction and abstraction refinement of Markov automata. In: QAPL. EPTCS, vol. 154, pp. 48–63 (2014)

9. Braitling, B., Ferrer Fioriti, L.M., Hatefi, H., Wimmer, R., Becker, B., Hermanns, H.: Abstraction-based computation of reward measures for Markov automata. In: D'Souza, D., Lal, A., Larsen, K.G. (eds.) VMCAI 2015. LNCS, vol. 8931, pp. 172–189. Springer, Heidelberg (2015). https://doi.org/10.1007/978-3-662-46081-8_10

10. Brázdil, T., Hermanns, H., Krcál, J., Kretínský, J., Rehák, V.: Verification of open interactive Markov chains. In: FSTTCS. LIPIcs, vol. 18, pp. 474–485. Schloss Dagstuhl - Leibniz-Zentrum fuer Informatik (2012)

11. Budde, C.E., D'Argenio, P.R., Hartmanns, A., Sedwards, S.: A statistical model checker for nondeterminism and rare events. In: Beyer, D., Huisman, M. (eds.) TACAS 2018. LNCS, vol. 10806, pp. 340–358. Springer, Cham (2018). https://doi.org/10.1007/978-3-319-89963-3_20

12. Budde, C.E., Dehnert, C., Hahn, E.M., Hartmanns, A., Junges, S., Turrini, A.: JANI: quantitative model and tool interaction. In: Legay, A., Margaria, T. (eds.) TACAS 2017. LNCS, vol. 10206, pp. 151–168. Springer, Heidelberg (2017). https://doi.org/10.1007/978-3-662-54580-5_9

13. Butkova, Y., Hartmanns, A., Hermanns, H.: A Modest approach to modelling and checking Markov automata. In: Parker, D., Wolf, V. (eds.) QEST 2019. LNCS, vol. 1785, pp. 52–69. Springer, Cham (2019). https://doi.org/10.1007/978-3-030-30281-8_4

14. Butkova, Y., Hatefi, H., Hermanns, H., Krčál, J.: Optimal continuous time Markov decisions. In: Finkbeiner, B., Pu, G., Zhang, L. (eds.) ATVA 2015. LNCS, vol. 9364, pp. 166–182. Springer, Cham (2015). https://doi.org/10.1007/978-3-319-24953-7_12

15. Butkova, Y., Wimmer, R., Hermanns, H.: Long-run rewards for Markov automata. In: Legay, A., Margaria, T. (eds.) TACAS 2017. LNCS, vol. 10206, pp. 188–203. Springer, Heidelberg (2017). https://doi.org/10.1007/978-3-662-54580-5_11

16. Butkova, Y., Wimmer, R., Hermanns, H.: Markov automata on discount!. In: German, R., Hielscher, K.-S., Krieger, U.R. (eds.) MMB 2018. LNCS, vol. 10740, pp. 19–34. Springer, Cham (2018). https://doi.org/10.1007/978-3-319-74947-1_2

17. D'Argenio, P.R., Hartmanns, A., Sedwards, S.: Lightweight statistical model checking in nondeterministic continuous time. In: Margaria, T., Steffen, B. (eds.) ISoLA 2018. LNCS, vol. 11245, pp. 336–353. Springer, Cham (2018). https://doi.org/10.1007/978-3-030-03421-4_22

18. Dehnert, C., Junges, S., Katoen, J.-P., Volk, M.: A Storm is coming: a modern probabilistic model checker. In: Majumdar, R., Kunčak, V. (eds.) CAV 2017. LNCS, vol. 10427, pp. 592–600. Springer, Cham (2017). https://doi.org/10.1007/978-3-319-63390-9_31

19. Eisentraut, C.: Principles of Markov automata. Ph.D. thesis, Saarland University, Saarbrücken, Germany (2017)

20. Eisentraut, C., Hermanns, H., Katoen, J.-P., Zhang, L.: A semantics for every GSPN. In: Colom, J.-M., Desel, J. (eds.) PETRI NETS 2013. LNCS, vol. 7927, pp. 90–109. Springer, Heidelberg (2013). https://doi.org/10.1007/978-3-642-38697-8_6

21. Eisentraut, C., Hermanns, H., Schuster, J., Turrini, A., Zhang, L.: The quest for minimal quotients for probabilistic and Markov automata. Inf. Comput. **262**(Part), 162–186 (2018)

22. Eisentraut, C., Hermanns, H., Zhang, L.: Concurrency and composition in a stochastic world. In: Gastin, P., Laroussinie, F. (eds.) CONCUR 2010. LNCS, vol. 6269, pp. 21–39. Springer, Heidelberg (2010). https://doi.org/10.1007/978-3-642-15375-4_3

23. Eisentraut, C., Hermanns, H., Zhang, L.: On probabilistic automata in continuous time. In: LICS, pp. 342–351. IEEE Computer Society (2010)

24. Fehnker, A., Chaudhary, K.: Twenty percent and a few days – optimising a Bitcoin majority attack. In: Dutle, A., Muñoz, C., Narkawicz, A. (eds.) NFM 2018. LNCS, vol. 10811, pp. 157–163. Springer, Cham (2018). https://doi.org/10.1007/978-3-319-77935-5_11

25. Fränzle, M., Hahn, E.M., Hermanns, H., Wolovick, N., Zhang, L.: Measurability and safety verification for stochastic hybrid systems. In: HSCC, pp. 43–52. ACM (2011)

26. Garavel, H., Lang, F., Mateescu, R., Serwe, W.: CADP 2011: a toolbox for the construction and analysis of distributed processes. STTT **15**(2), 89–107 (2013)

27. Guck, D., Han, T., Katoen, J.-P., Neuhäußer, M.R.: Quantitative timed analysis of interactive Markov chains. In: Goodloe, A.E., Person, S. (eds.) NFM 2012. LNCS, vol. 7226, pp. 8–23. Springer, Heidelberg (2012). https://doi.org/10.1007/978-3-642-28891-3_4

28. Guck, D., Hatefi, H., Hermanns, H., Katoen, J.P., Timmer, M.: Analysis of timed and long-run objectives for Markov automata. Logical Methods Comput. Sci. **10**(3) (2014)

29. Guck, D., Timmer, M., Hatefi, H., Ruijters, E., Stoelinga, M.: Modelling and analysis of Markov reward automata. In: Cassez, F., Raskin, J.-F. (eds.) ATVA 2014. LNCS, vol. 8837, pp. 168–184. Springer, Cham (2014). https://doi.org/10.1007/978-3-319-11936-6_13

30. Hahn, E.M., Hartmanns, A., Hermanns, H., Katoen, J.P.: A compositional modelling and analysis framework for stochastic hybrid systems. Formal Methods Syst. Des. **43**(2), 191–232 (2013)

31. Hartmanns, A.: A Modest Markov automata tutorial (artifact). 4TU.Centre for Research Data (2019). https://doi.org/10.4121/uuid:5a73169e-b494-411b-b3a8-051e62efba9e

32. Hartmanns, A., Hermanns, H.: The Modest Toolset: an integrated environment for quantitative modelling and verification. In: Ábrahám, E., Havelund, K. (eds.) TACAS 2014. LNCS, vol. 8413, pp. 593–598. Springer, Heidelberg (2014). https://doi.org/10.1007/978-3-642-54862-8_51

33. Hartmanns, A., Hermanns, H.: Explicit model checking of very large MDP using partitioning and secondary storage. In: Finkbeiner, B., Pu, G., Zhang, L. (eds.) ATVA 2015. LNCS, vol. 9364, pp. 131–147. Springer, Cham (2015). https://doi.org/10.1007/978-3-319-24953-7_10

34. Hartmanns, A., Klauck, M., Parker, D., Quatmann, T., Ruijters, E.: The quantitative verification benchmark set. In: Vojnar, T., Zhang, L. (eds.) TACAS 2019. LNCS, vol. 11427, pp. 344–350. Springer, Cham (2019). https://doi.org/10.1007/978-3-030-17462-0_20

35. Hatefi, H.: Finite horizon analysis of Markov automata. Ph.D. thesis, Saarland University, Germany (2017). scidok.sulb.uni-saarland.de/volltexte/2017/6743/

36. Hatefi, H., Hermanns, H.: Model checking algorithms for Markov automata. Electron. Commun. EASST **53** (2012)

37. Hatefi, H., Wimmer, R., Braitling, B., Fioriti, L.M.F., Becker, B., Hermanns, H.: Cost vs. time in stochastic games and Markov automata. Formal Asp. Comput. **29**(4), 629–649 (2017)

38. Haverkort, B.R.: Performance of Computer Communication Systems - A Model-Based Approach. Wiley, Hoboken (1998)

39. Hermanns, H.: Interactive Markov Chains: The Quest for Quantified Quality. LNCS, vol. 2428, pp. 35–55. Springer, Heidelberg (2002). https://doi.org/10.1007/3-540-45804-2_3
40. Hoare, C.A.R.: Communicating Sequential Processes. Prentice-Hall, Upper Saddle River (1985)
41. Jackson, J.R.: Jobshop-like queueing systems. Manag. Sci. **10**(1), 131–142 (1963)
42. Kwiatkowska, M.Z., Norman, G., Segala, R., Sproston, J.: Automatic verification of real-time systems with discrete probability distributions. Theor. Comput. Sci. **282**(1), 101–150 (2002)
43. Legay, A., Sedwards, S., Traonouez, L.-M.: Scalable verification of Markov decision processes. In: Canal, C., Idani, A. (eds.) SEFM 2014. LNCS, vol. 8938, pp. 350–362. Springer, Cham (2015). https://doi.org/10.1007/978-3-319-15201-1_23
44. Milner, R.: Communication and Concurrency. Prentice-Hall, Upper Saddle River (1989)
45. Nakamoto, S.: Bitcoin: A peer-to-peer electronic cash system (2009). bitcoin.org
46. Puterman, M.L.: Markov Decision Processes: Discrete Stochastic Dynamic Programming. Wiley Series in Probability and Statistics, Wiley, Hoboken (1994)
47. Quatmann, T., Junges, S., Katoen, J.-P.: Markov automata with multiple objectives. In: Majumdar, R., Kunčak, V. (eds.) CAV 2017. LNCS, vol. 10426, pp. 140–159. Springer, Cham (2017). https://doi.org/10.1007/978-3-319-63387-9_7
48. Rabe, M.N., Schewe, S.: Finite optimal control for time-bounded reachability in CTMDPs and continuous-time Markov games. Acta Inf. **48**(5–6), 291–315 (2011)
49. Segala, R.: Modeling and verification of randomized distributed real-time systems. Ph.D. thesis, Massachusetts Institute of Technology, Cambridge (1995)
50. Timmer, M.: Efficient modelling, generation and analysis of Markov automata. Ph.D. thesis, University of Twente, Enschede (2013)
51. Timmer, M., Katoen, J.-P., van de Pol, J., Stoelinga, M.I.A.: Efficient modelling and generation of Markov automata. In: Koutny, M., Ulidowski, I. (eds.) CONCUR 2012. LNCS, vol. 7454, pp. 364–379. Springer, Heidelberg (2012). https://doi.org/10.1007/978-3-642-32940-1_26
52. Timmer, M., Katoen, J.P., van de Pol, J., Stoelinga, M.: Confluence reduction for Markov automata. Theor. Comput. Sci. **655**, 193–219 (2016)

Explainable AI Planning (XAIP): Overview and the Case of Contrastive Explanation (Extended Abstract)

Jörg Hoffmann[1]([✉]) and Daniele Magazzeni[2]

[1] Saarland Informatics Campus, Saarland University, Saarbrücken, Germany
hoffmann@cs.uni-saarland.de
[2] King's College London, London, UK
daniele.magazzeni@kcl.ac.uk

Abstract. Model-based approaches to AI are well suited to explainability in principle, given the explicit nature of their world knowledge and of the reasoning performed to take decisions. AI Planning in particular is relevant in this context as a generic approach to action-decision problems. Indeed, explainable AI Planning (XAIP) has received interest since more than a decade, and has been taking up speed recently along with the general trend to explainable AI. In the lecture, we provide an overview, categorizing and illustrating the different kinds of explanation relevant in AI Planning; and we outline recent works on one particular kind of XAIP, contrastive explanation. This extended abstract gives a brief summary of the lecture, with some literature pointers. We emphasize that completeness is neither claimed nor intended; the abstract may serve as a brief primer with literature entry points.

1 Explainable AI Planning: Overview

The need for explainable AI (XAI) first became prominent in Machine Learning, where the lack of understandable decision rationales is particularly daunting. Model-based techniques are fundamentally better suited to providing explanations, yet their explainability has traditionally not received much interest. This has changed with the XAI trend. In particular, research on explainable AI planning (XAIP) has received increasing interest in recent years. One culminating point of this trend is the nascent series of XAIP workshops[1] at the International Conference on Automated Planning and Scheduling (ICAPS).

As is natural for a nascent area, at this time the XAIP landscape is still in the making. XAIP has attracted interest from researchers with widely different backgrounds and points of view, and it is too early to give a conclusive systematization into sub-topics and issues of interest. A roadmap for XAIP was

---

[1] See the 2019 edition at https://kcl-planning.github.io/XAIP-Workshops/ICAPS_2019.

© Springer Nature Switzerland AG 2019
M. Krötzsch and D. Stepanova (Eds.): Reasoning Web 2019, LNCS 11810, pp. 277–282, 2019.
https://doi.org/10.1007/978-3-030-31423-1_9

proposed by Fox et al. [10], categorizations have been attempted [16], and a systematization of possible objectives has just been published [6]. XAIP includes topics ranging from epistemic logic to machine learning, and techniques ranging from domain analysis to plan generation and goal recognition. Nevertheless, some major themes have emerged, that we refer to here as *plan explanation*, *contrastive explanation*, *human factors*, and *model reconciliation*.

Plan explanation is the oldest branch of XAIP. It aims at helping humans to understand the inner workings of a plan suggested by the system (e. g., [1,2,13,17,20,24,27]). This involves, in particular, the transformation of planner output into forms that humans can easily understand; the description of causal and temporal relations between individual plan steps; and the design of interfaces, in particular suitable dialogue systems, supporting human interaction and understanding.

In contrastive explanation, the aim is to answer user questions of the kind "Why do you suggest to do A here? (rather than B which seems more appropriate to me)". This is a frequent form of question as highlighted by a recent analysis [19] of lessons to be learned for XAI from social sciences. Answers to such questions take the form of reasons why A is preferable over B. Contrastive explanation is the major focus of this lecture, so we discuss it in more detail in Sects. 2 and 3.

Human factors research naturally has to be a major component of XAIP, whose ultimate aim is to communicate with human users. Manuela Veloso and her team investigate verbalizations describing the robot experience and intentions to human users [22]. Other work [29] focuses on a human's interpretation of plans. Learning is used to create a model of the interpretation, which is then used to measure the explicability and predictability of plans. A recent proposal is to combine cognitive measures with epistemic planning [21]. Many works, also ones cited here as belonging to other themes, include human factors research to varying degrees.

In *model reconciliation*, the focus is on the agent vs. human having different world models. The explanation must then identify and reconcile the relevant model differences. This has been intensively investigated in the last years [4, 15,23,28], with mature results and outreach to the robotics [7] and multi-agent communities [12].

There are of course various works on XAIP, or relating to XAIP, that do not fit into this categorization. To name but a few examples: Göbelbecker et al. [11] proposed a framework for "excuses", which can be viewed as explanations why a planning task is unsolvable; Smith [25] put forward the challenge of *planning as an iterative process*, which among others requires explanation facilities; and some work has considered particular forms of communication like lying [5].

2   Contrastive Explanations

As mentioned, an important type of questions in Explainable Planning are *contrastive* questions, of the form "Why action A instead of action B?". These questions arise when the planner is suggesting something different from what the user would expect. In such a scenario, one way to address this type of question is to allow the user to compare the plan suggested by the planner with what she/he was expecting. These are contrastive explanations that can highlight the differences between the decisions that have been made by the planner and what the user would expect, as well as to provide further insight into the model and the planning process. A detailed analysis of contrastive explanations in AI has been proposed by Tim Miller in [18].

Some recent work introduced contrastive explanations for Explainable Planning. In particular, in [14] contrastive questions are compiled into constraints that form a hypothetical model. Such a hypothetical model can be used to generate the hypothetical plan that the user would expect and from here the contrastive explanation can be presented to the user. The work focuses on temporal planning and presents domain-independent compilations.

Another related line of work focuses on providing contrastive explanations *as a service* [3]. Here the idea is to create a wrapper around an existing planner and use automatic compilations of user questions into models. In this way, the explanations are generated using the same planner already used by the user, and this increases the user confidence in the explanations provided.

In the lecture we give an overview of recent progress on using contrastive explanations for Explainable Planning.

3   Contrastive Explanation of Plan Space Through Plan-Space Dependencies

We finally consider a line of work, conducted by the authors, starting from the idea to answer questions "Why does the plan π start with action A rather than B?" by generating a new plan π' starting with B and highlighting undesirable properties of π'. A weakness of this approach is that there may be differences between π and π' unrelated to the use of A vs. B. Many comparison aspects (e. g. which other actions are used, or which "soft" objectives are satisfied) may be affected by arbitrary decisions in the planner's search. Therefore, the idea is to replace the *existential* answer generating a single alternative plan π' with a *universal* answer pertaining to *all* possible such alternatives.

This can be done at the level of *plan properties*: Boolean functions on plans that capture aspects of plans the user cares about (whether or not the plan starts with a particular action, whether or not a particular soft objective is satisfied, etc). Given a set of plan properties, one can determine dependencies across these properties, i. e., plan-space entailments: a plan property p entails another property p' if every plan that satisfies p also satisfies p'. A user question "Why does the current plan π satisfy p rather than q?" can then be answered in terms of the properties q' not true in π but entailed by q: things that will necessarily change when satisfying q.

We put forward, and explain in the lecture, a generic framework for this kind of analysis, as well as an instantiation and experiments in the context of oversubscription planning [8,26] where resources are insufficient to achieve all goals, and plan properties of obvious interest are those goals achieved by a plan. A first paper on this approach is published at XAIP'19 and serves as a reference for the reader interested in details [9].

Acknowledgments. This material is based upon work supported by the Air Force Office of Scientific Research under award number FA9550-18-1-0245. Jörg Hoffmann's research group has received support by DFG grant 389792660 as part of TRR 248 (see https://perspicuous-computing.science). Daniele Magazzeni's research group has received support by EPSRC grant EP/R033722/1: Trust in Human-Machine Partnerships.

References

1. Bercher, P., et al.: Plan, repair, execute, explain - how planning helps to assemble your home theater. In: Chien, S., Do, M., Fern, A., Ruml, W. (eds.) Proceedings of the 24th International Conference on Automated Planning and Scheduling (ICAPS 2014). AAAI Press (2014)
2. Bidot, J., Biundo, S., Heinroth, T., Minker, W., Nothdurft, F., Schattenberg, B.: Verbal plan explanations for hybrid planning. In: Proceedings MKWI (2010)
3. Cashmore, M., Collins, A., Krarup, B., Krivic, S., Magazzeni, D., Smith, D.: Explainable planning as a service. In: ICAPS-19 Workshop on Explainable Planning (2019)
4. Chakraborti, T., Sreedharan, S., Zhang, Y., Kambhampati, S.: Plan explanations as model reconciliation: moving beyond explanation as soliloquy. In: Sierra, C. (ed.) Proceedings of the 26th International Joint Conference on Artificial Intelligence (IJCAI 2017). AAAI Press/IJCAI (2017)
5. Chakraborti, T., Kambhampati, S.: (how) can ai bots lie? In: Proceedings of the 2nd Workshop on Explainable Planning (XAIP 2019) (2019)
6. Chakraborti, T., Kulkarni, A., Sreedharan, S., Smith, D.E., Kambhampati, S.: Explicability? Legibility? Predictability? Transparency? Privacy? Security? The emerging landscape of interpretable agent behavior. In: Proceedings of the 29th International Conference on Automated Planning and Scheduling (ICAPS 2019). AAAI Press (2019)
7. Chakraborti, T., Sreedharan, S., Grover, S., Kambhampati, S.: Plan explanations as model reconciliation. In: Proceedings of the 14th ACM/IEEE International Conference on Human-Robot Interaction (HRI 2019), pp. 258–266 (2019)
8. Domshlak, C., Mirkis, V.: Deterministic oversubscription planning as heuristic search: abstractions and reformulations. J. Artif. Intell. Res. **52**, 97–169 (2015)
9. Eifler, R., Cashmore, M., Hoffmann, J., Magazzeni, D., Steinmetz, M.: Explaining the space of plans through plan-property dependencies. In: Proceedings of the 2nd Workshop on Explainable Planning (XAIP 2019) (2019)
10. Fox, M., Long, D., Magazzeni, D.: Explainable planning. In: Proceedings of IJCAI 2017 Workshop on Explainable AI (2017)
11. Göbelbecker, M., Keller, T., Eyerich, P., Brenner, M., Nebel, B.: Coming up with good excuses: what to do when no plan can be found. In: Brafman, R.I., Geffner, H., Hoffmann, J., Kautz, H.A. (eds.) Proceedings of the 20th International Conference on Automated Planning and Scheduling (ICAPS 2010), pp. 81–88. AAAI Press (2010)

12. Kambhampati, S.: Synthesizing explainable behavior for human-AI collaboration. In: Proceedings of the 18th International Conference on Autonomous Agents and MultiAgent Systems (AAMAS 2019), pp. 1–2 (2019)
13. Khan, O.Z., Poupart, P., Black, J.P.: Minimal sufficient explanations for factored Markov decision processes. In: Gerevini, A., Howe, A., Cesta, A., Refanidis, I. (eds.) Proceedings of the 19th International Conference on Automated Planning and Scheduling (ICAPS 2009). AAAI Press (2009)
14. Krarup, B., Cashmore, M., Magazzeni, D., Miller, T.: Model-based contrastive explanations for explainable planning. In: ICAPS 2019 Workshop on Explainable Planning (2019)
15. Kulkarni, A., Zha, Y., Chakraborti, T., Vadlamudi, S.G., Zhang, Y., Kambhampati, S.: Explicable planning as minimizing distance from expected behavior. In: Proceedings of the 18th International Conference on Autonomous Agents and MultiAgent Systems, AAMAS 2019, Montreal, QC, Canada, 13–17 May 2019, pp. 2075–2077 (2019)
16. Langley, P., Meadows, B., Sridharan, M., Choi, D.: Explainable agency for intelligent autonomous systems. In: Singh, S., Markovitch, S. (eds.) Proceedings of the 31st AAAI Conference on Artificial Intelligence (AAAI 2017). AAAI Press, February 2017
17. McGuinness, D.L., Glass, A., Wolverton, M., da Silva, P.P.: Explaining task processing in cognitive assistants that learn. In: Proceedings of the 20th International Florida Artificial Intelligence Research Society Conference (FLAIRS 2007), pp. 284–289 (2007)
18. Miller, T.: Contrastive explanation: a structural-model approach. CoRR, abs/1811.03163 (2018)
19. Miller, T.: Explanation in artificial intelligence: insights from the social sciences. Artif. Intell. **267**, 1–38 (2019)
20. Nothdurft, F., Behnke, G., Bercher, P., Biundo, S., Minker, W.: The interplay of user-centered dialog systems and AI planning. In: Proceedings of the 16th Annual Meeting of the Special Interest Group on Discourse and Dialogue (SIGDAL2015), pp. 344–353 (2015)
21. Petrick, R., Dalzel-Job, S., Hill, R.: Combining cognitive and affective measures with epistemic planning for explanation generation. In: Proceedings of the 2nd Workshop on Explainable Planning (XAIP 2019) (2019)
22. Rosenthal, S., Selvaraj, S.P., Veloso, M.M.: Verbalization: narration of autonomous robot experience. In: Kambhampati, S. (ed.) Proceedings of the 25th International Joint Conference on Artificial Intelligence (IJCAI 2016). AAAI Press/IJCAI (2016)
23. Sarath, S., Alberto, O., Prasad, M., Subbarao, K.: Model-free model reconciliation. In: ICAPS-19 Workshop on Explainable Planning (2019)
24. Seegebarth, B., Müller, F., Schattenberg, B., Biundo, S.: Making hybrid plans more clear to human users - A formal approach for generating sound explanations. In: Bonet, B., McCluskey, L., Silva, J.R., Williams, B. (eds.) Proceedings of the 22nd International Conference on Automated Planning and Scheduling (ICAPS 2012). AAAI Press (2012)
25. Smith, D.: Planning as an iterative process. In: Hoffmann, J., Selman, B. (eds.) Proceedings of the 26th AAAI Conference on Artificial Intelligence (AAAI 2012), Toronto, ON, Canada, pp. 2180–2185. AAAI Press, July 2012
26. Smith, D.E.: Choosing objectives in over-subscription planning. In: Koenig, S., Zilberstein, S., Koehler, J. (eds.) Proceedings of the 14th International Conference on Automated Planning and Scheduling (ICAPS 2004), Whistler, Canada, pp. 393–401. Morgan Kaufmann (2004)

27. Sohrabi, S., Baier, J.A., McIlraith, S.A.: Preferred explanations: theory and generation via planning. In: Burgard, W., Roth, D. (eds.) Proceedings of the 25th National Conference of the American Association for Artificial Intelligence (AAAI 2011), San Francisco, CA, USA. AAAI Press, July 2011
28. Sreedharan, S., Chakraborti, T., Kambhampati, S.: Handling model uncertainty and multiplicity in explanations via model reconciliation. In: Proceedings of the Twenty-Eighth International Conference on Automated Planning and Scheduling, ICAPS 2018, Delft, The Netherlands, 24–29 June 2018, pp. 518–526 (2018)
29. Zhang, Y., Sreedharan, S., Kulkarni, A., Chakraborti, T., Zhuo, H., Kambhampati, S.: Plan explicability and predictability for robot task planning. In: Proceedings of ICRA (2017)

Author Index

Printed in the United States
By Bookmasters